New Carbon Materials from Biomass and Their Applications

New Carbon Materials from Biomass and Their Applications

Editors

Jorge Bedia
Carolina Belver

MDPI • Basel • Beijing • Wuhan • Barcelona • Belgrade • Manchester • Tokyo • Cluj • Tianjin

Editors
Jorge Bedia
Chemical Engineering
Universidad Autónoma
de Madrid
Madrid
Spain

Carolina Belver
Chemical Engineering
Universidad Autónoma
de Madrid
Madrid
Spain

Editorial Office
MDPI
St. Alban-Anlage 66
4052 Basel, Switzerland

This is a reprint of articles from the Special Issue published online in the open access journal *Applied Sciences* (ISSN 2076-3417) (available at: www.mdpi.com/journal/applsci/special_issues/carbon_biomass_applications).

For citation purposes, cite each article independently as indicated on the article page online and as indicated below:

LastName, A.A.; LastName, B.B.; LastName, C.C. Article Title. *Journal Name* **Year**, *Volume Number*, Page Range.

ISBN 978-3-0365-1435-2 (Hbk)
ISBN 978-3-0365-1436-9 (PDF)

© 2022 by the authors. Articles in this book are Open Access and distributed under the Creative Commons Attribution (CC BY) license, which allows users to download, copy and build upon published articles, as long as the author and publisher are properly credited, which ensures maximum dissemination and a wider impact of our publications.

The book as a whole is distributed by MDPI under the terms and conditions of the Creative Commons license CC BY-NC-ND.

Contents

About the Editors vii

Preface to "New Carbon Materials from Biomass and Their Applications" ix

Jorge Bedia and Carolina Belver
Special Issue on New Carbon Materials from Biomass and Their Applications
Reprinted from: *Appl. Sci.* **2021**, *11*, 2453, doi:10.3390/app11062453 1

Roberta Ferrentino, Riccardo Ceccato, Valentina Marchetti, Gianni Andreottola and Luca Fiori
Sewage Sludge Hydrochar: An Option for Removal of Methylene Blue from Wastewater
Reprinted from: *Appl. Sci.* **2020**, *10*, 3445, doi:10.3390/app10103445 5

Virpi Siipola, Stephan Pflugmacher, Henrik Romar, Laura Wendling and Pertti Koukkari
Low-Cost Biochar Adsorbents for Water Purification Including Microplastics Removal
Reprinted from: *Appl. Sci.* **2020**, *10*, 788, doi:10.3390/app10030788 27

Joan J. Manyà, David García-Morcate and Belén González
Adsorption Performance of Physically Activated Biochars for Postcombustion CO_2 Capture from Dry and Humid Flue Gas
Reprinted from: *Appl. Sci.* **2020**, *10*, 376, doi:10.3390/app10010376 45

Elsayed Mousa, Mania Kazemi, Mikael Larsson, Gert Karlsson and Erik Persson
Potential for Developing Biocarbon Briquettes for Foundry Industry
Reprinted from: *Appl. Sci.* **2019**, *9*, 5288, doi:10.3390/app9245288 63

Elena Diaz, Francisco Javier Manzano, John Villamil, Juan Jose Rodriguez and Angel F. Mohedano
Low-Cost Activated Grape Seed-Derived Hydrochar through Hydrothermal Carbonization and Chemical Activation for Sulfamethoxazole Adsorption
Reprinted from: *Appl. Sci.* **2019**, *9*, 5127, doi:10.3390/app9235127 79

Ping Yu, Qiansheng Li, Lan Huang, Genhua Niu and Mengmeng Gu
Mixed Hardwood and Sugarcane Bagasse Biochar as Potting Mix Components for Container Tomato and Basil Seedling Production
Reprinted from: *Appl. Sci.* **2019**, *9*, 4713, doi:10.3390/app9214713 93

Sari Tuomikoski, Riikka Kupila, Henrik Romar, Davide Bergna, Teija Kangas and Hanna Runtti et al.
Zinc Adsorption by Activated Carbon Prepared from Lignocellulosic Waste Biomass
Reprinted from: *Appl. Sci.* **2019**, *9*, 4583, doi:10.3390/app9214583 107

Lan Huang, Ping Yu and Mengmeng Gu
Evaluation of Biochar and Compost Mixes as Substitutes to a Commercial Propagation Mix
Reprinted from: *Appl. Sci.* **2019**, *9*, 4394, doi:10.3390/app9204394 123

Bogdan Saletnik, Marcin Bajcar, Grzegorz Zaguła, Aneta Saletnik, Maria Tarapatskyy and Czesław Puchalski
Biochar as a Stimulator for Germination Capacity in Seeds of Virginia Mallow (*Sida hermaphrodita* (L.) *Rusby*)
Reprinted from: *Appl. Sci.* **2019**, *9*, 3213, doi:10.3390/app9163213 135

Juana P. Moiwo, Alusine Wahab, Emmanuel Kangoma, Mohamed M. Blango, Mohamed P. Ngegba and Roland Suluku
Effect of Biochar Application Depth on Crop Productivity Under Tropical Rainfed Conditions
Reprinted from: *Appl. Sci.* **2019**, *9*, 2602, doi:10.3390/app9132602 149

JoungDu Shin, SangWon Park and SunIl Lee
Optimum Method Uploaded Nutrient Solution for Blended Biochar Pellet with Application of Nutrient Releasing Model as Slow Release Fertilizer
Reprinted from: *Appl. Sci.* **2019**, *9*, 1899, doi:10.3390/app9091899 165

Xuhui Li, Kunquan Li, Chunlei Geng, Hamed El Mashad, Hua Li and Wenqing Yin
Biochar from Microwave Pyrolysis of Artemisia Slengensis: Characterization and Methylene Blue Adsorption Capacity
Reprinted from: *Appl. Sci.* **2019**, *9*, 1813, doi:10.3390/app9091813 177

Bogdan Saletnik, Grzegorz Zaguła, Marcin Bajcar, Maria Tarapatskyy, Gabriel Bobula and Czesław Puchalski
Biochar as a Multifunctional Component of the Environment—A Review
Reprinted from: *Appl. Sci.* **2019**, *9*, 1139, doi:10.3390/app9061139 187

Xiaoyuechuan Ma, Shusheng Pang, Ruiqin Zhang and Qixiang Xu
Process Simulation and Economic Evaluation of Bio-Oil Two-Stage Hydrogenation Production
Reprinted from: *Appl. Sci.* **2019**, *9*, 693, doi:10.3390/app9040693 207

Xue Yang, Shiqiu Zhang, Meiting Ju and Le Liu
Preparation and Modification of Biochar Materials and their Application in Soil Remediation
Reprinted from: *Appl. Sci.* **2019**, *9*, 1365, doi:10.3390/app9071365 223

Ping Lu, Kebing Wang and Juhui Gong
Optimization of *Salix* Carbonation Solid Acid Catalysts for One-Step Synthesis by Response Surface Method
Reprinted from: *Appl. Sci.* **2019**, *9*, 1518, doi:10.3390/app9081518 249

About the Editors

Jorge Bedia

Jorge Bedia works in Industrial Engineering and obtained his PhD in Chemical Engineering at the University of Malaga (Spain). Currently, he is an Associate Professor at the Chemical Engineering Department of the Autonomous University of Madrid (Spain), and has been since 2013. His research interests include: (i) the synthesis and characterization of carbon-based materials for adsorption and catalysis; (ii) the synthesis and applications of MOFs; (iii) water purification by advanced oxidation processes, especially photocatalysis; (iv) the gas-phase hydrodechlorination of chlorinated volatile organic compounds; and (v) separation and catalytic processes with ionic liquids and supported ionic liquids. Dr. Bedia is a co-author of around 100 refereed journal publications (more than 3800 citations, H factor of 39), 3 book chapters and 2 Spanish patents. He has presented over 150 works to national and international conferences with over 20 oral presentations. He has been involved in more than 20 research projects from different entities—European, national, integrated or cooperative with other countries (Russia, Mexico, Germany, Peru, and the USA)—and has had several research contracts with private companies.

Carolina Belver

Carolina Belver is an Associate Professor in the Department of Chemical Engineering at Universidad Autonoma de Madrid (Spain). She received her B. Sc. in Chemistry from Universidad Autonoma de Madrid in 1997 and a Ph.D. in 2004 from University of Salamanca (Spain). She received competitive national postdoctoral contracts at the Catalysis and Petrochemistry Institute (Spain) (2005–2007) and Material Science Institute of Madrid (2008–2011). Dr. Belver's specialization and research deals with the design, processing, and evaluation of novel heterostructures for applications in environmental remediation, mainly focused on heterogeneous catalysis, photocatalysis and adsorption. She is the co-author of more than 85 referred publications and 13 chapter-books (more than 3800 citations; H factor of 36), and in 2013 she received a Fulbright Fellowship recognizing her accomplishments in her career as a researcher. Dr. Belver is Executive Editor of Chemical Engineering Journal, and a member of the Editorial Advisory Boards of the journals *Materials Science for Energy*, *Catalysts* and *Applied Surface Science*.

Preface to "New Carbon Materials from Biomass and Their Applications"

Carbon-based materials, such as chars, activated carbons, one-dimensional carbon nanotubes, and two-dimensional graphene nanosheets, have shown great potential for a wide variety of applications. These materials can be synthesized from any precursor with a high proportion of carbon in its composition. Although fossil fuels have been extensively used as precursors, their unstable cost and supply have led to the synthesis of carbon materials from biomass. Biomass covers all forms of organic materials, including plants both living and in waste form and animal waste products. It appears to be a renewable resource because it yields value-added products prepared using environmentally friendly processes. The applications of these biomass-derived carbon materials include electronic, electromagnetic, electrochemical, environmental and biomedical applications. Thus, novel carbon materials from biomass are a subject of intense research, with strong relevance to both science and technology. The main aim of this book is to present the most relevant and recent insights in the field of the synthesis of biomass-derived carbons for sustainable applications, including adsorption, catalysis and/or energy storage applications.

Jorge Bedia and Carolina Belver
Editors

Editorial

Special Issue on New Carbon Materials from Biomass and Their Applications

Jorge Bedia * and Carolina Belver *

Departamento de Ingeniería Química, Universidad Autónoma de Madrid, Cantoblanco, 28049 Madrid, Spain
* Correspondence: jorge.bedia@uam.es (J.B.); carolina.belver@uam.es (C.B.)

Citation: Bedia, J.; Belver, C. Special Issue on New Carbon Materials from Biomass and Their Applications. *Appl. Sci.* **2021**, *11*, 2453. https://doi.org/10.3390/app11062453

Received: 19 February 2021
Accepted: 4 March 2021
Published: 10 March 2021

Publisher's Note: MDPI stays neutral with regard to jurisdictional claims in published maps and institutional affiliations.

Copyright: © 2021 by the authors. Licensee MDPI, Basel, Switzerland. This article is an open access article distributed under the terms and conditions of the Creative Commons Attribution (CC BY) license (https://creativecommons.org/licenses/by/4.0/).

1. Introduction

Carbon-based materials, such as chars, activated carbons, one-dimensional carbon nanotubes, and two-dimensional graphene nanosheets, have shown great potential for a wide variety of applications. These materials can be synthesized from any precursor with a high proportion of carbon in its composition. Although fossil fuels have been extensively used as precursors, their unstable cost and supply have led to the synthesis of carbon materials from biomass [1]. More importantly, pollution and CO_2 emissions and their impact on climate change are real issues that should concern the scientific community now and lead to the replacement of fossil resources with more environmentally friendly sources. Biomass covers all forms of organic material, including plants, both living and in waste form, and animal waste products. To take full advantage of the renewable characteristics of biomass waste, it should be processed through sustainable processes following the green chemistry principles, such as low energy consumption, high atom efficiency, or use of less hazardous chemicals, among others. The applications of these biomass-derived carbon materials include electronic, electromagnetic, electrochemical, environmental, and biomedical applications. Thus, novel carbon materials from biomass are a subject of intense research, with strong relevance to both science and technology.

2. New Carbon Materials from Biomass and Their Applications

This special issue includes relevant works about the synthesis of carbon-based materials from biomass waste and their use in different applications, with special attention to biochar. In this sense, Ferrentino et al. [2] synthesized several hydrochars from municipal sewage sludge that showed high adsorption capacities of methylene blue dye. A subsequent KOH treatment enhanced adsorption capacity, which seems to be controlled by a complex result of various phenomena, including physi- and chemisorption and acid–base and redox equilibria. The adsorption of this same dye was analyzed by Li et al. [3], who used *Artemisia selengensis* to produce biochar by microwave pyrolysis. The results indicated that the increase of the pyrolysis temperature results in a decrease of the yield but an increase of the dye adsorption. The use of biochars as adsorbents for water purification was also analyzed by Siipola et al. [4]. In this case, the biochars from pine and spruce bark were subsequently activated with steam at 800 °C and studied in the removal of phenol, and microplastics retention and cation exchange capacity were employed as key test parameters. The work concluded that ultra-high porosities are not necessary for satisfactory water purification, supporting the economic feasibility of bio-based adsorbent production. Similarly, Diaz et al. [5] reported the preparation of low-cost activated grape seed-derived hydrochar through hydrothermal carbonization and chemical activation with different chemical activating agents (KOH, $FeCl_3$, and H_3PO_4) for sulfamethoxazole adsorption. The hydrochars showed low porosity, however their activation with KOH resulted in highly porous activated carbons (2200 $m^2 \cdot g^{-1}$). The adsorption capacity was determined by the porous texture, achieving a high saturation capacity of 650 $mg \cdot g^{-1}$. Tuomikoski et al. [6] detailed the zinc adsorption using activated carbon prepared from sawdust waste.

Carbonization and activation were performed in a single stage using steam as a physical activation agent at 800 °C. The adsorption capacity towards zinc was tested and compared favorably to those of the materials reported in the literature. Not only liquid-phase but also gas-phase adsorption applications are included in this special issue. Manyà et al. [7] analyzed the dynamic CO_2 capture on physically activated biochars. Those were obtained from vine shoots and wheat straw pellets through an initial slow pyrolysis and further activation with CO_2 up to different degrees of burn-off. The adsorbent prepared from the vine shoots-derived biochar with the most hierarchical pore size distribution exhibited a good and stable performance under dry conditions. However, the presence of relatively high concentrations of water vapor in the feeding gas clearly interfered with the CO_2 adsorption mechanism, leading to significantly shorter breakthrough times.

Besides biochar, many of the publications collected in this special issue were devoted to soil enrichment. Yu et al. [8] reported the use of mixed hardwood and sugarcane bagasse biochar as potting mix components for container tomato and basil seedling production. The study concluded that 70% mixed hardwood biochar could be amended with peat moss for tomato and basil seedling production without negative effects on plant biomass. Similarly, Saletnik et al. [9] investigated the stimulation and conditioning of seeds with biochar and the effects observed in the germination and emergence of Virginia mallow seedlings. The biochars, applied as conditioner added to water in the process of seed hydration, improved the germination capacity. The beneficial effects of biochar application were also reflected in the increased mass of Virginia mallow seedlings. Shin el al. [10] used nutrient uploaded biochar pellets as slow release fertilizers. It was observed that the cumulative ammonium nitrogen in the blended biochar pellets was slow released over the 77 days of precipitation period, but nitrite nitrogen was rapidly released, i.e., within 15 days of precipitation. Accumulated phosphate phosphorus concentrations were not much different, and slowly released until the final precipitation period. These findings indicated that blended biochar pellets can be used as slow-release fertilizers for agricultural practices. The effect of biochar application depth on crop productivity under tropical rainfed conditions was researched by Moiwo et al. [11]. The study determined the effect of biochar application depth on the productivity of NERICA-4 upland rice cultivars under tropical rainfed conditions, concluding that the biochar can enhance crop productivity. Finally, Huang et al. [12] evaluated the use of biochar and compost mixes as substitutes to a commercial propagation mix. High percentages (70% or 80%, by volume) of biochars with vermicompost or chicken manure compost were evaluated to substitute a commercial propagation mix. The combination of biochars with vermicompost improves the results, however the mixture with chicken manure compost was not recommended.

The use of biomass waste as a potential environmentally friendly energy source is also a relevant research field. In this sense, Mousa et al. [13] studied the use of biocarbon briquettes for foundry industry. The foundry industry, like many others, is currently facing challenges to reduce the environmental impacts from application of fossil fuels. Replacing foundry coke with alternative renewable carbon sources can lead to significant decreases in fossil fuel consumption and fossil CO_2 emission. The work was aimed at the design, optimization, and development of briquettes containing biocarbon for an efficient use in cupola furnace.

Biomass-derived carbons are also being extensively analyzed as part of catalysts and/or catalyst's support in different catalytic reactions. Lu et al. [14] synthesized solid acid catalysts by one-step carbonization and sulfonation of *Salix psammophila* in the presence of concentrated sulfuric acid. The catalysts were used in the esterification reaction between oleic acid and methanol to prepare the biodiesel, achieving a conversion of 94.15% in the optimized conditions. Ma et al. [15] reports the process simulation and economic evaluation of a two-stage bio-oil hydrogenation process. An Aspen Plus process simulation model was developed for the two-stage bio-oil hydrogenation demonstration plant, which was used to evaluate the effect of catalyst coking on the bio-oil upgrading process and the economic performance of the process. The model was also used to investigate the effect of catalyst deactivation caused by coke deposition in the mild stage.

Finally, we would like to mention two review contributions. The first one by Yang et al. [16] reviewed the preparation and modification of biochar materials and their application in soil remediation, which was extensively treated in this special issue. The wide application of biochar is due to its abilities to remove pollutants, remediate contaminated soil, and reduce greenhouse gas emissions. The study analyzed the influence of preparation methods, process parameters, and modification methods on the physicochemical properties of biochar, as well as the mechanisms of biochar in the remediation of soil pollution. The biochar applications in soil remediation in the past years were summarized, such as the removal of heavy metals and persistent organic pollutants (POPs), and the improvement of soil quality. The review also details the potential risks of biochar applications and the future research directions. The second review by Saletnik et al. [17] reported the use of biochar for environmental applications. The article reviews the information related to the broad uses of carbonization products. It also discusses the legal aspects and quality standards applicable to these materials, with special attention to the lack of uniform legal and quality conditions, which would allow for much better use of biochar. The review also aims to highlight the high potential for use of biochar in different environments. The presented text attempts to emphasize the importance of biochar as an alternative to classic products used for energy, for environmental and agricultural purposes.

Funding: This research received no external funding.

Institutional Review Board Statement: Not applicable.

Informed Consent Statement: Not applicable.

Data Availability Statement: Not applicable.

Acknowledgments: We would like to thank the authors, reviewers, and the editorial team of Applied Sciences for the work devoted to this special issue. We would like also to thank Tamia Qing for all her help advising and invaluable work during all the process.

Conflicts of Interest: The authors declare no conflict of interest.

References

1. Bedia, J.; Peñas-Garzón, M.; Gómez-Avilés, A.; Rodriguez, J.; Belver, C. A Review on the Synthesis and Characterization of Biomass-Derived Carbons for Adsorption of Emerging Contaminants from Water. *C* **2018**, *4*, 63. [CrossRef]
2. Ferrentino, R.; Ceccato, R.; Marchetti, V.; Andreottola, G.; Fiori, L. Sewage Sludge Hydrochar: An Option for Removal of Methylene Blue from Wastewater. *Appl. Sci.* **2020**, *10*, 3445. [CrossRef]
3. Li, X.; Li, K.; Geng, C.; El Mashad, H.; Li, H.; Yin, W. Biochar from Microwave Pyrolysis of Artemisia Slengensis: Characterization and Methylene Blue Adsorption Capacity. *Appl. Sci.* **2019**, *9*, 1813. [CrossRef]
4. Siipola, V.; Pflugmacher, S.; Romar, H.; Wendling, L.; Koukkari, P. Low-Cost Biochar Adsorbents for Water Purification Including Microplastics Removal. *Appl. Sci.* **2020**, *10*, 788. [CrossRef]
5. Diaz, E.; Manzano, F.J.; Villamil, J.; Rodriguez, J.J.; Mohedano, A.F. Low-Cost Activated Grape Seed-Derived Hydrochar through Hydrothermal Carbonization and Chemical Activation for Sulfamethoxazole Adsorption. *Appl. Sci.* **2019**, *9*, 5127. [CrossRef]
6. Tuomikoski, S.; Kupila, R.; Romar, H.; Bergna, D.; Kangas, T.; Runtti, H.; Lassi, U. Lassi Zinc Adsorption by Activated Carbon Prepared from Lignocellulosic Waste Biomass. *Appl. Sci.* **2019**, *9*, 4583. [CrossRef]
7. Manyà, J.J.; García-Morcate, D.; González, B. Adsorption Performance of Physically Activated Biochars for Postcombustion CO_2 Capture from Dry and Humid Flue Gas. *Appl. Sci.* **2020**, *10*, 376. [CrossRef]
8. Yu, P.; Li, Q.; Huang, L.; Niu, G.; Gu, M. Mixed Hardwood and Sugarcane Bagasse Biochar as Potting Mix Components for Container Tomato and Basil Seedling Production. *Appl. Sci.* **2019**, *9*, 4713. [CrossRef]
9. Saletnik, B.; Bajcar, M.; Zaguła, G.; Saletnik, A.; Tarapatskyy, M.; Puchalski, C. Biochar as a Stimulator for Germination Capacity in Seeds of Virginia Mallow (*Sida hermaphrodita* (L.) Rusby). *Appl. Sci.* **2019**, *9*, 3213. [CrossRef]
10. Shin, J.; Park, S.; Lee, S. Optimum Method Uploaded Nutrient Solution for Blended Biochar Pellet with Application of Nutrient Releasing Model as Slow Release Fertilizer. *Appl. Sci.* **2019**, *9*, 1899. [CrossRef]
11. Moiwo, J.P.; Wahab, A.; Kangoma, E.; Blango, M.M.; Ngegba, M.P.; Suluku, R. Effect of Biochar Application Depth on Crop Productivity Under Tropical Rainfed Conditions. *Appl. Sci.* **2019**, *9*, 2602. [CrossRef]
12. Huang, L.; Yu, P.; Gu, M. Evaluation of Biochar and Compost Mixes as Substitutes to a Commercial Propagation Mix. *Appl. Sci.* **2019**, *9*, 4394. [CrossRef]

13. Mousa, E.; Kazemi, M.; Larsson, M.; Karlsson, G.; Persson, E. Potential for Developing Biocarbon Briquettes for Foundry Industry. *Appl. Sci.* **2019**, *9*, 5288. [CrossRef]
14. Lu, P.; Wang, K.; Gong, J. Optimization of Salix Carbonation Solid Acid Catalysts for One-Step Synthesis by Response Surface Method. *Appl. Sci.* **2019**, *9*, 1518. [CrossRef]
15. Ma, X.; Pang, S.; Zhang, R.; Xu, Q. Process Simulation and Economic Evaluation of Bio-Oil Two-Stage Hydrogenation Production. *Appl. Sci.* **2019**, *9*, 693. [CrossRef]
16. Yang, X.; Zhang, S.; Ju, M.; Liu, L. Preparation and Modification of Biochar Materials and their Application in Soil Remediation. *Appl. Sci.* **2019**, *9*, 1365. [CrossRef]
17. Saletnik, B.; Zaguła, G.; Bajcar, M.; Tarapatskyy, M.; Bobula, G.; Puchalski, C. Biochar as a Multifunctional Component of the Environment—A Review. *Appl. Sci.* **2019**, *9*, 1139. [CrossRef]

Article

Sewage Sludge Hydrochar: An Option for Removal of Methylene Blue from Wastewater

Roberta Ferrentino [1], Riccardo Ceccato [2], Valentina Marchetti [1], Gianni Andreottola [1] and Luca Fiori [1,*]

[1] Department of Civil, Environmental and Mechanical Engineering, University of Trento, via Mesiano 77, 38123 Trento, Italy; roberta.ferrentino@unitn.it (R.F.); marchetti.vale@libero.it (V.M.); gianni.andreottola@unitn.it (G.A.)
[2] Department of Industrial Engineering, University of Trento, via Sommarive 9, 38123 Trento, Italy; riccardo.ceccato@unitn.it
* Correspondence: luca.fiori@unitn.it; Tel.: +39-0461282692

Received: 20 April 2020; Accepted: 12 May 2020; Published: 16 May 2020

Featured Application: Producing a material—a product—from sewage sludge—a waste—resulting from wastewater treatment plants (WWTPs) is a possible sustainable solution to reduce the amount of sewage sludge to be disposed of. Thus, sewage sludge hydrochar (simply produced as detailed in the paper) having good adsorption capabilities could be used, for instance, in municipal and industrial WWTPs for water remediation.

Abstract: Municipal sewage sludge was subjected to a hydrothermal carbonization (HTC) process for developing a hydrochar with high adsorption capacity for water remediation in terms of dye removal. Three hydrochars were produced from municipal sewage sludge by performing HTC at 190, 220 and 250 °C, with a 3 h reaction time. Moreover, a portion of each hydrochar was subjected to a post-treatment with KOH in order to increase the adsorption capacity. Physicochemical properties of sludge samples, raw hydrochars and KOH-modified hydrochars were measured and batch adsorption studies were performed using methylene blue (MB) as a reference dye. Data revealed that both raw and modified hydrochars reached good MB removal efficiency for solutions with low MB concentrations; on the contrary, MB in high concentration solutions was efficiently removed only by modified hydrochars. Interestingly, the KOH treatment greatly improved the MB adsorption rate; the modified hydrochars were capable of capturing above 95% of the initial MB amount in less than 15 min. The physicochemical characterization indicates that alkali modification caused a change in the hydrochar surface making it more chemically homogeneous, which is particularly evident for the 250 °C hydrochar. Thus, the adsorption process can be regarded as a complex result of various phenomena, including physi- and chemi-sorption, acid–base and redox equilibria.

Keywords: hydrothermal carbonization; HTC; sewage sludge; hydrochar; methylene blue; adsorption; water remediation; value-added product; waste-to-products

---

1. Introduction

Sewage sludge is an unavoidable waste of municipal wastewater treatment activity. The production of sludge in municipal wastewater treatment plants (WWTPs) has increased due to more stringent legislation and a growing number of new plants, becoming a critical issue [1]. Several methods can be adopted for sludge management such as landfill disposal, incineration and, where possible and allowed by the legislation, utilization in agriculture. However, each of these options has important limitations [2]. This has prompted the search for more cost effective and environmentally sustainable technologies able to promote the innovative and beneficial use of sewage sludge. Thus, the conversion of this class of waste products, as well as of organic wastes in general, into value-added products such as energy carriers, biomaterials, bioplastics and fertilizers is on the rise. A very appealing technology is hydrothermal carbonization (HTC) which, in recent years, has been proposed as a promising process that allows obtaining a solid carbonaceous material with different possible utilizations. This process is attractive due to its simplicity, low-cost and its energy and CO_2 containment efficiency. One of the major advantages of HTC over other technologies is its capability to convert wet biomass into solid products without the need for energy-intensive drying before and/or during the process [3], as HTC uses water as the reaction medium. High water content raw substrates such as animal manures, the organic fraction of municipal solid waste, sewage sludge, aqua culture and algal residues could be used as feedstock for the HTC process [4].

The HTC process is performed applying mild temperatures (180–260 °C) under saturated water vapor pressure for several hours [5,6]. Thus, the feedstock is subjected to a thermochemical process that includes simultaneous and sequential reactions of hydrolysis, dehydration, decarboxylation, condensation, polymerization and aromatization. However, the detailed reaction mechanisms are as yet unknown due to the complexity of the residual biomasses used as feedstock and the coexistence of several reactions in series and in parallel [7]. Reaction mechanisms involving complex biomasses can be investigated using a lumped components approach [8]. The resulting carbon rich solid product, referred to as hydrochar, is presently evaluated for use in a wide range of applications due to its properties and the diversity of materials used as feedstock. Proposed uses of hydrochar include its utilization as adsorbent material, carbon based smart material, energy vector and soil amendment [9]. Consequently, hydrochar is regarded as a valuable material for various industrial, environmental and agricultural applications. Regarding the use as an adsorbent material, hydrochar usually exhibits increased functional groups on its surface when compared with the raw biomass, which gives the hydrochar a high hydrophobicity, chemical affinity and potentialities towards adsorption applications [10,11]. For instance, Hammud et al. [12] performing HTC at 225 °C converted pine needles into hydrochar to remove malachite green dye from water and reached an adsorption capacity of 52.91 mg g^{-1}. Similarly, Wei et al. [13] used municipal sewage sludge as raw material to prepare an adsorbent by HTC (180 °C) and investigated its adsorption capacity for crystal violet. Results showed that the prepared hydrochar had a relatively high specific surface area, well-developed porosity and abundant surface organic functional groups that were beneficial for contaminant removal. Moreover, to enhance the adsorption capability of hydrochar, modification turned out to be an effective method [14]. By "modification" we mean a treatment in which the feedstock (hydrochar in this case) is treated with a chemical agent (e.g., KOH) via impregnation, which is not followed by a heat treatment at 500–850 °C in nitrogen typical of chemical activation [11,15–17]. For instance, Regmi et al. [18] tested the sorption capacities of KOH-modified hydrochar from switchgrass for removing copper and cadmium from aqueous solutions: they found a removal of about 100% within 24 h of contact time while the raw hydrochar only removed 16% of copper and 5.6% of cadmium. Similarly, Sun et al. [14] showed that alkali modification improved the sorption ability of hydrochars produced by HTC of sawdust, wheat straw and corn stalk. Moreover, Spataru et al. [19] tested a low-cost adsorbent derived from HTC of waste activated sludge after KOH treatment at room temperature for the removal of orthophosphate from the effluent of a municipal WWTP. The study demonstrated that modified hydrochar achieved more than 97% orthophosphate removal. Thus, literature studies demonstrate that the alkali modification process may positively affect hydrochar surface, thus improving its sorption capacity [14].

To date, no study has examined the adsorption capacity of municipal sewage sludge-derived hydrochar [13] compared to alkali modified hydrochar and, in addition, considering different HTC treatment temperatures. Thus, the aim of the present study is to investigate the adsorption capacity of sewage sludge hydrochar (raw or KOH-modified) obtained at different HTC operating conditions for dye removal from aqueous solutions using methylene blue (MB) as the adsorbate.

Importantly, these materials present some elements of complexity in comparison with other classes of organic-derived adsorbents: (i) the lower treatment temperatures give rise to products displaying lower specific surface area values than activated carbon-based adsorbents, usually ranging from 800 to above 1100 m^2·g^{-1}; (ii) the presence in sewage sludges of alkaline and heavy metals cannot be neglected.

Thus, the feasibility of this application must take into account these two aspects: (1) the adsorption mechanism has to involve not only the surface morphology of the samples, but also some chemical reactions have to occur; (2) the presence of metal ions can provide acid–base or oxy-reductive processes with respect to the adsorbate: actually, MB is a positive-charged dye and it also displays reducing behavior in solutions (it can be used as an indicator in redox titrations).

With all these features in mind, the aim of this work is reached by performing the following main steps:

- preparation of three hydrochars from municipal sewage sludge by performing HTC at 190, 220 and 250 °C with a reaction time of 3 h;
- modification with KOH in order to increase the adsorption capacity of hydrochars;
- comparison of the physicochemical properties of raw and modified hydrochars using elemental analysis, thermogravimetry (TGA), nitrogen physisorption analysis, Fourier-transform infrared spectroscopy (FTIR), inductively coupled plasma spectroscopy (ICP) and flow injection mercury system (FIMS);
- comparison of the potential application of raw and modified hydrochars as adsorbents for MB removal by performing batch adsorption studies, namely adsorption isotherms and adsorption kinetics tests.

Even if the HCT of sewage sludge has been previously addressed by several authors, as far as we know, there are no studies performing such a detailed evaluation considering sewage sludge as feedstock for the HTC process and the derived hydrochar, possibly KOH-modified, as adsorbent material for MB removal. Therefore, the novelty of this study is on the investigation of the adsorption capacity of the derived hydrochars, the main adsorption mechanisms involved and the assessment of the best production conditions that enhance the adsorption capacity of sewage sludge hydrochars.

2. Materials and Methods

2.1. Materials

The sewage sludge, used as the feedstock, was a mixture of diluted sludge exiting the anaerobic digester (digestate), dry matter content 2.7%, and the same digestate downstream of the addition of polyelectrolyte and passage in centrifuge (palatable sludge), dry matter content 22%. The mixture was prepared in order to feed the HTC reactor with a stream sufficiently rich in dry matter content and, at the same time, where the biomass was completely submerged into a liquid phase: the resulting mixture had a dry matter content of 12%. The sludge mixture dry matter content was chosen to simulate the implementation in full-scale applications [20]. As a matter of fact, a sludge with a dry matter content equal to 12% can be easily pumped to the HTC reactor. Both digestate and palatable sludge were collected from the local WWTP of Trento North, Italy.

The preparation of MB solutions at different concentrations was conducted by adding the required amount of MB powder to deionized water. The activated carbon AquaSorb™ BP2, with a reported specific surface area of 900 m^2·g^{-1}, was also used to test the MB adsorption capacity so to compare the results obtained with those of the sewage sludge hydrochars and modified hydrochars.

2.2. Preparation and Modification of Hydrochar

HTC of sewage sludge was performed in a stainless steel AISI 316 batch reactor of 2 L internal volume. The reactor, built in house, was designed for temperatures and pressures, respectively, up to

300 °C and 140 bar. The reactor top flange is connected to two pipes that allow nitrogen gas purging in order to ensure not oxidizing conditions. The reactor is equipped with a pressure transmitter, two pressure gauges and four thermocouples positioned at different heights within it. The pressure and temperature transmitters send data to software that allows the control of the reactor temperature and the monitoring of the four temperatures and the pressure. At the end of the HTC run, the reactor is cooled down to room temperature and then depressurized. Further details about the HTC reactor could be found in Merzari et al. [21].

1.7 kg of sewage sludge (0.8 kg of palatable sludge and 0.9 kg of digestate) were introduced into the reactor for each HTC run, in order to have a feedstock with a dry matter content equal to 12%. HTC tests were performed in duplicate at three different temperatures: 190, 220 and 250 °C, while the reaction time was set at 3 h for all the tests. At the end of each HTC run, the solid-phase material was separated from the liquid phase by vacuum filtration and later dried at 105 °C for 24 h to remove residual moisture. The obtained dry samples were designated as 190HC, 220HC and 250HC on the basis of their treatment temperature.

To produce the KOH-modified hydrochar, 5 g of hydrochar powder were mixed with 500 mL of a 2M KOH solution and then stirred for 1 h at room temperature. The modified hydrochars were separated from the liquid phase by vacuum filtration, then washed with deionized water and finally dried for 24 h at 105 °C. The obtained dry samples were labeled as 190MHC, 220MHC and 250MHC, on the basis of the treatment conditions as for the raw hydrochars, where M refers to the alkali modification process.

The hydrochar mass yield is here defined, as is the usual case [22], as the percentage ratio between the mass (on a dry basis, d.b.) of the solid (hydrochar) remaining after HTC and the mass (on a d.b.) of the raw sample before thermal treatment. Similarly, a mass yield is defined also for the KOH modification as the mass (on a d.b.) remaining after KOH treatment and the mass (on a d.b.) of the raw hydrochar before such treatment.

2.3. Materials Characterization

Elemental composition (carbon, C; hydrogen, H; and nitrogen, N) of the samples was assessed by using a LECO 628 analyzer (LECO, Moenchengladbach, Germany). Each analysis was performed in duplicate using about 0.1 g of sample per trial. Considering the ash value resulting from proximate analysis, the oxygen (O) content was calculated based on the mass difference [23].

In order to evaluate moisture (M), volatile matter (VM) and ash (A) contents of the samples, thermogravimetric analysis (TGA) was carried out by means of a Mettler Q5000 V3 (Columbus, OH, USA) thermobalance [24] and using about 30 mg of dried sample per trial. The thermal program, chosen from literature [25], can be considered as a modification of the ASTM reference methods E871, E872 and E830. To release residual moisture content, the sample was heated under air to 105 °C, temperature which was held for 30 min before heating at 16 °C/min from 105 to 900 °C (hold time: 7 min) in nitrogen to determine the volatile matter content. Ash was determined by switching to air while holding the temperature of 900 °C for an additional 30 min.

The surface morphological structure of the hydrochars was examined using a scanning electron microscopy (SEM) system (JEOL JSM-7001F Field Emission SEM, JEOL, Tokyo, Japan). The functional groups of the samples were examined by using Fourier transform infrared spectroscopy (FT-IR, Avatar 330, Nicolet, Waltham, MA, USA). Wavenumbers between 4000 and 400 cm^{-1} were covered using 64 scans in the investigation range with a resolution of 4 cm^{-1}. Moreover, the functional groups were examined using also FTIR-ATR (attenuated total reflectance) spectroscopy using a Spectrum One (Perkin Elmer, Boston, MA, USA) by averaging 16 scans with a resolution of 4 cm^{-1} in the wavenumber range between 4000 and 650 cm^{-1}, employing a ZnSe crystal. The load applied to squeeze the powdered samples towards the diamond was 130 ± 1 N.

The surface morphology and pore structure characteristics of the samples were determined by N_2 physisorption measurements performed at 77 K using a surface area porosimeter (ASAP 2010,

Micromeritics, Norcross, GA, USA). All the samples were degassed below 1.3 Pa at 25 °C prior to the measurement.

The specific surface area was calculated following multipoint N_2-Brunauer–Emmett–Teller (BET) adsorption method, in the interval $0.05 \leq (P/P_o) \leq 0.33$. Pore size distribution curves were determined using the Brunauer–Joyner–Halenda (BJH) method applied both on the adsorption and the desorption branches of the isotherms.

The chemical speciation analysis of alkali and heavy metals and other elements undetectable by ultimate analysis was carried out in duplicate on raw sludges, hydrochars and modified hydrochars. Heavy metals concentrations were determined by using an optical ICP system (Perkin Elmer 7300 DV) and following UNI EN 13,657 and UNI EN ISO 11885:2009 methods. Mercury concentrations were determined by using a flow injection mercury system (FIMS) (Perkin Elmer FIMS 100). Total nitrogen concentrations were determined following CNR-IRSA Method [26].

2.4. Batch Adsorption Study

Adsorption isotherms of MB onto both raw and KOH-modified hydrochars were determined by adding 10–16 mg of absorbent to 25 mL stoppered glass bottles containing 4 mL of MB solution with various initial concentrations ranging between 10 and 300 mg L^{-1}. The bottles were agitated with a magnetic stirrer at 250 rpm and were kept at 20 ± 2 °C for 48 h. The pH of the MB solutions was evaluated using a portable pH meter (pH 3110 ProfiLine with SenTix41 probe, WTW, Milan, Italy). Then, the obtained solutions were centrifuged at 3000 rpm for 5 min and the MB concentrations were measured from the calibration curve of MB solutions at 665 nm using a UV-visible spectrophotometer (Model V-3250, Jasco Europe, Lecco, Italy). The adsorption capacity at equilibrium q_e (mg g^{-1}) and the percentage removal of MB were evaluated by Equations (1) and (2), respectively [27,28]:

$$q_e = (C_o - C_e) V/W \tag{1}$$

$$\% \text{ removal} = ((C_o - C_e)/C_0) \, 100\% \tag{2}$$

where C_e and C_0 are, respectively, the equilibrium and initial MB concentration (mg L^{-1}), V is the solution volume (L) and W is the adsorbent mass used (g).

To investigate the mechanisms of adsorption, three well-known adsorption isotherms, Langmuir, Freundlich and Tempkin-type curves, were adopted. The Langmuir sorption isotherm is applied to equilibrium sorption assuming monolayer sorption on a surface with a finite number of identical sites. The linear Langmuir equation is expressed as Equation (3) [29]:

$$q_e = (K_L \, q_m \, C_e)/(1 + K_L \, C_e) \tag{3}$$

where K_L is the Langmuir constant (L mg^{-1}) related to the affinity of binding sites and the free energy of sorption, and q_m is the maximum adsorption capacity when a monolayer forms on the hydrochar (mg g^{-1}).

The Freundlich equation describes heterogeneous surface energy systems and is expressed as Equation (4) [30]:

$$q_e = K_F \, C_e^{(1/n)} \tag{4}$$

where K_F and $1/n$ are the Freundlich constants, determined from the plot of q_e versus C_e. The parameters K_F and $1/n$ are related to the sorption capacity and the sorption intensity of the system. The magnitude of the term $1/n$ gives an indication of the affinity of the sorbent/adsorbate systems [31].

The Tempkin equation is given by the following Equation (5) [32]:

$$q_e = B \ln (K \, C_e) \tag{5}$$

where B = (R T)/b, T is the absolute temperature in Kelvin, R is the universal gas constant (8.314 J mol^{-1} K^{-1}), b and K are Tempkin constants related to the heat of sorption (J mol^{-1}) and to the equilibrium binding (L g^{-1}), respectively.

The kinetics of MB adsorption on the raw and KOH-modified hydrochars were examined by adding 90 mg of each hydrochar to 150 mL glass bottles containing 50 mL of MB solution with a concentration of 100 mg L^{-1}. The glass bottles were then shaken at 250 rpm under a magnetic stirrer for 24 h. At different time intervals, the samples were centrifuged at 3000 rpm for 5 min and the residual MB concentration was measured by the UV-visible spectrophotometer (Model V-3250, Jasco Europe, Lecco, Italy) through the calibration curve.

The kinetics of the MB adsorption process was evaluated using pseudo-first order and pseudo-second-order kinetic models. The kinetic equations for these models are expressed by Equations (6) [33] and (7) [34], respectively:

$$\ln(q_e - q_t) = \ln q_e - k_1 t \quad (6)$$

$$t/q_t = t/q_e + 1/(k_2 q_e^2) \quad (7)$$

where q_e (mg g^{-1}) and q_t (mg g^{-1}) denote the MB sorption at equilibrium and time t (min), respectively, and k_1 (min^{-1}) and k_2 (g mg^{-1} min^{-1}) represent, respectively, the pseudo-first and pseudo-second order adsorption rate constants.

3. Results and Discussion

3.1. Characterization

Mass yields, elemental and proximate compositions and specific surface area of the various samples are reported in Table 1. Results reveal that the mass yield and the C content of the raw hydrochars are consistent with literature studies [35]. There was a significant decrease in mass yield at increasing HTC reaction temperature: when the reaction temperature rose from 190 to 250 °C, the hydrochar yield dropped down from about 83% to 63% and the C percentage increased from 26.9% to 36.3%.

Table 1. Mass yields, main physicochemical characteristics and elemental and proximate compositions of all the samples.

Samples	Mass Yield [%]	Elemental Composition (wt%, db)				Ash [%]	VM [% db]	FC [% db]	Atomic Ratio		Surface Area [m^2 g^{-1}]
		C [%]	H [%]	N [%]	O^1 [%]				H/C	O/C	
Digestate	-	25.6	4.0	3.6	21.9	45.0	50.2	4.8	1.88	0.64	-
Palatable	-	35.9	5.4	5.8	24.4	28.4	65.6	5.9	1.81	0.51	-
Mixture	-	34.4^2	5.2^2	5.5^2	24.1	30.8	63.4	5.7	1.81	0.53	-
190HC	83.3	26.9	5.1	3.0	20.5	44.5	55.2	0.3	2.28	0.57	31.00
220HC	76.3	28.2	4.0	1.8	18.5	47.5	52.4	0.1	1.70	0.49	8.82
250HC	62.9	36.3	5.0	5.0	8.7	45.0	54.9	0.1	1.65	0.17	11.85
190MHC	51.5	29.4	4.1	1.9	15.8	48.8	51.1	0.1	1.67	0.40	0.29
220MHC	71.2	33.6	4.4	3.0	9.5	49.5	50.0	0.5	1.57	0.20	2.74
250MHC	84.7	30.6	4.0	2.1	11.2	52.1	45.3	2.6	1.57	0.27	13.36

^{1} oxygen content is estimated as follows: O = 100 − (C+H+N+ash). ^{2} calculated as the average of the values of digestate and palatable (weighted average considering the relevant dry matter content) the absolute errors for mass yield are ≤ ±2.4, the absolute errors for C, H, N are, respectively, ≤ ±1.0, ±0.2, ±0.3.

The KOH treatment reduced the mass of the hydrochar with a clear trend: the higher the temperature at which the hydrochar was produced, the higher the KOH treatment yield. This behavior is in agreement with previous results obtained coupling HTC with classical KOH chemical activation [11] and testifies the higher stability of hydrochars obtained at higher temperatures.

Considering the elemental composition (H, C, N, O), the O percentage is maximal for the three raw sludges and decreases with the HTC reaction temperature, as expected. Interestingly, the KOH treatment significantly reduced the O content in the samples obtained at 190 and 220 °C. Once more, this testifies the higher stability of hydrochars obtained at higher temperatures. The reduction in O can be due to the loss of volatile matters comprising some surface functional groups containing O [35].

Conversely, results related to the content of N and H do not show a clear trend in both raw and modified hydrochars.

The carbon content data are particularly interesting and testify to the peculiarity of this type of substrate, which behaves differently than the more standard biomass (e.g., agro-industrial biowaste, ligno-cellulosic biomass, organic fraction of municipal solid waste). If for standard biomasses the HTC leads to an increase in the C percentage in the solid, here, on the contrary, the C percentage in some cases even decreased, passing for example from the value of 34.4% in the sludge mixture to the value of 26.9% in the hydrochar obtained at 190 °C. This atypical trend is explained by the remarkable ash content that characterizes the sewage sludge. HTC actually concentrates the ashes which, for the two samples mentioned above, passed from the value of 30.8% to the value of 44.5%. For sewage sludge, the C concentration effect is observable if the percentage values are considered on a dry ash free (daf) basis, less observable or not observable at all if the percentage values are considered on a dry basis. Thus, on a daf basis, the sludge mixture and the hydrochar obtained at 190 °C had a comparable carbon content (about 49%), and the carbon content rose for all the other hydrochars, both raw and KOH-modified, and it was in the range of 54%–67%. Therefore, for these types of substrate, the carbon enrichment takes place at the net of their ash content: the concentration effect of minerals masks the carbon concentration effect which is typical of HTC.

Regarding proximate analysis data, the ash content shows a significant increase after HTC, as expected. Moreover, a further increase is observed for the alkali-modified samples: among all the samples, the highest ash content (about 52%) was measured for the 250MHC.

The volatile matter of the sludge mixture reduced from about 63% to 52%–55% with the carbonization treatment, justifying the corresponding increase in mineral ashes [36]. Results show that there is no significant difference in the VM content of the three raw hydrochars. Furthermore, the KOH treatment reduced even more the VM content of the hydrochars, which resulted in a range from about 51% (190MHC) to 45% (250MHC). Accordingly, the ash content after the KOH treatment increased up to 52% (250MHC). However, the decrease in VM was not accompanied by an increase in fixed carbon content. This indicates that VM was also converted into other products, such as CO_2 or liquids [37,38]. Despite the reported data of FC content, which honestly appears quite controversial (actually, in the general case HTC increases FC, which is not the case here), the same thermogravimetric analysis testifies that both raw hydrochar and modified hydrochar samples show an increased thermal stability if compared to raw sludge samples, as expected. This is clear when considering the derivative thermogravimetric (DTG) curves in N_2: Figure A1 in Appendix A. The temperature at which the decomposition rate was maximal was around 280 °C for digestate, palatable and mixture sludge, while it was in the range 450–480 °C for raw and modified hydrochars.

Considering H/C and O/C atomic ratios, general trends appear even if some data are out of trend, likely due to the heterogeneity of this kind of substrates. H/C and O/C atomic ratios decreased after HTC, and they decreased even more after KOH post-treatment. The higher the temperature of the HTC process, the lower are the H/C and O/C ratios. Thus, increasing the reaction severity resulted in a decrease of H/C and O/C ratios.

The BET specific surface area values of hydrochars were 31.00, 8.82 and 11.85 $m^2\ g^{-1}$ for 190HC, 220HC and 250HC, respectively. Thus, sample 190HC had the highest surface area value. However, after treatment with KOH, samples 190 MHC and 220MHC showed very low BET surface areas in comparison with raw hydrochars, accounting for 0.29 and 2.74 $m^2\ g^{-1}$ respectively. Similar results and trends were observed also for cold alkali modification of other hydrochars as reported in previous studies [10,14,28]. A low apparent BET surface area can be related to pore blockage due to organic compounds that were not transferred to the liquid phase during KOH treatment [10]. On the contrary, the BET surface area of sample 250MHC slightly increased after the treatment with KOH reaching the value of 13.36 $m^2\ g^{-1}$.

Figure 1 shows adsorption/desorption branches for nitrogen isotherms at 77 K of raw (Figure 1a) and modified (Figure 1b) hydrochars. The desorption branch of sample 190MHC is not reported because it is superimposed onto the adsorption one, due to the low BET surface area value. All hydrochars present gas physisorption isotherms of type II b according to the IUPAC classification [39], typical of materials with slit macro-pores with an average diameter of around 100 nm and attributable to the formation of aggregates of discoidal particles. Macro-pores were present in small amount in all samples, thus confirming the low values of BET surface area observed. Adsorption/desorption isotherms are very close, indicating a narrow and homogenous pore distribution. Moreover, the presence of the hysteresis loop confirms the prevalent presence of macro-pores, except for the 190MHC sample. Furthermore, the shape of the hysteresis loop is of type H3, located at the high relative pressure range, displaying two steep, parallel branches, without a plateau. This shape indicates the presence of flat particles and slit pores. After the KOH treatment, hysteresis loops are wider, especially for the sample 220MHC, due to an amplification of the pore distribution curve. Moreover, in the case of sample 220MHC the shape of the hysteresis loop can be considered as intermediate between type H3 and H4, indicating a higher reduction in the number of pores compared to the other modified hydrochars. As expected, specific surface area values are much lower in comparison with reported ones for activated carbon materials, obtained from different organic sources [40,41]. However, the presence of various functional groups on the sample surfaces, the alkaline treatment and the presence of metal ions could modify in a more efficient way the interactions between adsorbent and MB adsorbate, as already reported for low-temperature treated organic-based adsorbents [42]. SEM micrographs (Figure A2 in Appendix A) show a smooth surface without pores in the micrometer size range for all the samples investigated.

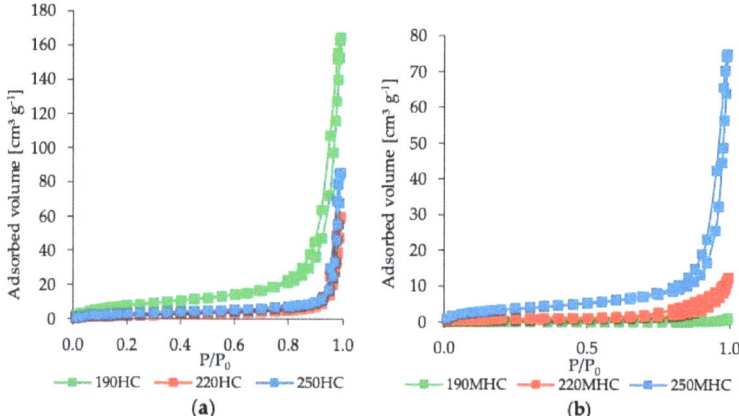

Figure 1. Nitrogen adsorption/desorption isotherms at 77 K of (**a**) raw and (**b**) modified hydrochars.

The FTIR analysis results of raw and modified hydrochars are illustrated in Figure 2. This analysis allows determining of the main functional groups present on the hydrochars' structures, which could attribute to them different adsorption capacities. The FTIR spectra of raw (Figure 2a) and modified (Figure 2b) hydrochars are quite similar in terms of their functional groups, since peaks fall in the same wavenumber interval. The bands within the range of 3600–3000 cm^{-1} correspond to O–H groups such as alcohols, phenols and carboxylic acids. These groups can provide hydrochars with good cation exchange capacity [6,43], just like MB, that is positively charged. Moreover, the peak has a fairly wide width meaning that there is a complex interaction among the same surrounding functional groups. The bands at 2920–2850 cm^{-1} are attributable to aliphatic C–H stretching bonds, as due to -CH$_2$ (methylene) or -CH$_3$ (methyl) groups, without aromatic components. The bands within the range of 1650–1530 cm^{-1} could correspond to C–O groups of amide groups and carboxylates. From a qualitative point of view, samples obtained at 190 °C do not show enhanced structural changes before and after the basic treatment, unless a decrease in intensity of the band located at about 1630 cm^{-1}, maybe due to a decrease in the adsorbed water content. For the samples produced at higher temperatures, a general decrease in intensity for the signals related to oxygen-containing functional groups was observed with the alkaline treatment, as a result of the partial removal of these groups from the sample surfaces, leading to a more homogeneous outer structure, as also evidenced by SEM analysis, confirming data from the elemental analysis.

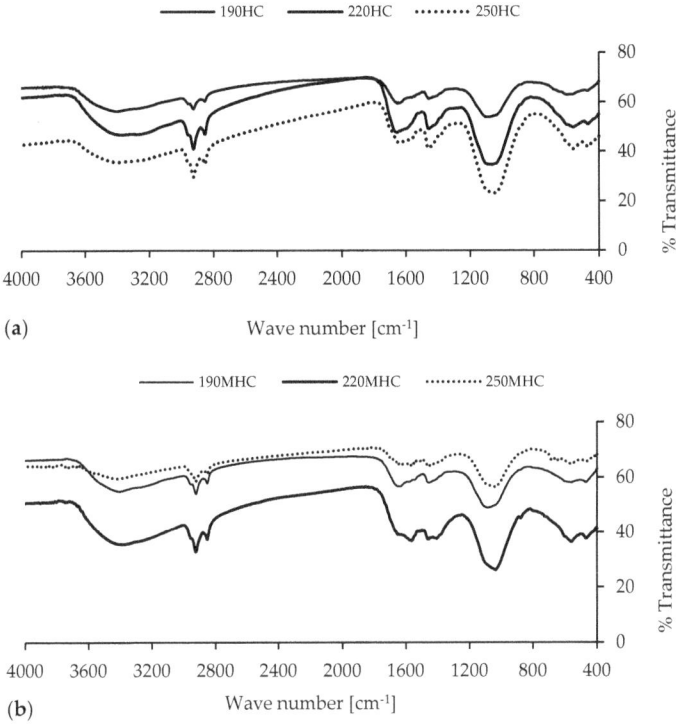

Figure 2. Fourier transform infrared spectroscopy (FTIR) spectra of (**a**) raw and (**b**) modified hydrochars.

The HTC treatment has a significant effect on the amount of heavy metals in the sewage sludge hydrochars [44]. The total contents of heavy metals, alkali earth metals and nutrient elements in the various samples is presented in Figure 3. In Table A1 in Appendix A the concentration values are reported. As evident from Figure 3, considerable amounts of N, Ca, Fe and P were present in the raw sludge mixture, followed by Al, Mg and K. Cu, Na and Zn were present in lower concentration. Cd, Cr, Ni, Hg and Pb are not reported in Figure 3 as their concentration is lower than 30 mg kg^{-1}. Since heavy metals mostly remain in the solid phase after HTC treatment, their concentration in the hydrochars increased [45] with the formation of more adsorption points to adsorb heavy metal ions [46]. Considering alkali metals, the HTC treatment concentrated both Ca and Mg in the hydrochar, while Na and K concentration decreased due to their solubility (and thus partial dissolution) into the HTC process water [47]. The KOH treatment did not substantially vary the concentration of metals in the hydrochars with some exceptions: the Mg and Na concentrations slightly increased while the increase in K concentration was extremely high. As a matter of fact, some K deriving from the KOH used to modify the hydrochars remained in the hydrochars themselves. The presence of metal ions with basic properties (K, Ca, Al) can improve the efficiency of acid–base interactions with MB, displaying acid behavior, thus an improvement of MB removal is expected in more basic environments [42]. Moreover, it should not be neglected that the presence of Fe ions in raw and modified hydrochars could improve the MB removal efficiency, due to the onset of a redox equilibrium between Fe ions and MB itself.

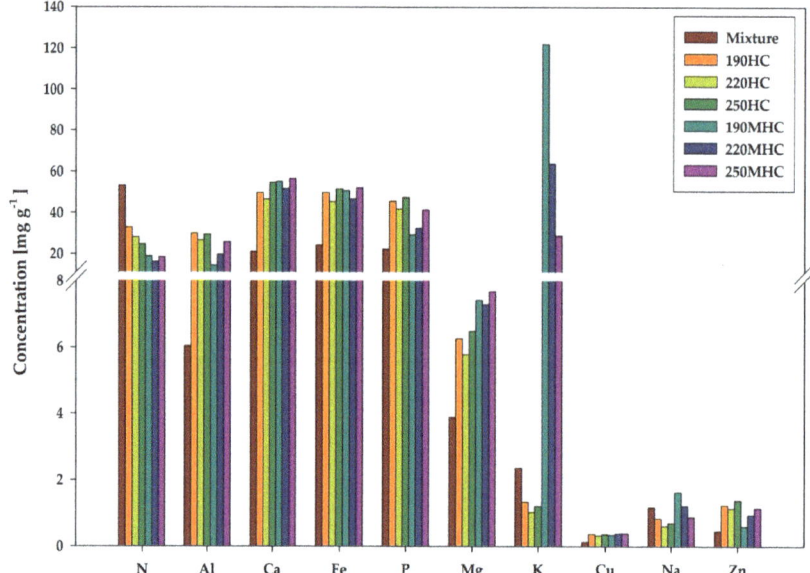

Figure 3. Heavy metals, alkali metals and nutrient content in mixture sludge, raw and modified hydrochars.

Furthermore, our data also prove that the HTC of sewage sludge is a process which enriches the resulting solids in P [48]. The produced hydrochar has thus a higher concentration of P compared to the raw sludge mixture; P which, together with other elements such as N, K, Na, Ca and Mg, is an essential nutrient for plants growth.

It is worthy commenting on the N content data calculated via ICP with that calculated via CHN analysis through the elemental analyzer. The data for the sludge mixture is statistically identical (ICP: 5.3% ± 0.1%; CHN analysis: 5.5% ± 0.3%). When considering raw and modified hydrochars, such data consistency is found only for three samples out of six: for three hydrochars, the values measured by the two techniques are statistically different. This could be due to some extent to the different analytical approach but, in the opinion of the authors, this is mainly due to the natural heterogeneity of the samples, also considering that each sample derives from the mixing of two different sludges having different chemical characteristics.

3.2. Adsorption Isotherms

The equilibrium adsorption isotherm is studied in detail, since it can provide information about the surface properties of hydrochars and their adsorption behavior. Adsorption equilibrium is a dynamic concept, i.e., it is achieved when the rate of (dye) adsorption is equal to the rate of (dye) desorption [49].

Figure 4 depicts the MB adsorption isotherms onto raw hydrochars, modified hydrochars and activated carbons.

Considering raw hydrochars (Figure 4a), isotherms are of the same type; they are characterized by a large increase in the amount adsorbed at low concentrations, while the amount adsorbed stabilizes around a limit value, different for the various hydrochars, when the concentration is higher. For all the treatment temperatures the q_e increased quickly for C_e lower than 30 mg L^{-1}, while the increase in q_e

slowed down for C_e values higher than 50 mg L^{-1}. The amount of adsorbed MB increased when the HTC treatment temperature increased from 190 to 220 °C; on the contrary, the amount of adsorbed MB decreased for the hydrochar obtained at 250 °C. However, the hydrochar that revealed the best adsorption capacity is the 190HC because it shows the lower C_e concentration (95 mg L^{-1}) compared to samples 220HC (150 mg L^{-1}) and 250HC (190 mg L^{-1}), when the equilibrium is reached using the most concentrated initial MB solution (295 mg L^{-1}). These findings, which may appear to be contradictory, are explained by the fact that the quantity of hydrochar used in the various tests was slightly variable (Section 2.4 and Tables A2 and A3 in Appendix A).

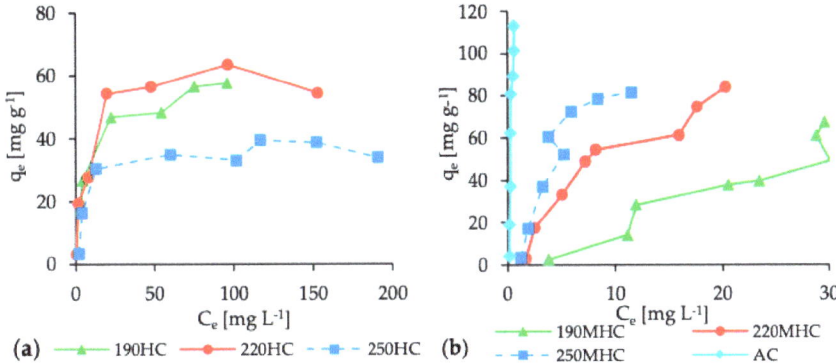

Figure 4. Methylene blue (MB) adsorption isotherms on (**a**) raw hydrochars and (**b**) modified hydrochars and activated carbons, AC.

The q_e increased quickly for low values of C_e for modified hydrochars (Figure 4b), too. However, the isotherm trends are quite different from those of raw hydrochars. Thus, it is clear that the KOH treatment produced strong changes in the adsorption characteristics of the hydrochars. Actually, for all the modified hydrochars the C_e values, measured after 48 h of test, are much lower (maximum C_e value: 30 mg L^{-1}) than those of raw hydrochars (maximum C_e value: 190 mg L^{-1}) and this corresponds to a higher amount of MB adsorbed. Considering all the hydrochars, the sample that shows the best adsorption capacity is 250MHC, i.e., that produced by HTC of sewage sludge at 250 °C for 3 h and post-treatment with KOH. In Figure 4b the adsorption isotherm trend for activated carbons is depicted, too. As expected, the q_e increased very quickly, reaching a value of 113 mg g^{-1} at the highest C_0 concentration, and C_e values were very reduced and always lower than 0.65 mg L^{-1}.

The pH values of the MB solutions were evaluated. The pH value slightly increased after the addition of MB to the deionized water, passing from a value of 5.5, referring to the deionized water, to values in the range 6.2–6.6. The trend of the pH variation corresponding to a MB concentration ranging between 10 and 300 mg L^{-1} has been reported in Figure A3 in Appendix A.

The percentages of MB removed by hydrochars, referring to different C_0 of the MB solutions, are reported in Figure 5.

Figure 5a shows the MB removal of raw hydrochars. Results reveal a high MB removal efficiency for initial concentrations of MB ranging between 10 and 100 mg L^{-1} accounting for an average value of 90%, 95% and 87% for samples 190HC, 220HC and 250HC, respectively. The removal efficiency decreased below 87% (190HC), 89% (220HC) and 65% (250HC) for C_0 values higher than 150 mg L^{-1}.

However, the reduction in MB removal efficiency was slow for samples 190HC and 220HC reaching 68% and 48%, respectively, for C_0 equal to 300 mg L^{-1}, while sample 250HC showed a quite quick reduction in the MB removal efficiency accounting for 48% already in correspondence of a C_0 value of 200 mg L^{-1} and further, to 35% for a C_0 equal to 300 mg L^{-1}. Considering the KOH modified hydrochars (Figure 5b), results show an increasing MB removal efficiency for initial concentrations of the MB solutions ranging between 10 and 100 mg L^{-1} accounting for 88%, 95% and 97% for sample

190MHC, 220MHC and 250MHC, respectively, at C_0 equal to 100 mg L^{-1}. Further, these removal efficiencies remained stable for C_0 values up to 300 mg L^{-1} confirming that the treatment with KOH caused a significant change in the structure of the hydrochars produced at all the HTC temperatures investigated and enhanced the adsorption capacity for removing MB from the aqueous solutions.

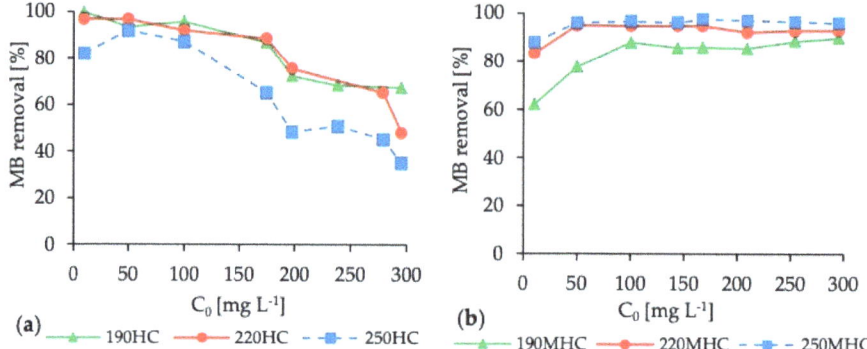

Figure 5. MB removal percentage corresponding to different C_0 values of MB solutions by (**a**) raw hydrochars and (**b**) modified hydrochars. Removal efficiency by activated carbons was always greater than 99% and thus not reported.

As expected, results of the MB adsorption isotherms onto activated carbon revealed a high MB removal efficiency (≥99%) for all the initial concentrations of MB investigated. The equilibrium adsorption data were analyzed using Langmuir, Freundlich and Tempkin adsorption isotherm models. The linear MB adsorption plots for raw and modified hydrochars by the three models are shown in Figure A4 in Appendix A.

Table 2 summarizes the parameters of Langmuir, Freundlich and Tempkin isotherms and the correlation coefficients; the parameters were obtained from the intercepts and slopes of the straight lines resulting from the linear regression of the data points.

Table 2. Parameters used in Langmuir, Freundlich and Tempkin equations.

Samples	Langmuir Equation			Freundlich Equation			Tempkin Equation		
	q_m (mg g^{-1})	K_L (L mg^{-1})	R^2	K_F (L mg^{-1})	n	R^2	B	K (L g^{-1})	R^2
190HC	70.51	0.05	0.9923	5.78	5.81	0.9151	12.49	1.20	0.9589
220HC	54.29	0.38	0.9676	9.74	2.44	0.7733	8.78	6.12	0.7122
250HC	37.64	0.14	0.9845	5.83	2.52	0.7659	6.43	2.57	0.7801
190MHC	247.06	0.01	0.1237	3.60	1.24	0.8000	34.51	0.17	0.8201
220MHC	140.13	0.06	0.8694	10.74	1.45	0.9412	29.35	0.70	0.9525
250MHC	203.16	0.07	0.4208	13.89	1.21	0.8156	36.36	0.98	0.8823

For raw hydrochars, results show that the correlation coefficients for Langmuir isotherms are significantly higher than those for Freundlich and Tempkin isotherms. The Langmuir isotherms describe closely the MB adsorption by raw hydrochars, suggesting that adsorption is localized on a monolayer and all the adsorption sites are homogeneous [50]. The maximum value of adsorption capacity is 70.51 mg g^{-1}, related to the sample subjected to 190 °C HTC process. After the treatment with KOH of this sample, the adsorption capacity increased to 247.06 mg g^{-1} according to the linear regression. However, after KOH treatment the correlation coefficient decreases strongly for the 190 °C samples (from 0.9923 to 0.1237), and this behavior is also observed for the 220 and 250 °C samples, meaning that the modified hydrochars do not follow Langmuir isotherms. On the contrary, Tempkin isotherms exhibit higher correlation coefficients for MB sorption by the modified hydrochars indicating the physicochemical nature of the sorption process with a multilayer coverage of active sites on the adsorbent surface [51]. Results show an R^2 equal to 0.8201, 0.9525 and 0.8823 for samples 190MHC, 220MHC and 250MHC, respectively. The equilibrium data were also fitted to the Freundlich equation. Data show good correlation coefficients, in particular for the modified hydrochars; however, they were lower than Langmuir (for HC) and Tempkin (for MHC) values meaning that the MB adsorption does not follow Freundlich model closely.

3.3. Adsorption Kinetics

Kinetics data for the adsorption of MB onto both the raw and modified hydrochars versus contact time are presented in Figure 6a and b, respectively.

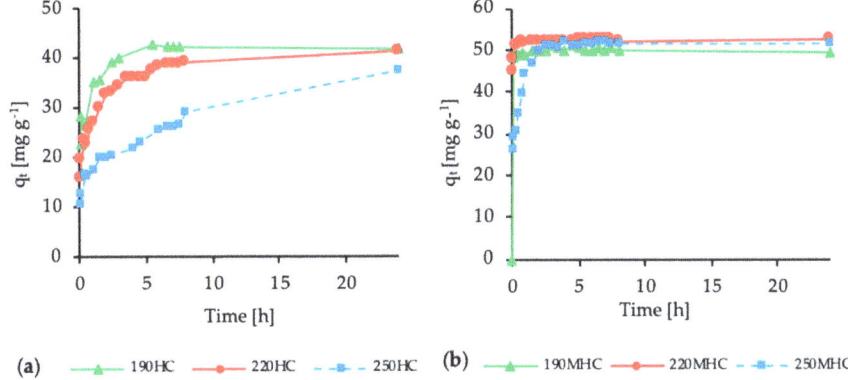

Figure 6. MB adsorption kinetics on (**a**) raw and (**b**) modified hydrochars.

Sorption of MB onto HC (Figure 6a) increased rapidly during the initial 2 and 4 h of contact time for 190HC and 220HC, respectively. For both samples, sorption increased a little more until 6 h of contact time and then slowly, reaching an asymptotic value: 41.8 mg g^{-1} and 41.2 mg g^{-1} for 190HC and 220HC, respectively. The trend for sample 250HC differs significantly from those discussed above. The adsorption of MB onto sample 250HC occurs very slowly with a q_t that increases from 10.2 to 29.4 mg g^{-1} during 8 h of contact time, finally reaching a value of 37.5 mg g^{-1} at 24 h. Thus, this observation is consistent with the adsorption isotherm results, where sample 250HC showed a lower adsorption capacity compared to samples 190HC and 220HC.

Considering the modified hydrochars (Figure 6b), the adsorption of MB onto MHC increased very rapidly. Samples 190MHC and 220MHC reached both a q_t value equal to 48.0 mg g^{-1} after only 15 min of contact time and this value remained quite stable up to 24 h. Sample 250MHC shows an adsorption of MB slightly slower than the other two samples; the q_t value increased up to 50 mg g^{-1} during the initial 2 h of contact time and then remained stable until 24 h. Thus, at the maximum contact time tested all the samples show approximately the same adsorption capacity accounting for 50.0, 52.5 and 51.5 mg g^{-1} for 190MHC, 220MHC and 250MHC, respectively. It is evident that the KOH-treated hydrochars reach a higher value of q_t compared to the raw hydrochars. The adsorption of MB onto MHC, especially for samples 220MHC and 250MHC, is to some extent comparable to that of the activated carbon. In all cases, the adsorption of MB was very rapid. Moreover, the adsorption capacity of the activated carbon resulted equal to 64.3 mg g^{-1}, that is about 22% higher than that of the MHC samples.

The trends of 190HC, 220HC, 190MHC, 220MHC and 250MHC are similar; in all the samples the sorption of MB onto hydrochars increased during the first period of contact time and then slowed down significantly until an asymptotic value was reached, most probably related to an equilibrium stage. This can be explained considering that there are available active sites on the adsorbent that are reduced gradually with the adsorption progress [28]. The main difference between raw and modified hydrochars is the adsorption rate; what happens for the raw hydrochars within 8 h of contact time occurs, instead, for the modified hydrochars within 15 min or at most 2 h. This means that the treatment with KOH causes a change in the hydrochar structure and surface, speeding up the adsorption process.

The pseudo-first order and pseudo-second order kinetic models were used to evaluate the kinetics of the MB adsorption process. Data and related linear regression lines for the two models are reported in Figure A5 (see Appendix A) while Table 3 reports the calculated kinetics parameters.

Table 3. Kinetics parameters for the adsorption of MB onto raw and modified hydrochars.

Sample	Pseudo First-Order			Pseudo Second-Order		
	q_e (mg g^{-1})	k_1 (min^{-1})	R^2	q_e (mg g^{-1})	k_2 (g mg^{-1} min^{-1})	R^2
190HC	22.46	0.0125	0.8897	42.15	0.0041	0.9997
220HC	21.18	0.0053	0.9378	41.56	0.0008	0.9990
250HC	24.94	0.0021	0.8783	36.65	0.0003	0.9631
190MHC	4.79	0.0390	0.3351	49.54	−0.0080	0.9999
220MHC	13.03	0.0740	0.7815	52.56	0.0190	0.9999
250MHC	37.95	0.0310	0.9501	52.01	0.0030	0.9998

The correlation coefficients of the second order kinetic model are greater and closer to unity (R^2 > 0.95) than those of the first order kinetic model for both the raw and modified hydrochars obtained at all the HTC treatment temperatures. These results reveal that the pseudo second-order equation is preferable to describe the adsorption kinetics, and it is very accurate for five samples ($R^2 \geq 0.999$) out of six. Thus, the adsorption process involved chemical interactions between MB and the polar functional groups present at the surface of HC and MHC [34].

3.4. Effect of KOH Modification

As reported above, differences between raw and modified hydrochars are notable for all the HTC treatment temperatures. In general, the KOH treatment significantly enhanced the adsorption capacity of MB onto hydrochars, showing also that MB in high concentrations in the range of 200–300 mg L^{-1} could be easily adsorbed. These results are consistent with the FTIR analysis which has shown that the KOH treatment causes a change in the hydrochar surface making it more homogeneous, which contributes significantly to an increase in the adsorption capacity of the modified hydrochars. However, the KOH treatment does not have the same effect on all of the hydrochars. In particular, the alkali modification on the 250MHC sample provided a sort of "soft cleaning effect" on the hydrochar surface, as reported by SEM analysis, which is consistent with the slight increase in the surface area proved by the BET model. This means an increased number of surface active sites available for the adsorption corresponding to an increase in the MB adsorption capacity of MHCs. Conversely, the KOH treatment applied at hydrochars obtained at the lower temperatures was detrimental in terms of surface area, which dramatically decreased for the hydrochar obtained at 190 °C, passing from 31 for the raw hydrochar to 0.29 m^2 g^{-1} for the KOH-treated sample—a similar but less pronounced effect was observed for the 220 °C hydrochars. Considering adsorption tests, consistently the 190MHC performed the worst when compared to the other MHCs.

Conclusively, results reveal that the alkali treatment on hydrochars generates different effects depending on the HTC treatment temperature, to which different hydrochars characteristics and chemical stability correspond. A common feature is that for all the hydrochars, the KOH post-treatment increased the MB adsorption rate and, more generally, increased their performances as adsorbents.

4. Conclusions

Sewage sludge hydrochars were tested as adsorbents as such and after a KOH post-treatment step. Results indicate that both raw and alkali-modified hydrochars could be used as adsorbents for MB solutions and, more in general, for the treatment of wastewater containing dyes. Batch adsorption tests demonstrated good adsorption capacity of hydrochars with a higher MB removal efficiency for the alkali post-treated hydrochars. Among raw hydrochars, the sample produced at 190 °C achieved the best adsorption results and, correspondingly, was characterized by a higher surface area compared to the other hydrochars. After KOH post-treatment, the sample carbonized at 250 °C showed the highest MB removal: the alkali modification contributed to a homogenization of the hydrochar surface favoring the interaction between surface functional groups and MB molecules, thus enhancing the adsorption capacity of the sample. Comparing the adsorption results with all physicochemical data from various characterization techniques, a complex mechanism for MB removal can be suggested, where surface active sites, chemical interactions and acid–base or redox equilibria between adsorbents and adsorbates can play an important role. As a whole, this work demonstrates a relatively simple process, i.e., HTC possibly followed by a cold alkali post-treatment, that allows the conversion of sewage sludge to an adsorbent material usable for water remediation.

Author Contributions: Conceptualization, R.F., L.F. and G.A.; methodology, R.F., R.C., L.F. and G.A.; validation, R.F., R.C., L.F. and G.A.; formal analysis, R.F. and V.M.; investigation, R.F., V.M., R.C.; resources, L.F., G.A., R.C.; data curation, R.F., V.M., R.C., L.F. and G.A.; writing—original draft preparation, R.F. and L.F.; writing—review and editing, R.F., L.F., R.C., G.A.; supervision, L.F.; funding acquisition, L.F. and G.A. All authors have read and agreed to the published version of the manuscript.

Funding: This research was partially funded by ECOOPERA SpA https://www.ecoopera.coop/it/.

Acknowledgments: Authors want to thank the help of some technicians of the University of Trento: Mirko D'Incau for SEM acquisition, Roberta Villa for support in the adsorption tests, Wilma Waona for ultimate analysis data. Authors want to thank also prof. Luca Fambri for TGA and Matteo Faccini for his contribution in the experimental activity.

Conflicts of Interest: The authors declare no conflict of interest.

Appendix A

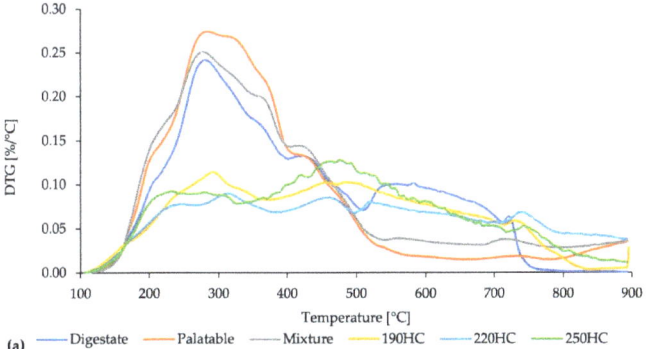

Figure A1. *Cont.*

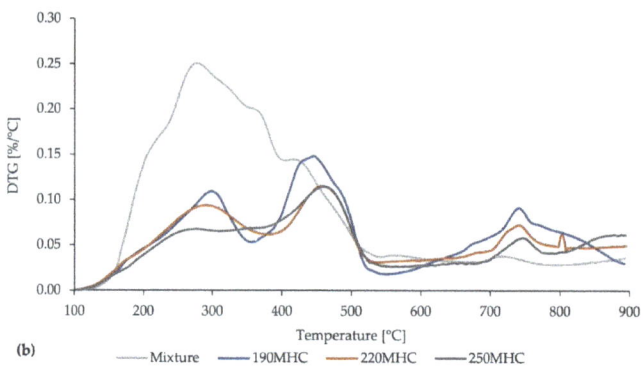

Figure A1. Derivative thermogravimetric (DTG) curves in N_2 atmosphere of (**a**) digestate, palatable, mixture sludge and hydrochars and (**b**) mixture sludge and modified hydrochars.

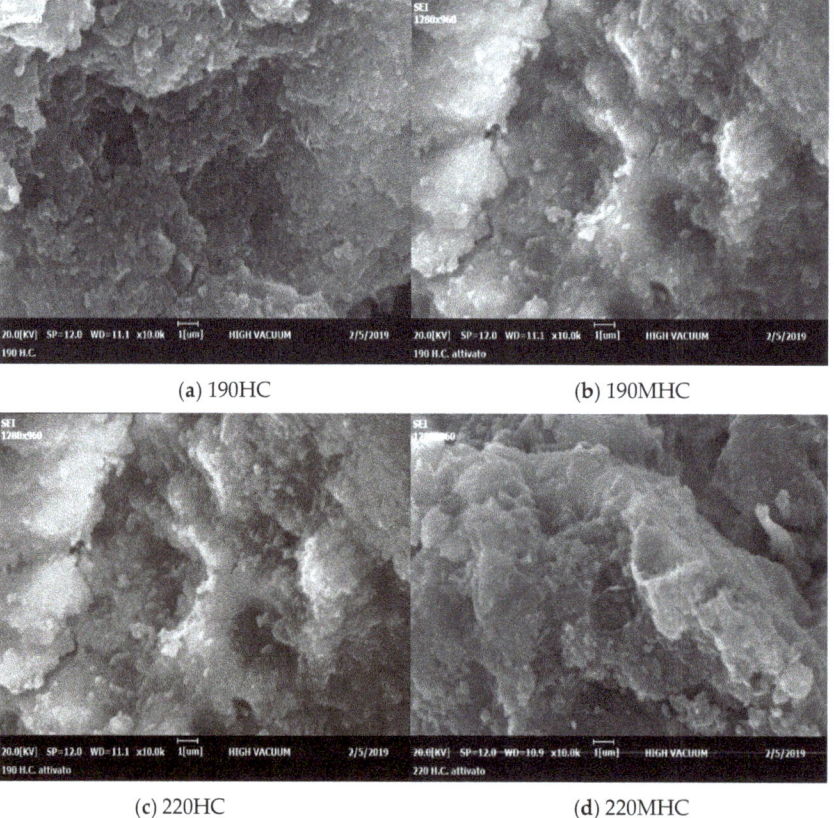

Figure A2. Cont.

(e) 250HC   (f) 250MHC

Figure A2. Scanning electron microscopy (SEM) images of (**a**) 190 °C, (**c**) 220 °C, (**e**) 250 °C raw hydrochars and (**b**) 190 °C, (**d**) 220 °C and (**f**) 250 °C modified hydrochars.

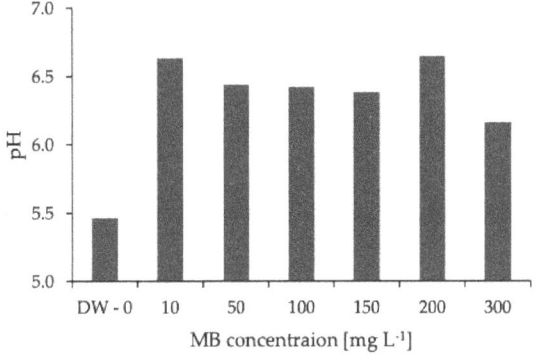

Figure A3. pH values at different MB concentrations. DW-0 represents deionized water without any MB. Standard deviation for the pH is < ±0.1.

Table A1. Content (mg kg^{-1}) of heavy metals, alkali metals and nutrients in the various samples. Absolute errors as follows: N < ±900, Al < ±800, Cd < ±0.1, Ca < ±2000, Cr < ±1, Fe < ±5000, P < ±1000, Mg < ±2000, Hg < ±0.1, Ni < ±1, Pb < ±2, K < ±700, Cu < ±10, Na < ±30, Zn < ±80.

	Digestate	Palatable	Mixture[1]	190HC	220HC	250HC	190MHC	220MHC	250MHC
N	46,400	54,300	53,200	32,800	28,000	24,700	18,800	16,000	18,500
Al	6500	6000	6000	30,000	26,600	29,400	14,200	19,900	25,800
Cd	0.9	1.0	1.0	1.6	1.4	1.5	1.5	2.1	1.6
Ca	26,000	20,000	21,000	50,000	47,000	55,000	55,000	52,000	56,000
Cr	18	19	19	28	26	33	31	27	38
Fe	27,000	24,000	24,000	50,000	45,000	51,000	50,000	47,000	52,000
P	25,000	22,000	22,000	46,000	42,000	47,000	29,000	32,000	41,000
Mg	6000	3500	4000	6000	6000	6500	7400	7300	7700
Hg	0.5	0.4	0.4	1.5	1.1	1.2	0.5	0.9	1.0
Ni	8	7.5	7.5	26	25	29	19	22	30
Pb	32	29	29	62	57	75	65	65	72
K	4900	1900	2400	1300	1000	1200	121,800	63,600	28,700
Cu	140	130	130	370	330	360	340	380	400
Na	2750	900	1170	840	610	700	1630	1210	890
Zn	490	460	470	1240	1140	1380	600	950	1160

[1] calculated as the average of the values of digestate and palatable (weighted average considering the relevant dry matter content).

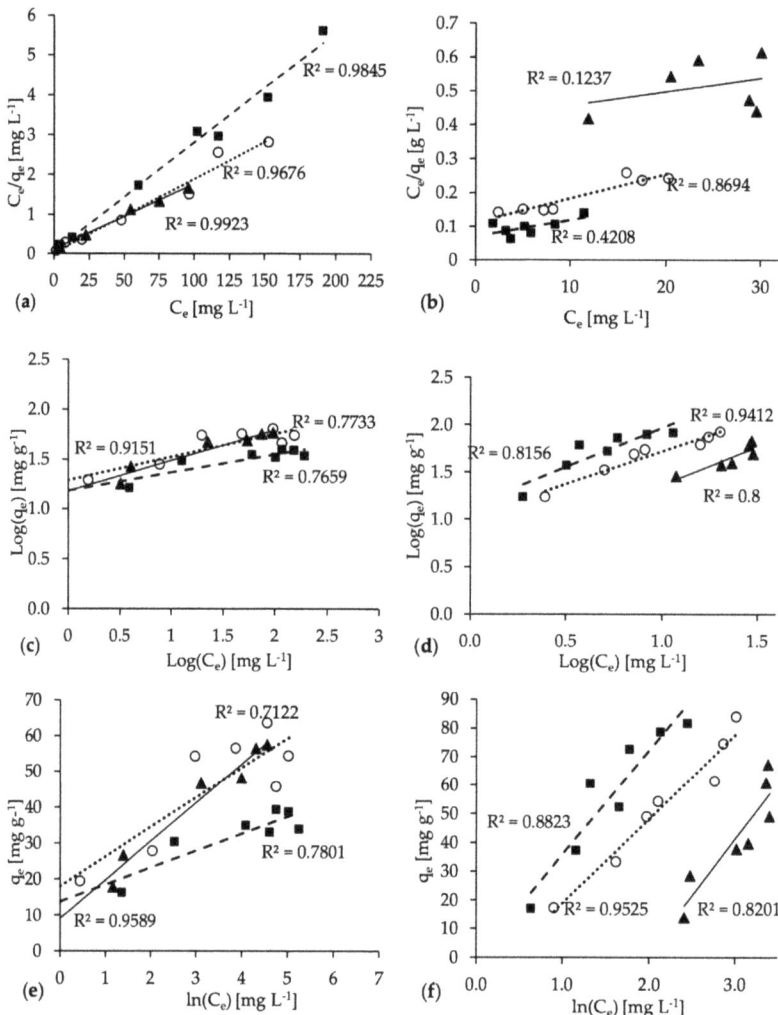

Figure A4. Linear Langmuir isotherm plots for the adsorption of MB on hydrochars (HC) (**a**) and modified hydrochars (MHC) (**b**); linear Freundlich isotherm plots for the adsorption of MB on HC (**c**) and MHC (**d**); linear Tempkin isotherm plots for the adsorption of MB on HC (**e**) and MHC (**f**) obtained at different HTC temperatures (triangle 190 °C, circle 220 °C and square 250 °C-initial MB concentrations: 10–300 mg L^{-1}; contact time: 48 h).

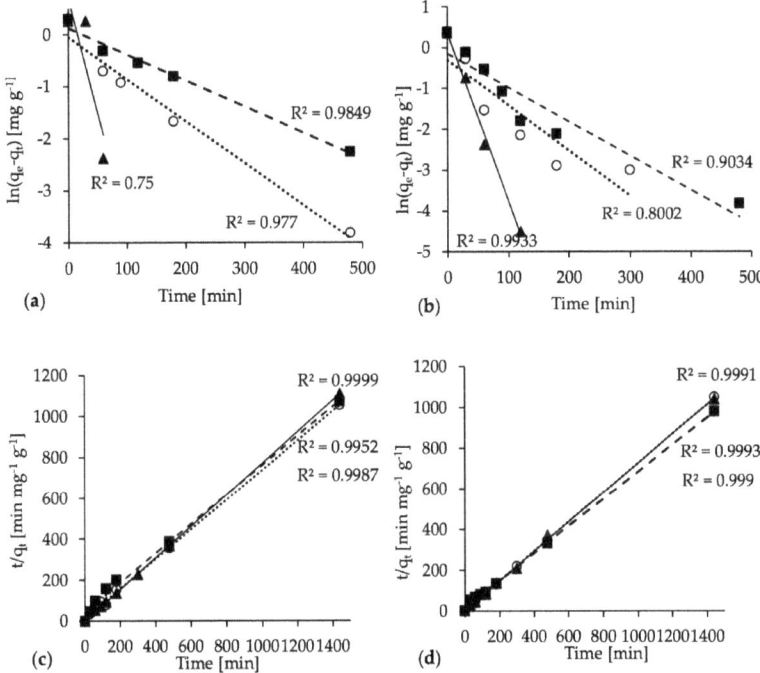

Figure A5. Pseudo first-order kinetics plots for adsorption of MB on HC (**a**) and MHC (**b**); pseudo second-order kinetics plots for adsorption of MB on HC (**c**) and MHC (**d**) (triangle 190 °C, circle 220 °C and square 250 °C-initial MB concentration: 100 mg L^{-1}; maximum contact time: 24 h).

Table A2. Experimental data of the isotherm adsorption batch tests for raw hydrochar (n.a.: not available).

C_0 [mg L^{-1}]	190HC				220HC				250HC			
	C_e [mg L^{-1}]	HC Mass [g]	q_e [mg g^{-1}]	MB Removal [%]	C_e [mg L^{-1}]	HC Mass [g]	q_e [mg g^{-1}]	MB Removal [%]	C_e [mg L^{-1}]	HC Mass [g]	q_e [mg g^{-1}]	MB Removal [%]
10	2	0.0150	3	80	0	0.0126	3	97	2	0.0101	3	82
50	3	0.0105	18	94	2	0.0100	19	97	4	0.0113	16	92
100	4	0.0144	27	96	8	0.0133	28	92	13	0.0115	30	87
174	22	0.0130	47	87	20	0.0114	54	89	60	0.0131	35	65
198	54	0.0119	48	73	48	0.0106	57	76	102	0.0116	33	48
239	75	0.0116	56	69	n.a.	n.a.	n.a.	n.a.	117	0.0124	39	51
279	n.a.	n.a.	n.a.	n.a.	96	0.0115	64	65	152	0.0131	39	45
295	96	0.0139	80	68	153	0.0105	54	48	191	0.0123	34	35

Table A3. Experimental data of the isotherm adsorption batch tests for modified hydrochar.

C_0 [mg L^{-1}]	190MHC				220MHC				250MHC			
	C_e [mg L^{-1}]	HC Mass [g]	q_e [mg g^{-1}]	MB Removal [%]	C_e [mg L^{-1}]	HC Mass [g]	q_e [mg g^{-1}]	MB Removal [%]	C_e [mg L^{-1}]	HC Mass [g]	q_e [mg g^{-1}]	MB Removal [%]
10	4	0.0105	2	62	2	0.0108	3	83	1	0.0107	3	88
50	11	0.0112	14	78	2	0.0110	17	95	2	0.0112	17	96
100	12	0.0124	28	88	5	0.0114	33	95	3	0.0104	37	97
144	21	0.0131	38	86	7	0.0112	49	95	5	0.0106	52	96
167	23	0.0145	40	86	8	0.0117	54	95	4	0.0108	60	98
209	30	0.0146	49	86	16	0.0126	61	92	6	0.0112	73	97
254	29	0.0148	61	89	18	0.0127	74	93	8	0.0125	79	97
295	30	0.0158	67	90	20	0.0131	84	93	11	0.0139	82	96

References

1. Ferrentino, R.; Langone, M.; Andreottola, G.; Rada, E.C. An anaerobic side-stream reactor in wastewater treatment: A review. *WIT Trans. Ecol. Environ.* **2014**, *191*, 1435–1446.
2. Saetea, P.; Tippayawong, N. Recovery of Value-Added Products from Hydrothermal Carbonization of Sewage Sludge. *ISRN Chem. Eng.* **2013**, *2013*, 1–6. [CrossRef]
3. Sharma, H.B.; Sarmah, A.K.; Dubey, B. Hydrothermal carbonization of renewable waste biomass for solid biofuel production: A discussion on process mechanism, the influence of process parameters, environmental performance and fuel properties of hydrochar. *Renew. Sustain. Energy Rev.* **2020**, *123*, 109761. [CrossRef]
4. Villamil, J.A.; de la Rubia, M.A.; Diaz, E.; Mohedano, A.F. *Technologies for Wastewater Sludge Utilization and Energy Production: Hydrothermal Carbonization of Lignocellulosic Biomass and Sewage Sludge*; Elsevier Inc.: Amsterdam, The Netherlands, 2020; ISBN 9780128162040.
5. Merzari, F.; Langone, M.; Andreottola, G.; Fiori, L. Methane production from process water of sewage sludge hydrothermal carbonization. A review. Valorising sludge through hydrothermal carbonization. *Crit. Rev. Environ. Sci. Technol.* **2019**, *49*, 1–42. [CrossRef]
6. Román, S.; Libra, J.; Berge, N.; Sabio, E.; Ro, K.; Li, L.; Ledesma, B.; Alvarez, A.; Bae, S. Hydrothermal carbonization: Modeling, final properties design and applications: A review. *Energies* **2018**, *11*, 216. [CrossRef]
7. Funke, A.; Ziegler, F. Hydrothermal carbonization of biomass: A summary and discussion of chemical mecha- nisms for process engineering. *Biofuels Bioprod. Biorefining* **2010**, *4*, 160–177. [CrossRef]
8. Lucian, M.; Volpe, M.; Fiori, L. Hydrothermal Carbonization Kinetics of Lignocellulosic Agro-Wastes: Experimental Data and Modeling. *Energies* **2019**, *12*, 516. [CrossRef]
9. Fang, J.; Zhan, L.; Ok, Y.S.; Gao, B. Minireview of potential applications of hydrochar derived from hydrothermal carbonization of biomass. *J. Ind. Eng. Chem.* **2018**, *57*, 15–21. [CrossRef]
10. Román, S.; Valente Nabais, J.M.; Ledesma, B.; González, J.F.; Laginhas, C.; Titirici, M.M. Production of low-cost adsorbents with tunable surface chemistry by conjunction of hydrothermal carbonization and activation processes. *Microporous Mesoporous Mater.* **2013**, *165*, 127–133. [CrossRef]
11. Purnomo, C.W.; Castello, D.; Fiori, L. Granular Activated Carbon from Grape Seeds Hydrothermal Char. *Appl. Sci.* **2018**, *8*, 331. [CrossRef]
12. Hammud, H.H.; Shmait, A.; Hourani, N. Removal of Malachite Green from water using hydrothermally carbonized pine needles. *RSC Adv.* **2015**, *5*, 7909–7920. [CrossRef]
13. Wei, J.; Liu, Y.; Li, J.; Yu, H.; Peng, Y. Removal of organic contaminant by municipal sewage sludge-derived hydrochar: Kinetics, thermodynamics and mechanisms. *Water Sci. Technol.* **2018**, *78*, 947–956. [CrossRef]
14. Sun, K.; Tang, J.; Gong, Y.; Zhang, H. Characterization of potassium hydroxide (KOH) modified hydrochars from different feedstocks for enhanced removal of heavy metals from water. *Environ. Sci. Pollut. Res.* **2015**, *22*, 16640–16651. [CrossRef]
15. Diaz, E.; Manzano, F.J.; Villamil, J.; Rodriguez, J.J.; Mohedano, A.F. Low-Cost Activated Grape Seed-Derived Hydrochar through Hydrothermal Carbonization and Chemical Activation for Sulfamethoxazole Adsorption. *Appl. Sci.* **2019**, *9*, 5127. [CrossRef]
16. Falco, C.; Marco-Lozar, J.P.; Salinas-Torres, D.; Morallón, E.; Cazorla-Amorós, D.; Titirici, M.M.; Lozano-Castelló, D. Tailoring the porosity of chemically activated hydrothermal carbons: Influence of the precursor and hydrothermal carbonization temperature. *Carbon N. Y.* **2013**, *62*, 346–355. [CrossRef]
17. Park, J.E.; Lee, G.B.; Hong, B.U.; Hwang, S.Y. Regeneration of Activated Carbons Spent by Waste Water Treatment Using KOH Chemical Activation. *Appl. Sci.* **2019**, *9*, 1–10.
18. Regmi, P.; Garcia Moscoso, J.L.; Kumar, S.; Cao, X.; Mao, J.; Schafran, G. Removal of copper and cadmium from aqueous solution using switchgrass biochar produced via hydrothermal carbonization process. *J. Environ. Manag.* **2012**, *109*, 61–69. [CrossRef]
19. Spataru, A.; Jain, R.; Chung, J.W.; Gerner, G.; Krebs, R.; Lens, P.N.L. Enhanced adsorption of orthophosphate and copper onto hydrochar derived from sewage sludge by KOH activation. *RSC Adv.* **2016**, *6*, 101827–101834. [CrossRef]
20. Lucian, M.; Fiori, L. Hydrothermal carbonization of waste biomass: Process design, modeling, energy efficiency and cost analysis. *Energies* **2017**, *10*, 211. [CrossRef]

21. Merzari, F.; Lucian, M.; Volpe, M.; Andreottola, G.; Fiori, L. Hydrothermal carbonization of biomass: Design of a bench-Scale reactor for evaluating the heat of reaction. *Chem. Eng. Trans.* **2018**, *65*, 43–48.
22. Volpe, M.; Wüst, D.; Merzari, F.; Lucian, M.; Andreottola, G.; Kruse, A.; Fiori, L. One stage olive mill waste streams valorisation via hydrothermal carbonisation. *Waste Manag.* **2018**, *80*, 224–234. [CrossRef] [PubMed]
23. Sun, K.; Ro, K.; Guo, M.; Novak, J.; Mashayekhi, H.; Xing, B. Sorption of bisphenol A, 17α-ethinyl estradiol and phenanthrene on thermally and hydrothermally produced biochars. *Bioresour. Technol.* **2011**, *102*, 5757–5763. [CrossRef] [PubMed]
24. Fiori, L.; Valbusa, M.; Lorenzi, D.; Fambri, L. Modeling of the devolatilization kinetics during pyrolysis of grape residues. *Bioresour. Technol.* **2012**, *103*, 389–397. [CrossRef]
25. García, R.; Pizarro, C.; Lavín, A.G.; Bueno, J.L. Biomass proximate analysis using thermogravimetry. *Bioresour. Technol.* **2013**, *139*, 1–4. [CrossRef]
26. CNR-IRSA. *Metodi Analitici Per i Fanghi*; CNR-IRSA: Montelibretti, Italy, 1985.
27. Guo, J.Z.; Li, B.; Liu, L.; Lv, K. Removal of methylene blue from aqueous solutions by chemically modified bamboo. *Chemosphere* **2014**, *111*, 225–231. [CrossRef]
28. Qian, W.; Luo, X.; Wang, X.; Guo, M.; Li, B. Removal of methylene blue from aqueous solution by modi fi ed bamboo hydrochar. *Ecotoxicol. Environ. Saf.* **2018**, *157*, 300–306. [CrossRef]
29. Langmuir, I. The constitution and fundamental properties of solids and liquids. *J. Am. Chem. Soc.* **1916**, *38*, 2221–2295. [CrossRef]
30. Freundlich, H. Uber Die Adsorption in Losungen. *Zeitschrift fur Physikalische Chemie* **1906**, *57*, 385–470. [CrossRef]
31. Malik, P.K. Use of activated carbons prepared from sawdust and rice-husk for adsoprtion of acid dyes: A case study of acid yellow 36. *Dye. Pigment.* **2003**, *56*, 239–249. [CrossRef]
32. Foo, K.Y.; Hameed, B.H. Preparation, characterization and evaluation of adsorptive properties of orange peel based activated carbon via microwave induced K 2CO 3 activation. *Bioresour. Technol.* **2012**, *104*, 679–686. [CrossRef]
33. Ho, Y.S.; Ng, J.C.Y.; McKay, G. Kinetics of pollutant sorption by biosorbents: Review. *Sep. Purif. Methods* **2000**, *29*, 189–232. [CrossRef]
34. Ho, Y.S.; McKay, G. Pseudo-second order model for sorption processes. *Process Biochem.* **1999**, *34*, 451–465. [CrossRef]
35. Zhai, Y.; Peng, C.; Xu, B.; Wang, T.; Li, C. Hydrothermal carbonisation of sewage sludge for char production with different waste biomass: Effects of reaction temperature and energy recycling. *Energy* **2017**, *127*, 167–174. [CrossRef]
36. He, C.; Giannis, A.; Wang, J. Conversion of sewage sludge to clean solid fuel using hydrothermal carbonization: Hydrochar fuel characteristics and combustion behavior. *Appl. Energy* **2013**, *111*, 257–266. [CrossRef]
37. Kang, S.; Li, X.; Fan, J.; Chang, J. Characterization of hydrochars produced by hydrothermal carbonization of lignin, cellulose, d-xylose, and wood meal. *Ind. Eng. Chem. Res.* **2012**, *51*, 9023–9031. [CrossRef]
38. Berge, N.D.; Ro, K.S.; Mao, J.; Flora, J.R.V.; Chappell, M.A.; Bae, S. Hydrothermal carbonization of municipal waste streams. *Environ. Sci. Technol.* **2011**, *45*, 5696–5703. [CrossRef]
39. Sing, K.S.W.; Everett, D.H.; Haul, R.A.W.; Moscou, L.; Pierotti, R.A.; Rouquérol, J.; Siemieniewska, T. Reporting physisorption data for gas/solid systems with special reference to the determination of surface area and porosity. *Pure Appl. Chem.* **1985**, *57*, 603–619. [CrossRef]
40. Hameed, B.H.; Ahmad, A.L.; Latiff, K.N.A. Adsorption of basic dye (methylene blue) onto activated carbon prepared from rattan sawdust. *Dye. Pigment.* **2007**, *75*, 143–149. [CrossRef]
41. El Qada, E.N.; Allen, S.J.; Walker, G.M. Adsorption of Methylene Blue onto activated carbon produced from steam activated bituminous coal: A study of equilibrium adsorption isotherm. *Chem. Eng. J.* **2006**, *124*, 103–110. [CrossRef]
42. Ahmed, T.; Noor, W.; Faruk, O.; Bhoumick, M.C.; Uddin, M.T. Removal of methylene blue (MB) from waste water by adsorption on jackfruit leaf powder (JLP) in continuously stirred tank reactor. *J. Phys. Conf. Ser.* **2018**, *1086*. [CrossRef]
43. Wiedner, K.; Rumpel, C.; Steiner, C.; Pozzi, A.; Maas, R.; Glaser, B. Chemical evaluation of chars produced by thermochemical conversion (gasification, pyrolysis and hydrothermal carbonization) of agro-industrial biomass on a commercial scale. *Biomass Bioenergy* **2013**, *59*, 264–278. [CrossRef]

44. Wang, X.; Li, C.; Zhang, B.; Lin, J.; Chi, Q.; Wang, Y. Migration and risk assessment of heavy metals in sewage sludge during hydrothermal treatment combined with pyrolysis. *Bioresour. Technol.* **2016**, *221*, 560–567. [CrossRef] [PubMed]
45. Huang, R.; Zhang, B.; Saad, E.M.; Ingall, E.D.; Tang, Y. Speciation evolution of zinc and copper during pyrolysis and hydrothermal carbonization treatments of sewage sludges. *Water Res.* **2018**, *132*, 260–269. [CrossRef] [PubMed]
46. Xiong, J.B.; Pan, Z.Q.; Xiao, X.F.; Huang, H.J.; Lai, F.Y.; Wang, J.X.; Chen, S.W. Study on the hydrothermal carbonization of swine manure: The effect of process parameters on the yield/properties of hydrochar and process water. *J. Anal. Appl. Pyrolysis* **2019**, *144*, 104692. [CrossRef]
47. Volpe, M.; Fiori, L. From olive waste to solid biofuel through hydrothermal carbonisation: The role of temperature and solid load on secondary char formation and hydrochar energy properties. *J. Anal. Appl. Pyrolysis* **2017**, *124*, 63–72. [CrossRef]
48. Wang, L.; Li, A.; Chang, Y. Hydrothermal treatment coupled with mechanical expression at increased temperature for excess sludge dewatering: Heavy metals, volatile organic compounds and combustion characteristics of hydrochar. *Chem. Eng. J.* **2016**, *297*, 1–10. [CrossRef]
49. Fu, J.; Chen, Z.; Wang, M.; Liu, S.; Zhang, J.; Zhang, J.; Han, R.; Xu, Q. Adsorption of methylene blue by a high-efficiency adsorbent (polydopamine microspheres): Kinetics, isotherm, thermodynamics and mechanism analysis. *Chem. Eng. J.* **2015**, *259*, 53–61. [CrossRef]
50. Chen, L.; Bai, B. Equilibrium, kinetic, thermodynamic, and in situ regeneration studies about methylene blue adsorption by the raspberry-like TiO$_2$@yeast microspheres. *Ind. Eng. Chem. Res.* **2013**, *52*, 15568–15577. [CrossRef]
51. Pathania, D.; Sharma, S.; Singh, P. Removal of methylene blue by adsorption onto activated carbon developed from Ficus carica bast. *Arab. J. Chem.* **2017**, *10*, S1445–S1451. [CrossRef]

© 2020 by the authors. Licensee MDPI, Basel, Switzerland. This article is an open access article distributed under the terms and conditions of the Creative Commons Attribution (CC BY) license (http://creativecommons.org/licenses/by/4.0/).

Article

Low-Cost Biochar Adsorbents for Water Purification Including Microplastics Removal

Virpi Siipola [1,*], Stephan Pflugmacher [2], Henrik Romar [3], Laura Wendling [1] and Pertti Koukkari [1]

1. VTT Technical Research Centre of Finland Ltd., P.O. Box 1000, 02044 Espoo, Finland; laura.wendling@vtt.fi (L.W.); pertti.koukkari@vtt.fi (P.K.)
2. Faculty of Biological and Environmental Sciences, University of Helsinki, Niemenkatu 73, 15140 Lahti, Finland; stephan.pflugmacher@helsinki.fi
3. Research Unit of Sustainable Chemistry, University of Oulu, P.O. Box 3000, 90014 Oulu, Finland; henrik.romar@chydenius.fi
* Correspondence: virpi.siipola@vtt.fi

Received: 19 December 2019; Accepted: 20 January 2020; Published: 22 January 2020

Abstract: The applicability of steam activated pine and spruce bark biochar for storm water and wastewater purification has been investigated. Biochar samples produced from the bark of scots pine (*Pinus sylvestris*) and spruce (*Picea* spp.) by conventional slow pyrolysis at 475 °C were steam activated at 800 °C. Steam activation was selected as a relatively inexpensive method for creating porous biochar adsorbents from the bark-containing sidestreams of the wood refining industry. A suite of standard analytical procedures were carried out to quantify the performance of the activated biochar in removing both cations and residual organics from aqueous media. Phenol and microplastics retention and cation exchange capacity were employed as key test parameters. Despite relatively low surface areas (200–600 m^2/g), the steam-activated biochars were highly suitable adsorbents for the chemical species tested as well as for microplastics removal. The results indicate that ultra-high porosities are not necessary for satisfactory water purification, supporting the economic feasibility of bio-based adsorbent production.

Keywords: biochar; activated carbon; steam activation; phenol adsorption; microplastics; bark

1. Introduction

As purification standards for residential and industrial wastewaters become increasingly restrictive and the negative environmental consequences of untreated urban stormwater runoff discharge to surface waterbodies become more apparent, there is an increasing need for more efficient adsorbents. Unprocessed stormwaters are typically captured and transferred directly to the environment via separate sewer systems, or periodically processed by municipal waste water treatment plants in large volumes where sewers are combined, resulting in capacity pressures on wastewater treatment infrastructure and decreased efficiency of resource recovery processes (e.g., [1]). Stormwater runoff contains organic residues in addition to micro- and nanoplastics, e.g., from vehicle tires [2]. Biochars and activated biochars produced from a variety of forestry and agricultural sidestreams have been extensively tested for both organic and inorganic contaminant sorption (see, e.g., [3]). Their suitability for water purification is well established but the primary focus of existing studies has been the development of high surface area carbons, with limited consideration of the economic feasibility of these carbons for various intended applications. Therefore, there is a need to establish the minimum requirements for bio-based adsorbents with respect to surface area, porosity, and surface chemistry for efficient water purification. For profitable production of a bio-adsorbent, both the raw material and the treatment process need to be low-cost.

Chemical activation can be used to produce ultrahigh surface areas and porosities because of extensive microporosity development. High micropore volume promotes adsorption of particularly small-sized metals and molecules. Still, a broader pore structure plays a vital role in adsorption processes. The small micropores that are accessible to nitrogen molecules in surface area measurements may not be accessible for contaminant molecules in solution. Meso- and macroscale pores are essential as vectors to areas deeper within the biochar particle, and their respective quantities are primarily dependent on the raw materials used in biochar production. In our previous study [4], 3D-modeling of pine bark biochar and phosphoric acid activated pine bark revealed that chemical activation did not affect the micrometer scale porosity of the activated biochar. Also significant are the elevated costs arising from the use of chemicals and intensive washing procedures [5]. Producing chemically activated carbons for wastewater treatment may be uneconomical because of the large quantities of chemicals needed and reuse of the used activated carbons may not be possible. The main applications for chemically activated carbons should be in higher value products such as supercapacitors, where the surface area and specific pore size distribution are critical parameters for their functionality [6,7].

Thermal treatment of biomass can be divided into three different paths: torrefaction, gasification, and pyrolysis [8]. The most important differences are the residence times and temperature gradients used, particle size of feedstock materials, and the distribution of products into gas, pyrolytic liquids, and solid materials. The pyrolysis of biomass can further be divided into fast, medium, and slow pyrolysis; of these the fast and slow pyrolysis are the ones mostly used. Fast pyrolysis with a residence time of seconds is used for the production of liquids whereas slow pyrolysis with residence times of minutes to hours is used to produce chars. Characteristics of the individual processes are summarized in Table 1.

Table 1. Summary of different pyrolysis processes with product distributions.

	Pyrolysis			Torrefaction	Gasification
	Fast	Intermediate	Slow		
Temperature	500	400	400	300	800–900
Residence time	<1 s	10–30 s	1–5 h	Hours-days	s-min
Liquid %	75	40 (2 phases)	35		1–5
Char %	12	40	35	85	<1
Gas %	13	20	30	15	95–99

The chars obtained from slow pyrolysis can undergo further physical or chemical treatment to generate activated carbons. Physical activations using CO_2 or steam eliminate the need for chemicals and subsequent washing procedures. The use of steam minimizes the activation chemical costs and promotes the formation of larger pores in the activated carbon (AC), although the resulting porosity is also dependent upon characteristics of the feedstock raw material [9]. The adsorption efficiency is related to the surface functionalities of the adsorbent carbon where, for example, a relatively large number of oxygen groups, enhance adsorption of cationic contaminants. The total number of surface functional groups in physically activated carbons are usually less than for chemically activated carbons because of the higher temperatures used. Critical views for their suitability for water treatment have been presented elsewhere [5]. Activated biochars possessing sufficient surface area and suitable porosity for tertiary wastewater purification that can be produced economically are of widespread interest [3,10,11].

The availability of bio-based feedstock is an essential variable for biochar and activated carbon (AC) production. Large quantities of lignocellulose sidestreams suitable for biochar production, such as sawdust and bark residues, are generated by the forest industry. The forest industry has traditionally used these sidestreams for energy production, but this use is increasingly limited because of the growing need to decrease carbon dioxide (CO_2) emissions from industries. There is a need to find alternative uses for these sidestreams, of which biochar is one possibility. Bio-based carbons are creating a new market segment in water treatment and metallurgy based on their potentially low cost compared with traditional fossil carbons subdued to emission trade. Both applications present "new" industrial utilizations with positive export potential for countries with significant forest products industries, both domestically and internationally.

Based on the earlier reports on economically feasible raw materials for biochar (e.g., [12,13]), a range of different wood-based wastes and sidestreams are suitable for biochar production. The steam activation method has been used to produce ACs from various biomasses, such as white spruce sawdust, canola, and wheat straw [14], switchgrass, hard and soft wood [15], oil palm stones [16], oil palm shells [17], and seed cakes [18]. Despite the large volumes generated by the forest products industry, tree bark has not been extensively tested for production of steam activated carbons. Mixed soft wood bark residue has been successfully converted into AC in a small-scale thermogravimetric experiment [19], producing surface areas between 455 and 613 m^2/g at different temperatures (600–985 °C). Poplar wood bark biochar has also been used for steam activation with similar surface areas of 547 and 555 m^2/g at 700 and 800 °C, respectively [20].

In the present study, we have investigated activated biochar production from two forest industry sidestreams, pine and spruce bark. The suitability of these steam-activated biochars for application to treatment of urban runoff and wastewater purification were investigated by examining the attenuation of selected metals, microplastics, and organic contaminants. Microplastics in stormwaters originate from microscopic plastic spheres or particles that are intentionally added to a product, or from disintegrating plastic and rubber materials. One of the major sources of microplastics is created by traffic through the abrasion of vehicle tires, brakes, and the road surface itself [2]. Vehicle-generated plastic particles can be mobilized by wind and passing traffic, becoming deposited in surface waters, soil, or sediment. Deposition of a large quantity of plastic particles to surface waters can cause significant damage to the aquatic environment and organisms [21–24]. Recent studies of microplastics removal have focused on agglomerate formation [22,25] or activated sludge [26]. Biochar and activated biochar also have the potential to retain microplastics. Microplastic particles can be immobilized between biochar particles or, in the case of nano-and micrometer-scale particles, retained within the pore structure. The present study examined microplastics removal by steam-activated biochar generated from pine and spruce bark.

The particular focus of the study was on the characteristics of the biochar products, e.g., the particle size and chemical composition, as forest residues may be comprised of highly inhomogeneous raw materials. The materials and methods section is followed by a detailed presentation and discussion of the results obtained that may affect the economics of biochar and AC production from the forest residues examined herein. The results indicate that the selected low-cost biomasses were suitable as adsorbents for all tested contaminants, and that sufficient adsorption capacities do not necessitate ultrahigh surface areas.

2. Materials and Methods

The selected methods were used for testing the differences in the produced biochars and AC after the slow pyrolysis or activation treatments. Elemental composition, surface area, and porosity were used to detect the differences in the chemical and physical properties. Potential material applicability was further examined in a series of laboratory trials, including phenol adsorption as an indicator of organic contaminant removal and cation exchange capacity (CEC) determination to estimate inorganic contaminate removal capacity. The microplastics (MP) removal capacity of produced biochars and AC was tested in a column experiment using various sizes and shapes of MP particles.

2.1. Raw Materials

Materials used in the experiments were scots pine (*Pinus sylvestrus*) bark and spruce (*Picea* spp.) bark. The pine bark biomass was acquired from Sweden and the spruce bark biomass from a Finnish sawmill. The samples contained small quantities of stem wood, which were not removed prior to carbonization. The bark samples were oven-dried at <70 °C to approximately 10% moisture content.

2.2. Slow Pyrolysis and Activation Treatments

Oven-dried bark samples were carbonized using slow pyrolysis in a 115-L reactor. The samples were distributed in the reactor on four levels of steel grids (Figure 1). The carbonization time and temperature were three hours and 475 °C, respectively.

Figure 1. The sample grid of the slow pyrolysis/activation reactor.

The produced biochars were steam activated using the same reactor as for slow pyrolysis. The biochars were weighed on steel vessels, which were placed on the steel grids. The particle size effect on activation results was studied via separation of the biochar particles into two different fractions. The larger particle size fraction consisted of biochar chunks up 10 cm in diameter formed directly from the biomass. The smaller particle size consisted of approximately 50% <5 mm particles and 50% <2 cm biochar particles, determined using standard sieves. The steam activations were performed using low (1.1 L/min) and high (5 L/min) N_2 gas flows with different water flow rates (Table 2) such that the volumetric quantity of steam was approximately 30–40% of the total gas volume injected in the oven (steam + nitrogen). The 30% steam activations were performed using low and high gas rates while in the 40% steam treatment only high N_2 flow was used. The steam was generated from deionized water and the water was pumped using a peristaltic pump. The water line was connected to the N_2 gas line, which circulated the heated reactor evaporating the water before entering the oven. The activation time was 3.5 h at 800 °C.

Table 2. Slow pyrolysis and steam activation conditions.

Treatment	Water, mL/min	N_2, L/h	Temperature, °C	Time, h
Slow pyrolysis	-	300	475	3.0
30% steam, low gas flow	0.28	66	800	3.5
30% steam, high gas flow	1.40	300	800	3.5
40% steam, low gas flow	1.97	300	800	3.5

2.3. Characterization Methods

All biomass and the produced biochars and ACs were analyzed for their elemental composition (C, H, N, S, and O) using a FLASH 2000 series analyzer (Thermo Scientific, Waltham, MA, USA). The ash content was determined gravimetrically after burning the samples at 550 °C for 23 h. The BET surface area and pore size distribution were determined via N_2 adsorption using a Micromeritics ASAP 2020 analyzer (Norcross, GA, USA). Prior to the surface area measurements, the samples were degassed at 2 µm Hg and 140 °C for 3 h to clean the surfaces. The N_2 adsorption tests were performed at isothermal conditions achieved by immersion of the sample tubes in liquid nitrogen. Nitrogen (N_2) was added in small doses, and the resulting isotherms were used for further calculations. The specific surface areas (SSA) we calculated using the BET [27] algorithm and pore size distributions were calculated using the density functional theory (DFT) [28]. The system applied facilitated measurement of pore sizes in the range of 1.5–300 nm in diameter even where smaller pores likely contribute to the adsorption at low pressure.

2.4. Adsorption Tests

The biochars and selected AC samples were tested for their organic contaminant adsorption capacity using phenol. Sub-samples of 0.1 g biochar or AC were agitated in 15 mL of phenol solution (100, 200, 500, 1000, and 2000 mg/L) for 24 h, after which the suspensions were filtered to 0.45 µm and analyzed spectrophotometrically (Shimadzu UV-1800) at 271 nm. The ACs were tested using two replicate samples and the calculated relative standard deviations (SD/mean*100, RSD) ranged from 0.1 to 18.1%. The highest RSDs (>10%) were found with the low 100 and 200 mg/L concentrations. The biochars were tested as single determinations.

The cation exchange capacity (CEC) and the exchangeable cations were determined for selected activated biochars as described in [29]. In addition, the concentration of released phosphorus was measured. Briefly, AC cations were exchanged for NH_4^+ by an overnight extraction (1:10 *w/v* ratio) using 0.5 M NH_4OAc (pH 7). After extraction, the biochars were centrifuged and resuspended twice with equal amounts of 0.5 M NH_4OAc (pH 7) to ensure saturation of exchange sites with NH_4^+. The three supernatants were combined and analyzed. Excess NH_4^+ was rinsed using deionized water. The adsorbed NH_4^+ was then exchanged by an overnight extraction using 1 M KCl. Concentrations of Al, Ca, Fe, K, Mg, Mn, Na, and P were determined in NH_4OAC extracts and the CEC from the quantity of exchangeable NH_4^+ in the KCl extracts. The standard deviations of the CEC measurements ranged from 0.3 to 1.1 mmol/kg.

For the microplastics experiments, three activated biochars with increasing surface areas and different pore size distributions were selected. A glass column was filled with 20 g of the respective biochar material. The filled column was washed with 5 L of tap water to remove fine biochar particles. Microplastic particles of various sizes and shapes were simulated using 2 g of spherical polyethylene (PE) microbeads (10 µm), 2 g of cylindrical, smooth PE pieces (2–3 mm) as well as 2 g fleece shirt fibers. Each column was eluated with 30 fractions of 50-mL tap water each. The fractions were filtered using pre-weighed glass fiber filters that were weighted again after drying for 3 d in a heated 25 °C closed cabinet. Each experiment was performed in triplicate. The biochar material was recycled by intensive washing with tap water and ultrasonication prior to the next use. The MP material recovered was assessed on glass fiber filters using a microscope (Pflugmacher et al. in prep).

3. Results

3.1. Yields, Surface Areas, and Porosities

The produced AC were characterized with respect to yield, elemental composition, surface area, and porosity (Tables 3 and 4). The surface area of pine bark biochar was most affected by increasing the gas flow rate, whereas greater amounts of steam yielded improved surface area results for spruce biochar. Raising the steam proportion to 40% did not induce higher surface area for pine bark biochar despite the larger activation burn-off. Mesoporosity development was greater for both bark biochars using the higher gas flow rate.

Table 3. The surface areas, porosities and activation burn-offs of the produced activated biochars.

Starting Material	Steam Amount	Particle Size	N_2 Gas Flow	Biochar					
				Burn-off	Surface Area	Total Pore Volume	Micropores <2 nm	Mesopores 2–50 nm	Macropores >50 nm
	%		L/h	%	m²/g	cm³/g	%	%	%
Pine bark	-	large	300	61.1	2.2	0.005	8.9	68.9	22.2
Spruce bark	-	large	300	63.1	12	0.016	18.8	81.3	0.0

	Steam Amount	Particle Size	N_2 Gas Flow	Activated Biochar					
				Activation Burn-off	Surface Area	Total Pore Volume	Micropores <2 nm	Mesopores 2–50 nm	Macropores >50 nm
	%		L/h	%	m²/g	cm³/g	%	%	%
Pine bark biochar	30	small	66	25	454	0.165	92.7	6.7	0.6
	30	small	300	28.7	603	0.240	79.6	20.4	0.0
	30	large	300	24.5	615	0.230	86.1	13.9	0.0
	40	small	300	31.6	539	0.200	86.5	13.5	0.0
	40	large	300	27.4	556	0.206	86.9	13.1	0.0
Spruce bark biochar	30	small	66	21.5	272	0.098	91.3	5.8	2.9
	30	large	66	22.3	233	0.084	89.3	8.9	2.4
	30	small	300	22	187	0.071	85.4	10.4	4.2
	30	large	300	20	185	0.072	84.5	9.9	5.6
	40	small	300	23.3	369	0.132	90.0	7.7	2.3
	40	large	300	21.6	222	0.084	86.5	9.9	3.6

Table 4. Elemental composition of the studied biomasses, biochars, and activated biochars.

Parameter	Pine Bark Biomass	Pine Bark Biochar	Pine Bark AC 30% Steam		Pine Bark AC 30% Steam		Pine Bark AC 40% Steam	
N_2 gas flow, L/h	-	-	66		300		300	
Particle size	large	large	small		small	large	small	large
Carbon, wt-%	53 ± 3	77 ± 2	81 ± 3		84 ± 2	90 ± 7	85 ± 2	91 ± 2
Nitrogen, wt-%	0.13 ± 0.02	0.43 ± 0.02	0.14 ± 0.01		0.11 ± 0.00	0.21 ± 0.06	0.17 ± 0.03	0.11 ± 0.02
Hydrogen, wt-%	5.8 ± 0.03	3.1 ± 0.1	0.69 ± 0.08		0.67 ± 0.01	0.70 ± 0.05	0.67 ± 0.03	0.74 ± 0.01
Sulfur, wt-%	0 ± 0	0 ± 0	0 ± 0		0 ± 0	0 ± 0	0 ± 0	0 ± 0
Oxygen, wt-%	41 ± 0	13 ± 0	2.4 ± 0.2		3.6 ± 0.4	3.5 ± 0.4	3.0 ± 0.2	2.8 ± 0.1
Ash, wt-%	1.4 ± 0.0	4.7 ± 0.7	5.8 ± 0.1		5.7 ± 0.3	5.6 ± 0.1	6.2 ± 0.1	6.1 ± 0.1

Parameter	Spruce Bark Biomass	Spruce Bark Biochar	Spruce Bark AC 30% Steam		Spruce Bark AC 30% Steam		Spruce Bark AC 40% Steam	
N_2 gas flow, L/h	-	-	66		300		300	
Particle size	large	large	small		large	small	large	large
Carbon, wt-%	47 ± 0.1	68 ± 3	65 ± 1	68 ± 3	63 ± 2	65 ± 4	57 ± 4	69 ± 8
Nitrogen, wt-%	0.42 ± 0.02	0.62 ± 0.05	0.53 ± 0.04	0.45 ± 0.03	0.40 ± 0.01	0.42 ± 0.04	0.46 ± 0.02	0.43 ± 0.08
Hydrogen, wt-%	5.7 ± 0.07	2.5 ± 0.1	0.70 ± 0.03	0.60 ± 0.01	0.60 ± 0.04	0.65 ± 0.03	0.61 ± 0.03	0.60 ± 0.08
Sulfur, wt-%	0 ± 0	0 ± 0	0 ± 0	0 ± 0	0 ± 0	0 ± 0	0 ± 0	0 ± 0
Oxygen, wt-%	42 ± 0	12 ± 0	4.7 ± 0.1	3.9 ± 0.3	5.3 ± 0.2	5.1 ± 0.5	5.2 ± 0.1	4.9 ± 0.6
Ash, wt-%	5.4 ± 0.1	11 ± 1	15 ± 0	24 ± 1	13 ± 1	15 ± 0	12 ± 0	18 ± 0

The surface areas of the activated biochars increased as a function of the burn-off, however the results were also influenced by pore formation. Large particles exposed to 30% steam, for example, had lower burn-off than small particle size AC (24.5% vs. 28.7%) but higher surface area because of the relatively greater quantity of micropores. The particle size effect was more substantial with pine bark, but the smaller particle size of both bark biochars yielded a slightly greater surface area, as the smaller particle size provides more reaction surface for the steam.

The total burn-off of pine bark biochar because of activation increased in a more linear fashion than that observed for spruce bark biochar (Figure 2). The surface areas of the spruce bark ACs were considerably lesser than those of pine bark ACs. This result may have been due to the high quantity of ash in the spruce bark (Table 4), which can cause pore blockage. The pore size distributions of pine and spruce bark ACs remained similar, consisting of primarily micropores. Increasing the activation time and temperature may have resulted in an increase in the quantity of larger pores.

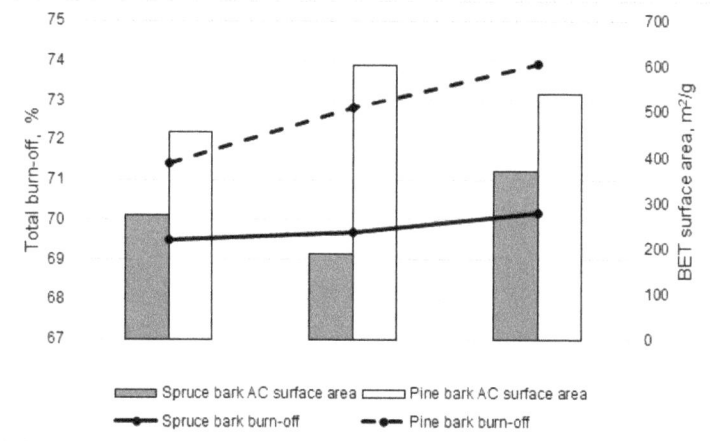

Figure 2. Total burn-off of small particle size activated carbons (ACs) and their surface areas.

Some of the observed differences between pine and spruce bark biochars and ACs may be an artefact of the non-homogenous nature of the biomasses. The bark residues may contain some amounts of stem wood, for example, that has a different pore structure than the bark.

3.2. Elemental Composition and Ash

Both raw materials were analyzed for their elemental and ash composition throughout the treatment chain (Table 4). In the untreated biomass, pine bark had higher carbon content than spruce, whereas nitrogen and oxygen were higher in the spruce bark. No differences in hydrogen content were observed and neither bark biomass contained sulphur.

Comparison of the elemental compositions of biochars and ACs revealed that spruce biochars contained a lesser quantity of carbon compared with the pine biochars. This result may indicate insufficient activation time for spruce biochar, resulting in incomplete carbonization and the inferior surface areas as compared with the pine biochar. Spruce bark ACs contained relatively greater quantities of oxygen and nitrogen than the pine bark ACs. The heteroatom contents of biochars are also biomass dependent. The ash content of spruce biochars and ACs were much greater than ash contents of pine bark carbons. The biochars were analyzed using X-ray fluorescence (XRF) for selected alkali metals (Mg, Ca, K), to determine the cause of the high ash content of the spruce biochar (data not shown). The XRF results showed that spruce biochar contained approximately 16 g/kg of calcium compared to 9 g/kg in the pine biochar. There were no other readily apparent differences between the alkali metal concentrations of the two biochars. The particle size effect was primarily observed in the carbon content, which was higher for the biochars of larger particle size.

3.3. Sorption Experiments

3.3.1. Removal of Organic Contaminants

The capacity of the activated biochars for attenuation of organic contaminants was examined using phenol. The experiments were conducted using biochars and small particle size ACs as there were no distinct differences between the particle sizes. The adsorption capacity increased with both raw materials as a function of surface area (Figure 3). As the material surface area increased to ca. 350 m^2/g the measured adsorption capacity increased in a more linear fashion. Phenol sorption by the biochars with low surface area was nearly equal to that of the higher surface area spruce bark ACs.

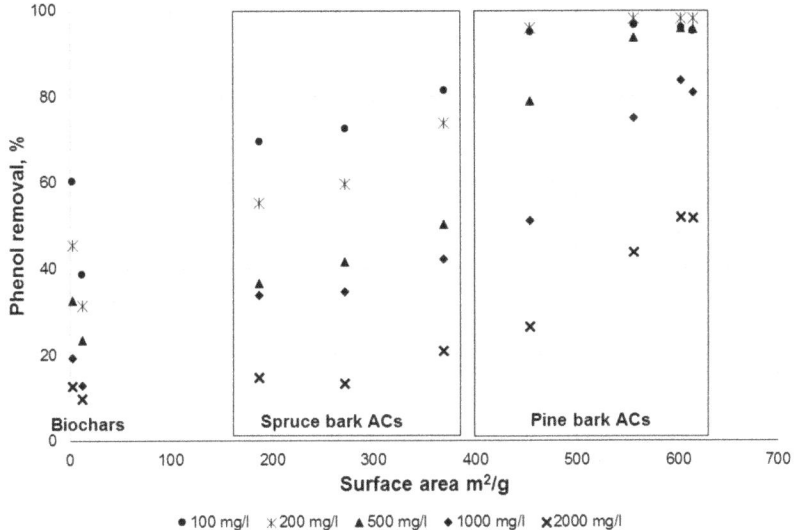

Figure 3. Phenol removal using biochars and activated biochars as a function of surface area. The rectangles are for clarity, presenting the locations of samples in this work, not general separation between spruce and pine bark activated carbons.

The higher surface area of pine bark ACs was associated with greater capacity for phenol removal from solution relative to the spruce bark AC. Nearly 100% phenol attenuation was observed at low solution concentrations (100 and 200 mg/L) and material surface areas ≥540 m^2/g. Phenol was efficiently removed from solutions of concentration 500 mg/L. The quantity of adsorbed phenol increased with solution concentration for both bark ACs, and the maximum removal using 2000 mg/L phenol concentration was approximately 50% for pine bark. The most efficient spruce bark AC removed about 20% of the phenol at the highest concentration tested (2000 mg/L) and 80% at the lowest concentration tested (100 mg/L).

The maximum phenol adsorption capacity obtained at the highest initial solution concentration was observed for pine bark activated with 30% steam and high gas flow, which also had the greatest surface area of the tested carbons (Figure 4, Table 5). The isothermal curves in Figure 4 show increasing adsorption with increasing phenol concentration. Pine bark ACs were efficient adsorbents at all phenol concentrations tested. Phenol adsorption with increasing aqueous concentration did not increase as sharply for spruce bark ACs as for pine bark ACs. The sharp rise of the pine bark curves (Figure 4) indicates a lesser quantity of competing ions for the adsorption sites [30], which is supported by the lower ash concentration of the pine bark ACs. Lower ash content delivers fewer dissolving (alkaline) ions to the solution.

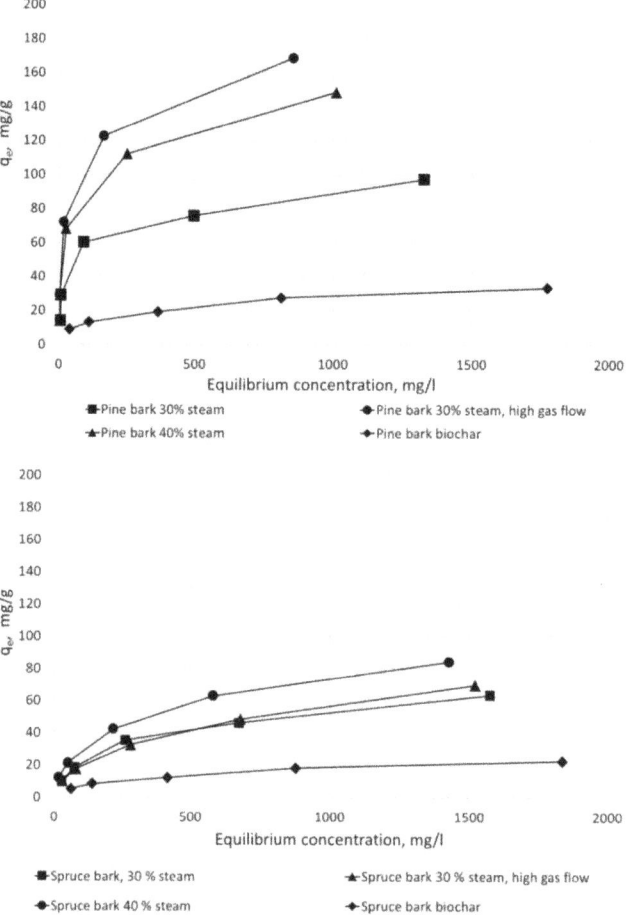

Figure 4. Phenol adsorption capacities of small particle size-activated biochars at equilibrium concentration.

Table 5. The maximum phenol adsorption capacities of the produced activated carbons.

Raw Material	Steam %	N_2 Gas Flow L/h	Particle Size	Surface Area m^2/g	Maximum Adsorption Capacity mg/g
Biochars					
Pine Bark	-	300	large	2.3	33
Spruce bark	-	300	large	12	23
Activated biochars					
Pine Bark	30	66	small	454	97
	30	300	small	603	169
	40	300	small	539	149
Spruce bark	30	66	small	272	64
	30	300	small	187	70
	40	300	small	369	84

3.3.2. Cation Exchange Capacity (CEC)

The cation exchange capacity (CEC) was determined using six cations, but the quantity of released phosphorus was also measured as it is one of the leading causes of eutrophication in natural waters. The quantities of extracted cations in mg/g are presented in Figure 5. The total quantity of exchangeable cations on spruce bark is approximately twice that of the pine bark. Still, of the exchangeable cations, only the amount of exchangeable calcium is substantially greater for spruce. The exchangeable Ca^{2+} result correlates with the much higher ash content of the spruce (12–15 wt-% spruce AC vs. 5–6 wt-% pine AC), as calcium oxides form a significant fraction of the ash components present in wood-based biochar [31]. XRF measurements also confirmed a clear difference between Ca^{2+} contents of the biomasses (data not shown). The CECs, corresponding to the relative quantities of charged surface sites, are greater for pine bark ACs (16–24 mmol/kg for pine, 9–11 mmol/kg for spruce) as these ACs had a relatively larger surface area compared with spruce ACs and, therefore, more surface groups for cations adsorption. The greater CEC of pine bark is due to the higher surface areas of the ACs.

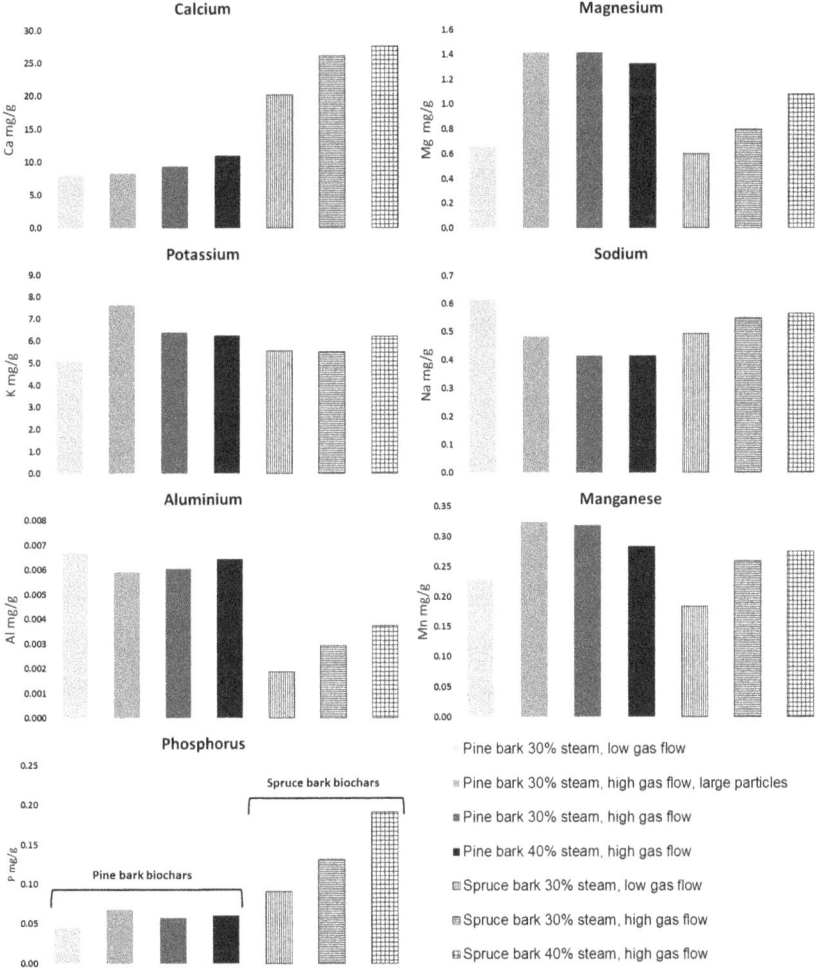

Figure 5. The amounts of exchangeable cations in the activated bark biochars.

A comparison of the individual cations showed K and Na concentrations in the same range for both biomasses. The levels of Mg, Al, and Mn were slightly higher with pine bark, whereas P concentration was higher for spruce. All Fe concentrations were below the detection limit of 0.015 mg/L. These results are consistent with other CEC studies performed for different biochar adsorbents (e.g., [32]).

3.3.3. Microplastics Retention

The selected ACs for MP retention possessed different surface areas and porosities. Pine bark ACs treated with 30 and 40% steam had surface areas of 603 and 556 m^2/g, respectively, and a moderate amount of mesoporosity. The third AC tested for microplastics retention was spruce bark AC with low surface area (187 m^2/g) but broader pore size distribution compared to the pine bark ACs (Table 3). All tested ACs had excellent retention performance for the larger MPs tested. The retention was 100% for the PE particles and nearly 100% for the fleece fibers, with only 1–4 fibers detected after elution. There were no differences between the tested ACs.

The 10-µm spherical microbead retention was not as efficient as for the larger particles, and some differences between the ACs could be detected (Figure 6). The 30 and 40% steam-activated pine biochars exhibited weaker retention of spherical microbead MPs compared to the spruce bark AC and the majority of the spherical microbead MPs eluted rapidly within the first 2–14 fractions. The 40% steam activated spruce biochar exhibited rapid elution of the spherical microbead MP particles, but slightly higher retention as compared with the pine bark AC. The experiments were performed in triplicate and the used biochars were washed and ultrasonicated between the tests. The MPs that could not be removed from the ACs by washing represent retained MP materials (Table 6). These results also support the superior performance of the spruce bark AC despite its lesser surface area.

Table 6. MP material that could not be removed from the activated biochars by washing.

Sample	Retention
Pine bark AC, 30% steam activation	0.165 ± 0.096 g
Pine bark AC, 40% steam activation	0.130 ± 0.040 g
Spruce bark AC, 40% steam activation	0.293 ± 0.046 g

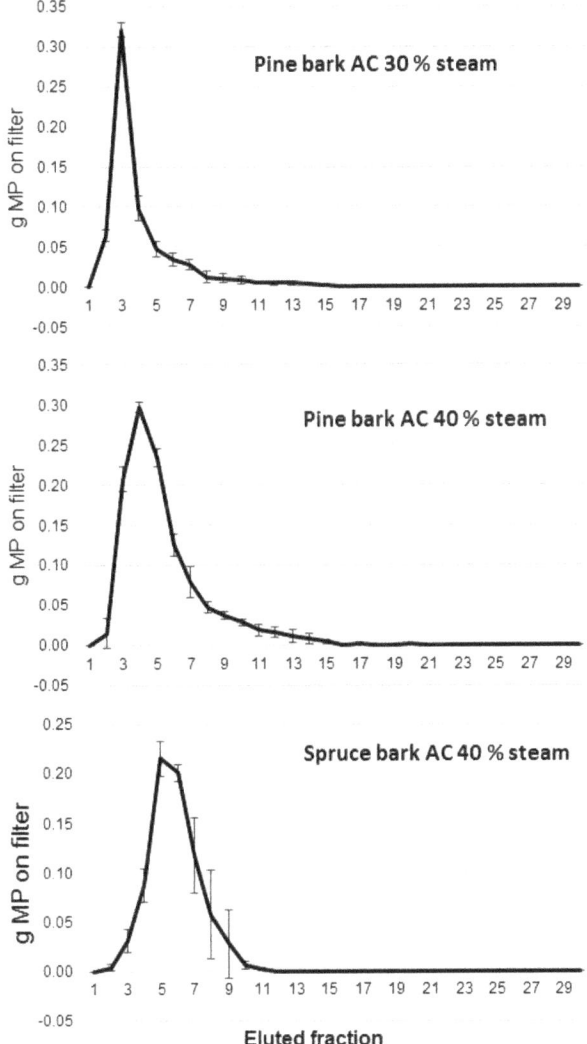

Figure 6. Retention of 10 μm microplastics particles on activated biochar.

4. Discussion

The objective of the research was to evaluate the suitability of two low-cost forest sidestreams, pine and spruce bark, for the production of biochar and activated carbons. Slow pyrolysis at 475 °C was used to generate the biochars. The activation method selected was steam activation (800 °C), which complemented the low-cost perspective of the work. Factors affecting the economic feasibility of biochar production were analyzed, including yield, particle size and key chemical, and physical characteristics. The produced biochars were also examined for selected water purification applications, including organic and inorganic pollutant removal and microplastics retention.

4.1. Raw Material Selection and Particle Size

The raw materials selected for these experiments were two soft-wood bark materials, which were not entirely homogeneous as some stem wood was also mixed in the samples. The effect of the stem wood was, however, likely minimal as the quantities of entrained stem wood were small. The main differences between stem wood and bark are different porosities and amounts of ash. The porosity of the biomass affects the porosity of the biochar and activated biochar [4]. Bark material also has higher ash content than stem wood [33,34]. Mineral impurities could have caused the high ash content of spruce, but the freshness and storage method of the material also affect the ash content [33]. The spruce bark biomass was more fresh and moist compared to the pine bark, which had been stored in dry conditions for a longer time period.

Both bark materials produced microporous ACs, but differences were found in the larger pore sizes. Spruce bark biochar had a higher amount of mesopores before activation, and the pores were enlarged during the activation into macropores. Pine bark biochar did not increase in macropore volume due to activation, but mesoporosity increased by ten-fold. It has been shown that steam activation produces more mesoporous carbons than CO_2 activation [35,36]. Steam reacts more readily with the carbon pore walls and begins to expand the existing pores, whereas CO_2 creates microporosity by reacting primarily with the active sites at the pore centers [35]. Longer activation times are needed with CO_2 if larger pores are required. The results reported herein are supported by those of Zhang et al. (2014) who used poplar wood bark and Cao et al. (2002), who detailed the outcomes of steam activation of poplar wood bark and a mixture of soft wood bark, respectively [19,37]. Both materials produced microporous ACs with similar surface areas (e.g., 555 m^2/g for poplar bark at 800 °C and 60 min). The observed differences in the results are due to the different raw material porosity, activation times, and activation temperatures. The development of mesoscale porosity is important to take into account when selecting raw material and activation method for water purification purposes. The adsorption occurs mainly in the micropores, but larger pores work as channels into the micropores. Bark materials are therefore suitable for AC production as they possess mesoporosity and the required microporosity can be generated by activation.

The biochar particle size before activation did not significantly affect the surface area and porosity. Although small particle size generally favors porosity development because of the larger reactive surface area for the activation reagent, crushing is an extra step in the manufacturing process. Surface area as high as 1361 m^2/g has been achieved for crushed walnut shells (particle size 1–2 mm), for example, using steam activation at 850 °C and 60 min [35]. When aiming to produce ACs with moderate surface areas, however, sufficient results can be achieved with larger biochar particles.

4.2. Adsorption Capacities of Chemical Compounds

The pine bark ACs showed excellent adsorption capacities for phenol. The obtained maximum adsorption capacity for pine bark AC (169 mg/g, Table 5) is comparable to capacities found in the literature for much higher surface area ACs (Table 7). These results along with selected previous work show that efficient phenol removal is not solely dependent on an ultrahigh surface area (Table 7). Phenol adsorption occurs in the micropores, but sufficient mesoporosity is also needed to create channels into the micropores. Steric effects inhibit diffusion of the phenol molecule deeper within the carbon pores [38]. Another parameter influencing results is the solution pH. At pH less than the pK_a value of phenol (9.89 at 298 K), phenol occurs in its non-dissociated form, which is the most active adsorbing form. Above pH 9, phenol dissociates into the phenolate anion and has to compete with other negatively charged ions (e.g., hydroxyl ions) for adsorption sites. Lower pH is, therefore, more favorable for phenol adsorption and respectively, in some cases removal of ash using an acid wash may be necessary [18,39]. The results of this experiment were affected as the high ash content of the ACs increased the solution pH above 9 for most of the AC samples, particularly the spruce ACs.

Table 7. Comparison of phenol adsorption results of current work to literature.

Raw Material	Activation Condition	Surface Area m²/g	Phenol Adsorption Capacity	Reference
Spruce and pine bark	Steam	185–615	64–169 mg/g (=0.7–1.8 mmol/g)	Current work
Cherrystone	KOH and ZnCl$_2$	170–465	Max ~70 mg/g for KOH and ZnCl$_2$	[39]
Eucalyptus seed	ZnCl$_2$	250–300	200 mg/g	[38]
Switchgrass (SG), hardwood (HW) and softwood (SW)	Steam	167–383	1.0–1.6 mmol/g	[15]
Rattan sawdust	KOH	1083	149 mg/g	[40]
Oil palm shells	Steam	988	166 mg/g	[17]
Rapeseed and raspberry seed cakes	Steam, CO$_2$	141–1179	Max ~250 mg/g steam, ~170 mg/g CO$_2$	[18]
Fir woods and pistachio shells	Steam, KOH	1009–1096	2.58 and 2.72 mol/kg steam, 2.74 and 3.03 mol/kg KOH	[41]

The CEC measures an adsorbent's capability to sorb positively charged ions and is, therefore, directly related to adsorbents' applicability for water treatment. The cation sorption capacity of biochars and ACs is dependent on various parameters such as surface charge density, chemical properties (atom/molecule radius, solubility, pK_a), and solution ion density [42]. It has also been found that ion exchange is the dominating mode by which biochars from fast pyrolysis adsorb metal ions [43]. Higher CEC estimates for biochars can be found in literature (e.g., [44]) compared to current work. Comparison is, however, difficult because of the high variability in the used CEC and carbonization methodology and feedstock materials [45]. Both bark ACs had acceptable CECs for application in water treatment, with pine bark ACs performing slightly better because of their higher surface area. The quantities of potentially environmentally harmful elements, such as aluminum and phosphorus, were low. The alkalinity caused by the high ash contents of both bark ACs favors metal adsorption.

4.3. Microplastics

The investigations of microplastics removal showed great potential for their recovery using activated biochar. The large particles were retained completely, but the micrometer-scale MP particles did not absorb as efficiently. The sorption mechanism of microplastics in biochars remains unknown, but the existence of much larger pores may facilitate micro- and nanoplastics retention. This is supported by the superior performance of the spruce bark AC in the present experiments, which had relatively low surface area but included macro-scale porosity. The retention mechanism of large particles is most likely physical attachment between the biochar particles. Thus, biochar surface roughness may be of benefit. These experiments were performed using the activated carbons, but the results indicate that non-activated biochar may as well be suitable for removal of larger MP particles. There is a need to develop more detailed knowledge regarding the mechanisms of MP retention by biochars. Biochars present an inexpensive means of removing MPs from waters with the added benefit of the simultaneous removal of other contaminants.

5. Conclusions

Both spruce and pine bark were suitable for biochar and activated biochar production for water treatment purposes, but some differences were observed between the two materials. The surface area of spruce bark AC remained less than the surface area of pine bark AC. One of the reasons for this was likely due to the high ash content of spruce bark, causing pore-clogging through carbonate formation. The lower carbon content of spruce bark also indicated that longer activation time or greater steam proportion may have yielded a higher degree of carbonization and greater porosity development. Pine bark activated using 30% steam and the higher rate of gas flow performed best in the phenol adsorption tests and had relatively greater CEC due to its larger surface area. Spruce ACs possessed a greater quantity of exchangeable cations, largely due to a high Ca content. The activated biochars tested herein efficiently removed the larger microplastics particles. The removal of 10-µm spherical microplastics was not as sufficient. Results indicated that higher meso-and macropore contents could be beneficial for the removal of the smallest MP particles. In conclusion, steam activation is a suitable method for activated biochar production. Surface areas in the range of 400–600 m^2/g are adequate for the efficient removal of contaminants from storm and wastewater. The results support the economic feasibility of steam-activated biochar for such water purification purposes. Carbonization provides an added value use for the ligno-cellulosic forest residues with environmentally benign applications. Removal of microplastics using biochar requires further research, particularly regarding the recovery of micrometer-scale MP size fractions as well as the identification of the retention mechanism. In addition, the economic feasibility of biochar production using different sidestreams or combinations thereof should be further considered in the biochar and activated biochar-related research.

Author Contributions: V.S. and P.K. conceptualized and designed the research topic; V.S. and P.K. supervised the investigation; V.S., S.P., H.R., and L.W. designed and executed the laboratory work; V.S. prepared the original manuscript draft; all authors contributed to reviewing and editing the manuscript. All authors have read and agreed to the published version of the manuscript.

Funding: This work has been funded by the Ministry of Agriculture and Forestry of Finland, agreement number: 754/03.02.06.00/2018.

Acknowledgments: The authors are grateful to Sari Tuomikoski (Oulu University, Finland) for performing the phenol adsorption tests and to Anssi Källi (VTT) for extensive support with the slow pyrolysis oven. We thank the laboratory staff at VTT for performing the analyses. We are grateful for the support by the FinnCERES Materials Bioeconomy Ecosystem.

Conflicts of Interest: The authors declare no conflict of interest.

References

1. Mohanty, S.K.; Valenca, R.; Berger, A.W.; Yu, I.K.M.; Xiong, X.; Saunders, T.M.; Tsang, D.C.W. Plenty of room for carbon on the ground: Potential applications of biochar for stormwater treatment. *Sci. Total Environ.* **2018**, *625*, 1644–1658. [CrossRef] [PubMed]
2. Sommer, F.; Dietze, V.; Baum, A.; Sauer, J.; Gilge, S.; Maschowski, C.; Gieré, R. Tire Abrasion as a Major Source of Microplastics in the Environment. *Aerosol Air Qual. Res.* **2018**, *18*, 2014–2028. [CrossRef]
3. Mohan, D.; Sarswat, A.; Ok, Y.S.; Pittman, C.U. Organic and inorganic contaminants removal from water with biochar, a renewable, low cost and sustainable adsorbent—A critical review. *Bioresour. Technol.* **2014**, *160*, 191–202. [CrossRef] [PubMed]
4. Siipola, V.; Tamminen, T.; Källi, A.; Lahti, R.; Romar, H.; Rasa, K.; Keskinen, R.; Hyväluoma, J.; Hannula, M.; Wikberg, H. Effects of biomass type, carbonization process, and activation method on the properties of bio-based activated carbons. *Bioresource* **2018**, *13*, 5976–6002.
5. Ahmed, M.B.; Zhou, J.L.; Ngo, H.H.; Guo, W.; Chen, M. Progress in the preparation and application of modified biochar for improved contaminant removal from water and wastewater. *Bioresour. Technol.* **2016**, *214*, 836–851. [CrossRef] [PubMed]
6. Fic, K.; Platek, A.; Piwek, J.; Frackowiak, E. Sustainable materials for electrochemical capacitors. *Mater. Today* **2018**, *21*, 437–454. [CrossRef]
7. Raymundo-Piñero, E.; Leroux, F.; Béguin, F. A High-Performance Carbon for Supercapacitors Obtained by Carbonization of a Seaweed Biopolymer. *Adv. Mater.* **2006**, *18*, 1877–1882. [CrossRef]
8. Bridgewater, A. Biomass fast pyrolysis. *Therm. Sci.* **2004**, *8*, 21–50. [CrossRef]
9. Arriagada, R.; García, R.; Molina-Sabio, M.; Rodriguez-Reinoso, F. Effect of steam activation on the porosity and chemical nature of activated carbons from Eucalyptus globulus and peach stones. *Microporous Mater.* **1997**, *8*, 123–130. [CrossRef]
10. Calisto, V.; Ferreira, C.I.A.; Oliveira, J.A.B.P.; Otero, M.; Esteves, V.I. Adsorptive removal of pharmaceuticals from water by commercial and waste-based carbons. *J. Environ. Manag.* **2015**, *152*, 83–90. [CrossRef]
11. Huggins, T.M.; Haeger, A.; Biffinger, J.C.; Ren, Z.J. Granular biochar compared with activated carbon for wastewater treatment and resource recovery. *Water Res.* **2016**, *94*, 225–232. [CrossRef]

12. Hakala, J.; Kangas, P.; Penttilä, K.; Alarotu, M.; Björnström, M.; Koukkari, P. *Replacing Coal Used in Steelmaking with Biocarbon from Forest Industry Side Streams*; VTT Technical Research Centre of Finland: Espoo, Finland, 2019.
13. Farrokh, N.T.; Suopajärvi, H.; Mattila, O.; Umeki, K.; Phounglamcheik, A.; Romar, H.; Sulasalmi, P.; Fabritius, T. Slow pyrolysis of by-product lignin from wood-based ethanol production—A detailed analysis of the produced chars. *Energy* **2018**, *164*, 112–123. [CrossRef]
14. Kwak, J.-H.; Islam, M.S.; Wang, S.; Messele, S.A.; Naeth, M.A.; El-Din, M.G.; Chang, S.X. Biochar properties and lead(II) adsorption capacity depend on feedstock type, pyrolysis temperature, and steam activation. *Chemosphere* **2019**, *231*, 393–404. [CrossRef]
15. Han, Y.; Boateng, A.A.; Qi, P.X.; Lima, I.M.; Chang, J. Heavy metal and phenol adsorptive properties of biochars from pyrolyzed switchgrass and woody biomass in correlation with surface properties. *J. Environ. Manag.* **2013**, *118*, 196–204. [CrossRef]
16. Lua, A.C.; Guo, J. Activated carbon prepared from oil palm stone by one-step CO2 activation for gaseous pollutant removal. *Carbon N. Y.* **2000**, *38*, 1089–1097. [CrossRef]
17. Jia, Q.; Lua, A.C. Effects of pyrolysis conditions on the physical characteristics of oil-palm-shell activated carbons used in aqueous phase phenol adsorption. *J. Anal. Appl. Pyrolysis* **2008**, *83*, 175–179. [CrossRef]
18. Smets, K.; De Jong, M.; Lupul, I.; Gryglewicz, G.; Schreurs, S.; Carleer, R.; Yperman, J. Rapeseed and raspberry seed cakes as inexpensive raw materials in the production of activated carbon by physical activation: Effect of activation conditions on textural and phenol adsorption characteristics. *Materials (Basel)* **2016**, *9*, 565. [CrossRef]
19. Cao, N.; Darmstadt, H.; Soutric, F.; Roy, C. Thermogravimetric study on the steam activation of charcoals obtained by vacuum and atmospheric pyrolysis of softwood bark residues. *Carbon N. Y.* **2002**, *40*, 471–479. [CrossRef]
20. Zhang, T.; Walawender, W.; Fan, L.; Fan, M.; Daugaard, D.; Brown, R. Preparation of activated carbon from forest and agricultural residues through CO activation. *Chem. Eng. J.* **2004**, *105*, 53–59. [CrossRef]
21. Andrady, A.L. Microplastics in the marine environment. *Mar. Pollut. Bull.* **2011**, *62*, 1596–1605. [CrossRef]
22. Herbort, A.F.; Sturm, M.T.; Fiedler, S.; Abkai, G.; Schuhen, K. Alkoxy-silyl Induced Agglomeration: A New Approach for the Sustainable Removal of Microplastic from Aquatic Systems. *J. Polym. Environ.* **2018**, *26*, 4258–4270. [CrossRef]
23. Ivar do Sul, J.A.; Costa, M.F. The present and future of microplastic pollution in the marine environment. *Environ. Pollut.* **2014**, *185*, 352–364. [CrossRef] [PubMed]
24. Li, J.; Liu, H.; Paul Chen, J. Microplastics in freshwater systems: A review on occurrence, environmental effects, and methods for microplastics detection. *Water Res.* **2018**, *137*, 362–374. [CrossRef] [PubMed]
25. Herbort, A.F.; Sturm, M.T.; Schuhen, K. A new approach for the agglomeration and subsequent removal of polyethylene, polypropylene, and mixtures of both from freshwater systems—A case study. *Environ. Sci. Pollut. Res.* **2018**, *25*, 15226–15234. [CrossRef] [PubMed]
26. Lares, M.; Ncibi, M.C.; Sillanpää, M.; Sillanpää, M. Occurrence, identification and removal of microplastic particles and fibers in conventional activated sludge process and advanced MBR technology. *Water Res.* **2018**, *133*, 236–246. [CrossRef]
27. Brunauer, S.; Emmett, P.H.; Teller, E. Adsorption of gases in multimolecular layers. *J. Am. Chem. Soc.* **1938**, *60*, 309. [CrossRef]
28. Lastoskie, C.; Gubbins, K.E.; Quirke, N. Pore size heterogeneity and the carbon slit pore: A density functional theory model. *Langmuir* **1993**, *9*, 2693–2702. [CrossRef]
29. Sumner, M.E.; Miller, W.P. Cation Exchange Capacity and Exchange Coefficients. In *Methods of Soil Analysis Part 3: Chemical Methods*; Sparks, D.L., Ed.; Soil Science Society of America: Madison, WI, USA, 1996; pp. 1201–1230.
30. Medellin-Castillo, N.A.; Padilla-Ortega, E.; Regules-Martínez, M.C.; Leyva-Ramos, R.; Ocampo-Pérez, R.; Carranza-Alvarez, C. Single and competitive adsorption of Cd(II) and Pb(II) ions from aqueous solutions onto industrial chili seeds (Capsicum annuum) waste. *Sustain. Environ. Res.* **2017**, *27*, 61–69. [CrossRef]
31. Tortosa Masiá, A.A.; Buhre, B.J.P.; Gupta, R.P.; Wall, T.F. Characterising ash of biomass and waste. *Fuel Process. Technol.* **2007**, *88*, 1071–1081. [CrossRef]

32. Heikkinen, J.; Keskinen, R.; Soinne, H.; Hyväluoma, J.; Nikama, J.; Wikberg, H.; Källi, A.; Siipola, V.; Melkior, T.; Dupont, C.; et al. Possibilities to improve soil aggregate stability using biochars derived from various biomasses through slow pyrolysis, hydrothermal carbonization, or torrefaction. *Geoderma* **2019**, *344*, 40–49. [CrossRef]
33. Lehtikangas, P. Quality properties of pelletised sawdust, logging residues and bark. *Biomass Bioenergy* **2001**, *20*, 351–360. [CrossRef]
34. Wang, L.; Barta-Rajnai, E.; Skreiberg, Ø.; Khalil, R.; Czégény, Z.; Jakab, E.; Barta, Z.; Grønli, M. Effect of torrefaction on physiochemical characteristics and grindability of stem wood, stump and bark. *Appl. Energy* **2018**, *227*, 137–148. [CrossRef]
35. González, J.F.; Román, S.; González-García, C.M.; Nabais, J.M.V.; Ortiz, A.L. Porosity Development in Activated Carbons Prepared from Walnut Shells by Carbon Dioxide or Steam Activation. *Ind. Eng. Chem. Res.* **2009**, *48*, 7474–7481. [CrossRef]
36. Molina-Sabio, M.; Gonzalez, M.T.; Rodriguez-Reinoso, F.; Sepúlveda-Escribano, A. Effect of steam and carbon dioxide activation in the micropore size distribution of activated carbon. *Carbon N. Y.* **1996**, *34*, 505–509. [CrossRef]
37. Zhang, J.; Zhang, W. Preparation and Characteristics of Activated Carbon from Wood Bark and Its Use for Adsorption of Cu (II). *Mater. Sci.* **2014**, *20*, 474–478. [CrossRef]
38. Rincón-Silva, N.G.; Moreno-Piraján, J.C.; Giraldo, L. Equilibrium, kinetics and thermodynamics study of phenols adsorption onto activated carbon obtained from lignocellulosic material (Eucalyptus Globulus labill seed). *Adsorption* **2016**, *22*, 33–48. [CrossRef]
39. Beker, U.; Ganbold, B.; Dertli, H.; Gülbayir, D.D. Adsorption of phenol by activated carbon: Influence of activation methods and solution pH. *Energy Convers. Manag.* **2010**, *51*, 235–240. [CrossRef]
40. Hameed, B.H.; Rahman, A.A. Removal of phenol from aqueous solutions by adsorption onto activated carbon prepared from biomass material. *J. Hazard. Mater.* **2008**, *160*, 576–581. [CrossRef]
41. Wu, F.-C.; Tseng, R.-L.; Juang, R.-S. Comparisons of porous and adsorption properties of carbons activated by steam and KOH. *J. Colloid Interface Sci.* **2005**, *283*, 49–56. [CrossRef]
42. Dias, J.M.; Alvim-Ferraz, M.C.M.; Almeida, M.F.; Rivera-Utrilla, J.; Sánchez-Polo, M. Waste materials for activated carbon preparation and its use in aqueous-phase treatment: A review. *J. Environ. Manag.* **2007**, *85*, 833–846. [CrossRef]
43. Mohan, D.; Pittman, C.U.; Bricka, M.; Smith, F.; Yancey, B.; Mohammad, J.; Steele, P.H.; Alexandre-Franco, M.F.; Gómez-Serrano, V.; Gong, H. Sorption of arsenic, cadmium, and lead by chars produced from fast pyrolysis of wood and bark during bio-oil production. *J. Colloid Interface Sci.* **2007**, *310*, 57–73. [CrossRef] [PubMed]
44. Yuan, J.H.; Xu, R.K.; Zhang, H. The forms of alkalis in the biochar produced from crop residues at different temperatures. *Bioresour. Technol.* **2011**, *102*, 3488–3497. [CrossRef] [PubMed]
45. Munera-Echeverri, J.L.; Martinsen, V.; Strand, L.T.; Zivanovic, V.; Cornelissen, G.; Mulder, J. Cation exchange capacity of biochar: An urgent method modification. *Sci. Total Environ.* **2018**, *642*, 190–197. [CrossRef] [PubMed]

© 2020 by the authors. Licensee MDPI, Basel, Switzerland. This article is an open access article distributed under the terms and conditions of the Creative Commons Attribution (CC BY) license (http://creativecommons.org/licenses/by/4.0/).

Article

Adsorption Performance of Physically Activated Biochars for Postcombustion CO_2 Capture from Dry and Humid Flue Gas

Joan J. Manyà *, David García-Morcate and Belén González

Aragón Institute of Engineering Research (I3A), Technological College of Huesca, University of Zaragoza, crta. Cuarte s/n, E-22071 Huesca, Spain; 657247@unizar.es (D.G.-M.); belenglez@unizar.es (B.G.)
* Correspondence: joanjoma@unizar.es

Received: 14 November 2019; Accepted: 30 December 2019; Published: 3 January 2020

Abstract: In the present study, the performance of four biomass-derived physically activated biochars for dynamic CO_2 capture was assessed. Biochars were first produced from vine shoots and wheat straw pellets through slow pyrolysis (at pressures of 0.1 and 0.5 MPa) and then activated with CO_2 (at 0.1 MPa and 800 °C) up to different degrees of burn-off. Cyclic adsorption-desorption measurements were conducted under both dry and humid conditions using a packed-bed of adsorbent at relatively short residence times of the gas phase (12–13 s). The adsorbent prepared from the vine shoots-derived biochar obtained by atmospheric pyrolysis, which showed the most hierarchical pore size distribution, exhibited a good and stable performance under dry conditions and at an adsorption temperature of 50 °C, due to the enhanced CO_2 adsorption and desorption rates. However, the presence of relatively high concentrations of water vapor in the feeding gas clearly interfered with the CO_2 adsorption mechanism, leading to significantly shorter breakthrough times. In this case, the highest percentages of a used bed were achieved by one of the other activated biochars tested, which was prepared from the wheat straw-derived biochar obtained by pressurized pyrolysis.

Keywords: postcombustion CO_2 capture; biomass-based adsorbents; cyclic breakthrough measurements; selectivity CO_2/N_2; humid conditions; hierarchical porosity

---

1. Introduction

Global warming is gaining wider recognition in the world. Fossil fuel power plants are responsible for approximately one-third of global CO_2 emissions to the atmosphere. Therefore, removing CO_2 from low-pressure flue gas (i.e., CO_2 capture in postcombustion) has been the focus of extensive research over the last few decades. As an alternative to the energy-intensive amine-based chemical absorption processes, CO_2 capture via adsorption on renewable biomass-derived carbons has gained increased interest, since these adsorbents are relatively cheap, require low energy for regeneration, and show a relatively good tolerance to moisture existing in flue gas [1].

An increasing number of studies have focused on producing activated carbons (ACs) from different biomass precursors (through physical or chemical activation) and assessing their performance in terms of CO_2 uptake at 10–15 kPa and selectivity towards CO_2 over N_2 [2–10]. At equilibrium conditions (i.e., adsorption isotherms of pure components), the biomass-derived ACs exhibit a relatively high CO_2 adsorption capacity, usually in the range of 1.0–2.0 mmol g^{-1} at 15 kPa and 25 °C [9]. Regarding the apparent CO_2/N_2 selectivity, which is defined as the ratio of molar uptakes divided by the ratio of partial pressures, values in the range of 8–14 (deduced from single component adsorption data at 25–50 °C) have been reported in previous studies [6,11,12]. It is generally accepted that the narrower micropores of ACs are the main responsible for the physical adsorption of CO_2 at low pressure. Therefore, and for a given AC, the CO_2 uptake at 10–15 kPa is primarily a function of its ultra-micropore volume (pore width below 0.7 nm).

From a more realistic point of view, studies focusing on the behavior of biomass-derived ACs under dynamic conditions are explicitly required, since cyclic breakthrough measurements are more representative of practical separations. Adsorbents have to exhibit relatively fast adsorption and desorption kinetics for their successful implementation in swing adsorption processes. In this sense, the number of studies available is relatively modest compared to the large amount of research already conducted on developing ACs from biomass precursors. In a very interesting study [13], the performance under dynamic conditions of olive stone and almond shells-derived ACs (produced by single-step activation with CO_2 at 800 °C) was tested. For a feed stream composed of a binary mixture of CO_2/N_2 (14/86 vol. %) at 50 °C and 120 kPa, the authors reported good CO_2 adsorption capacities (around 0.6 mmol g^{-1}) and apparent selectivities CO_2/N_2 (20–30) for fresh adsorbent. The fact that the selectivity increased for dynamic measurements was mainly explained by a certain decrease in the adsorbed amount of the weaker adsorbate (N_2) in the presence of the stronger one (CO_2). In other words, the apparent CO_2/N_2 selectivity estimated from the single-component adsorption isotherms could be lower than that measured under dynamic conditions for binary mixtures.

Shahkarami et al., [14] assessed the performance of several agricultural waste-based carbons, which were produced through slow or fast pyrolysis and then activated with KOH. The best results were obtained for a pinewood sawdust-derived AC, which exhibited high CO_2 uptake (1.8 mmol g^{-1} after nine cycles) and apparent selectivity CO_2/N_2 (29) at 25 °C and 101.3 kPa (for a binary mixture CO_2/N_2 15/85 vol. %). However, the temperature tested by Shahkarami et al., [14] was not within the typical range for postcombustion flue gas (40–60 °C). Moreover, the gas residence time selected by the authors (31 s) was relatively long. In this sense, capital costs for large-scale units operating at these conditions could be unacceptable. More representative operating conditions (adsorption temperatures of 45 and 60 °C, and gas residence times of 16–17 s) were considered in the study by Shafeeyan et al., [15], in which CO_2 adsorption capacities of 0.57 and 0.47 mmol g^{-1} (at 45 and 60 °C, respectively) were reported for a granular palm shell-derived N-doped AC.

The three above-mentioned studies, however, were conducted under dry conditions. Given that the real flue gas contains a certain amount of moisture (which is unavoidable in practice), the effect of water vapor on the CO_2 adsorption capacity and CO_2/N_2 selectivity under dynamic conditions should also be assessed. In this regard, there are relatively few experimental studies available in the literature that have explored the performance of biomass-derived ACs under humid conditions. Among them, it is worth mentioning the study by Xu et al., [16], where the performance of a commercial coconut shell-derived AC was tested during a vacuum swing adsorption (VSA) cycle under humid conditions. In the experiment conducted by Xu et al., the adsorption step was conducted at an absolute pressure of 120 kPa, at a bed temperature of 60 °C, and using a feed gas steam composed of CO_2 (12 vol. %), water (4.8 vol. %), and air (balance). Surprisingly, the authors reported an almost identical recovery and purity of CO_2 in the presence of water as compared to the dry condition. However, this study did not assess the performance of the tested adsorbent over sequential adsorption/desorption cycles.

Durán et al., [17] investigated the ability of a pine sawdust-based AC to selectively adsorb CO_2 from wet biogas under dynamic conditions (at an absolute pressure of 135 kPa, 30 °C, and using a quaternary gas mixture of $CH_4/CO_2/N_2/H_2O_{(v)}$ with a relative humidity of 55%). The authors observed that, under humid conditions, the CO_2 uptake and breakthrough time decreased in comparison with those measured during the breakthrough experiments under dry conditions. In fact, the breakthrough time was reduced by a factor of 1.85 for a molar ratio CO_2/CH_4 of 30/70 in the feed gas mixture. However, as a remarkable finding, Durán et al., [17] stated that the presence of water vapor on the bed can promote the CO_2 adsorption over CH_4, leading to a more efficient separation with respect to the dry case.

The specific aim of the present study is to explore the feasibility of using biomass-derived physically activated biochars as cost-effective adsorbents for CO_2 capture from humid postcombustion flue gas.

The performance of physically activated carbons based on vine shoots as well as wheat straw pellets was assessed via cyclic breakthrough experiments under dry and severe humid conditions (relative humidity of 100%). Both biomass sources were selected on the basis of their potential sustainability, since they are agricultural wastes (residual biomass) that do not compete with food, feed or timber production either directly or indirectly within a specific area.

2. Materials and Methods

2.1. Biochar-Based Adsorbents

Two types of agricultural wastes were used as carbon precursors: (1) vine shoots (VS) from a local vineyard (which were previously cut into pieces of 1.0–3.5 cm in length), and (2) free-binder wheat straw pellets (WS) from a Belgian company (7 mm OD and approximately 12 mm long). Proximate and ultimate analyses were conducted for both biomass sources (further details on these measurements are given in the Supplementary Materials).

The biomass feedstocks were pyrolyzed under nitrogen in a packed-bed reactor at a highest temperature (HTT) of 500 and 600 °C for WS and VS, respectively. Two different values of absolute pressures (0.1 and 0.5 MPa) were tested. Details on the pyrolysis device and experimental procedure are available elsewhere [18,19]. Briefly, approximately 400 g of raw biomass were heated at an average heating rate of 5 °C min^{-1} to the highest temperature with a soaking time of 60 min at this temperature. The volumetric flow rate of the carrier gas (N_2) within the reactor at the highest temperature was kept constant, regardless of the pressure applied, by properly adjusting the mass flow rate. Assuming an entire reactor's void-volume fraction of 0.9, the above-mentioned flow rate led to a as residence time of the carrier gas within the reactor of 100 s.

The obtained biochars (which were milled and sieved to a particle size distribution of 0.212–1.41 mm) were physically activated with CO_2 at 800 °C in a quartz tubular fixed-bed reactor (ID = 16 mm), which was placed in a vertical tube furnace (model EVA 12/300 from Carbolite Gero, UK). The reactor was filled with biochar at a bed height of 300 mm and then heated at 10 °C min^{-1} under a steady flow of N_2 (500 mL min^{-1} STP) at atmospheric pressure. Once the bed reached the desired temperature (800 °C), the gas flow was switched from pure N_2 to a mixture of CO_2 and N_2 (in the volume ratio of 20:80) at a total mass flow rate of 500 mL min^{-1} STP. Given that the porosity of raw biochars ranged from 0.65 to 0.75, a gas-hourly space velocity (GHSV) of 5600–7800 h^{-1} was estimated. A holding time of 1 h was applied in order to obtain a burn-off degree of 20–30%, depending on the reactivity of the given precursor. To properly determine the degree of burn-off, biochars were previously pyrolyzed (under N_2 atmosphere) at a highest temperature of 800 °C. Table 1 summarizes the preparation procedure for the ACs tested in the present study. The proximate and ultimate analyses of the activated biochars were also performed.

Table 1. Summary of the preparation procedure for the physically activated carbons (ACs) tested in the present study.

Activated Carbon	Biomass Source	Pyrolysis Conditions	Activation Conditions
AC_VS_600	Vine shoots	HTT = 600 °C; p = 101 kPa	With CO_2 at 800 °C and atmospheric pressure; 1 h holding time
AC_VS_600_P	Vine shoots	HTT = 600 °C; p = 500 kPa	
AC_WS_500	Wheat straw	HTT = 500 °C; p = 101 kPa	
AC_WS_500_P	Wheat straw	HTT = 500 °C; p = 500 kPa	

2.2. Static Gas Adsorption Measurements

For textural characterization purposes, N_2 adsorption/desorption isotherms at −196 °C and CO_2 adsorption isotherms at 0 °C were acquired using an ASAP 2020 gas sorption analyzer from Micromeritics (USA). Samples (around 120–180 mg) were previously degassed under dynamic vacuum conditions to constant weight at a temperature of 150 °C. From the N_2 adsorption/desorption isotherms at −196 °C, we estimated the apparent specific surface area (S_{BET}), micropore volume (V_{mic}), mesopore volume (V_{mes}), and pore size distribution (PSD) for pore sizes above 0.9–1.0 nm. The data from the CO_2 adsorption isotherms at 0 °C were used to estimate the ultra-micropore volume (V_{ultra}, for pore sizes lower than 0.7 nm) and the PSD for narrow micropores. Further details regarding the procedures used to estimate the above-mentioned parameters are given in the Supplementary Material.

The single-component adsorption isotherms (for both CO_2 and N_2), at temperatures of 25 and 50 °C, on the activated biochars (at pressures ranging from 0 to 101 kPa) were measured using the same above-mentioned device as well as the same degasification procedure.

2.3. Dynamic Breakthrough Experiments

Cyclic adsorption-desorption measurements were conducted in a custom-built device, whose schematic diagram is given in Figure 1. The adsorption column consisted of an AISI 316-L tubular reactor (250 mm length; 20.9 mm ID), which was heated by a PID controlled electric furnace. A K-type thermocouple was placed in the center of the packed bed. Experiments were performed using an initial mass of adsorbent that corresponded to a bed height of 220 mm. Small particle sizes (i.e., below 0.212 mm) were discarded to avoid excessive pressure drop. A small portion of fused quartz wool was placed at both ends of the column to prevent the loss of adsorbent.

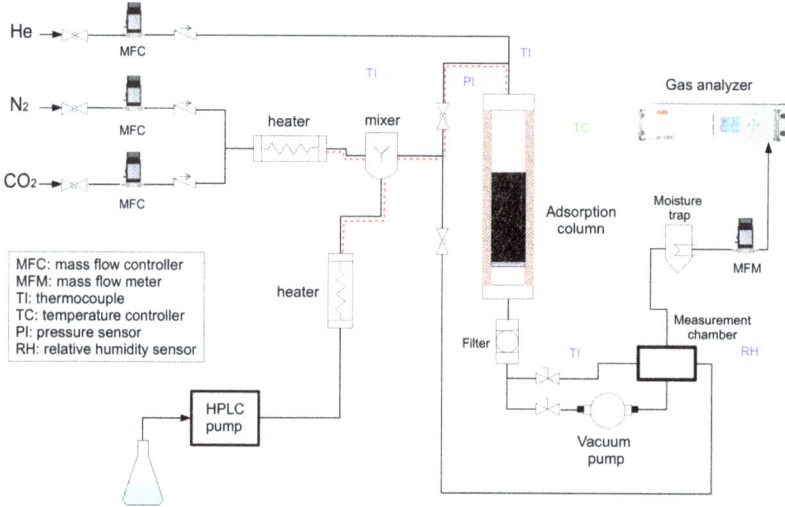

Figure 1. Schematic diagram of the packed-bed setup used for dynamic breakthrough tests.

The feed flow rates of pure N_2, CO_2, and helium were adjusted using mass flow controllers (Brooks, USA). A mass flow meter (Omega Engineering, UK) was used to measure the outlet gas flow rate. CO_2 concentrations were measured using a continuous gas analyzer (ABB model EL3020, Germany). During the adsorption step, the absolute pressure was kept constant at a value of 110–115 kPa. The desorption step was conducted under vacuum (VSA) using an oil rotary vane vacuum pump (Busch, Germany), which assured an absolute pressure of 10 kPa at suction flow rate up to 0.33 m^3 s^{-1} STP.

In a typical cyclic experiment under dry conditions, the adsorbent was initially outgassed by purging with helium at 150 °C for 4 h. Afterwards, the bed was cooled down to the desired temperature (25 or 50 °C). Adsorption was then started by switching the feeding gas from helium to a dry mixture of N_2 and CO_2 (with a CO_2 composition of 13.75–14.25 vol. %) at a total flow rate of 3.67 cm^3 s^{-1} STP. Under these operating conditions and assuming a void fraction of 0.65–0.75 (estimated from the measured pressure drop across the bed, which ranged from 165 to 430 Pa, and the Ergun equation), the gas residence time within the packed bed was 13.2 ± 1.3 and 12.2 ± 1.2 s at 25 and 50 °C, respectively. Once saturation was reached, the feeding gas was turned off and the column outlet was connected to the vacuum pump. During the desorption step, the temperature of the packed bed was kept at the same value than for the preceding adsorption step.

For the dynamic adsorption measurements under humid conditions (at a temperature of 50 °C), 0.0226 cm^3 min^{-1} of water was added to the dry feeding gas (at the same flow rate as mentioned above) using a HPLC pump (model 521 from Analytical Scientific Instruments, USA). The amount of added water corresponded to a relative humidity (RH) of the wet gas of approximately 100% at the operating conditions of the column. The water content of the wet feeding gas was 11.2 vol. %.

Since the dead volume of the system was relatively large, especially for the experiments conducted under humid conditions (in which the moisture trap and measurement chamber were also used), blank adsorption tests were carried out to estimate the time required to displace the dead volume. The blank tests were conducted in both dry and humid conditions for an empty and filled (with glass beads of 1.0 mm OD) adsorption column.

3. Results and Discussion

3.1. Properties of the Used Adsorbents

The results from proximate and ultimate analyses for the raw biomasses and activated biochars are listed in Table 2. Both raw biomass samples exhibited relatively high carbon contents (47.1 and 44.2 wt. % in dry and ash-free basis for VS and WS, respectively) and low ash contents (below 5 wt. %), indicating their potential to be used as precursors for activated carbons. After pyrolysis and further activation with CO_2, an expected higher carbon content (above 85 wt. % in daf basis), at the expense of both hydrogen and oxygen, was observed. However, the ash content of activated biochars was relatively high (from 9.00 to 16.5 wt. %). This can be explained by the loss of organic matter during the thermal treatments, leading to final mass yields of activated carbons (from the raw feedstock) within the range of 18–20 wt. % (dry basis), and the absence of any acid-leaching post-treatment. Despite the fact that the presence of ash can decrease the surface area available for adsorption, biomass-derived ash constituents (especially K) could have positive effect on CO_2 adsorption, as reported by Yin et al., [20]. At this point, it should be noted that the contents of potassium (as K_2O) in the ashes from the biomass sources used here were found to be considerably high: 18.4 wt. % for VS [18] and 53.2 wt. % for WS [19].

Table 2. Proximate and ultimate analyses of biomass sources (vine shoots (VS) and wheat straw pellets (WS)) and activated biochars.

	Sample					
	VS	WS	AC_VS_600	AC_VS_600_P	AC_WS_500	AC_WS_500_P
			Proximate (wt. %)			
Ash	4.43	4.23	9.00	13.9	15.6	16.5
Moisture	7.48	7.60	0.74	0.20	3.21	3.11
Volatile matter	75.7	74.8	13.2	10.2	12.9	12.2
Fixed carbon	12.4	13.4	77.1	75.5	68.3	68.2
			Ultimate (wt.%, daf basis)			
C	47.1	44.2	88.8	89.7	86.5	86.9
H	5.29	6.31	1.65	1.61	2.26	2.18
N	0.66	0.62	3.21	3.25	2.98	3.03
O [1]	47.0	48.9	6.33	5.44	8.26	7.89

[1] Calculated by difference.

On the other hand, Table 3 reports the key textural properties (i.e., apparent surface area and pore volumes at different pore size ranges) of activated biochars, which were deduced from the adsorption isotherms of both N_2 and CO_2 at −196 °C and 0 °C, respectively (see Figure 2). As expected, the N_2-based specific surface areas (S_{BET}) reported in Table 3 were relatively low due to the high content of ash and the predominantly microporous narrow structure in biomass-derived chars activated with CO_2. From Figure 2a, it can be observed a certain contribution of narrow macropores (isotherms do not reach a plateau at high relative pressures). In addition, a slight increase in the degree of burn-off resulted in a slightly better microporosity development (i.e., higher V_{mic} values). This trend, which is consistent with previous studies [21,22], could be explained by a progressive widening of the narrow micropores with the degree of burn-off. However, the mesopore volume did not seem to be correlated with the degree of burn-off, suggesting that the nature of the biomass feedstock and the pressure applied during the pyrolysis play a certain role in the resulting pore size distribution.

Table 3. Main textural properties and degrees of burn-off of activated biochars.

Activated Biochar	Apparent Specific Surface Area (m² g⁻¹)		Specific Pore Volume (cm³ g⁻¹)			Degree of Burn-Off (%) [2]
	S_{BET} [3]	S_{BET} [4]	V_{mic}	V_{mes}	V_{ultra}	
AC_VS_600	405	371	0.147	0.027	0.096	22.4 ± 1.9
AC_VS_600_P	536	416	0.161	0.014	0.094	26.2 ± 2.6
AC_WS_500	459	403	0.172	0.011	0.141	19.5 ± 1.1
AC_WS_500_P	514	416	0.191	0.016	0.094	28.9 ± 2.3

[2] Determined as the percentage of mass loss during the activation process of a given biochar, which was previously heated up to 800 °C under pure N_2 atmosphere. Standard deviation is also reported, since several activation runs were conducted to produce the required amount of activated biochar. [3] Determined from N_2 adsorption data at −196 °C. [4] Determined from CO_2 adsorption data at 0 °C.

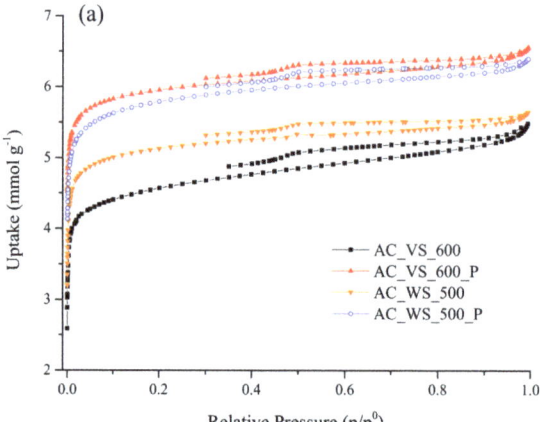

Figure 2. Cont.

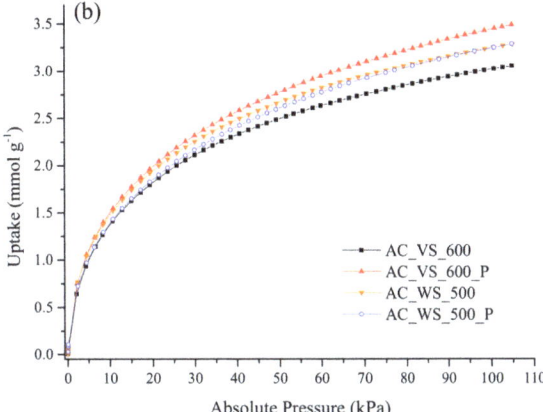

Figure 2. Adsorption isotherms of N_2 at −196 °C (**a**) and CO_2 at 0 °C (**b**) on the activated biochars (see Table 1 for sample designation).

3.2. Adsorption Isotherms

In light of the adsorption isotherms (for both CO_2 and N_2 at 25 and 50 °C) shown in Figure 3, the AC_WS_500 adsorbent appeared to be the best material in terms of CO_2 adsorption capacity at relatively low pressures. For instance, at an absolute pressure of 14 kPa, the highest CO_2 uptake at both temperatures was measured for this adsorbent (1.3 and 0.74 mmol g^{-1} at 25 and 50 °C, respectively; as reported in Table 4). This finding is consistent with the highest ultra-micropore volume (V_{ultra}) reported in Table 3 for this activated biochar (which also exhibited the lowest burn-off degree), since many studies have reported that the CO_2 uptake at absolute pressures of 5–15 kPa is mainly dependent on the availability of ultra-micropores [6,7,9,12].

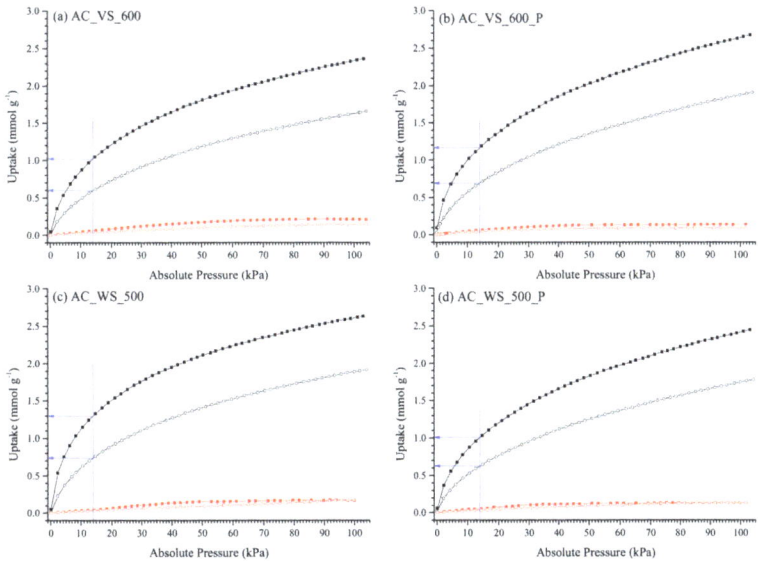

Figure 3. Adsorption isotherms of CO_2 and N_2 on the activated biochars at 25 and 50 °C: AC_VS_600 (**a**), AC_VS_600_P (**b**), AC_WS_500 (**c**), and AC_WS_500_P (**d**). Black filled squares: CO_2 at 25 °C; black open circles: CO_2 at 50 °C; red filled squares: N_2 at 25 °C; red open circles: N_2 at 50 °C.

Table 4. CO$_2$ adsorption capacities and apparent selectivities (deduced from single-component adsorption isotherms) for the activated biochars at 25 and 50 °C.

Activated Biochar	CO$_2$ Uptake at 14 kPa (mmol g^{-1})		CO$_2$ Uptake at 101.3 kPa (mmol g^{-1})		Apparent Selectivity CO$_2$/N$_2$ [5]	
	25 °C	50 °C	25 °C	50 °C	25 °C	50 °C
AC_VS_600	1.02	0.598	2.35	1.65	28.2	25.7
AC_VS_600_P	1.16	0.695	2.67	1.90	50.0	46.5
AC_WS_500	1.30	0.738	2.63	1.91	43.4	28.9
AC_WS_500_P	1.00	0.627	2.44	1.78	45.6	28.6

[5] Calculated from Equation (1).

The CO$_2$ adsorption capacities at low pressure (14 kPa) reported in Table 4 for the activated biochars can be considered as more than acceptable considering the relatively simple activation process (without the use of corrosive and/or costly reagents). At the more realistic temperature of 50 °C, the CO$_2$ uptakes at equilibrium (0.60–0.74 mmol g^{-1}) were within the range of those reported in previous studies for physically activated carbons derived from olive stones and almond shells [13], as well as for KOH-activated lignin-derived hydrochars [11].

Besides the CO$_2$ uptakes measured for the activated biochars, Table 4 also lists their apparent selectivities at 25 and 50 °C, which were estimated according to Hao et al., [11]:

$$S_{app} = \frac{q^*_{CO2} p_{N2}}{q^*_{N2} p_{CO2}} \quad (1)$$

In Equation (1), q^* is the uptake of CO$_2$ or N$_2$ from the respective pure component adsorption isotherms at an absolute pressure equal to the partial pressure of each gas in a binary mixture (p_{CO2} = 14 kPa and p_{N2} = 86 kPa). This relatively simple approach, which can provide a reasonable estimate of the binary selectivity from single-component adsorption data [11], represents a good alternative to the most commonly used method, which is based on the ideal adsorption solution theory (IAST). The IAST-based method requires as inputs the fitted adsorption isotherms for a given model (e.g., Toth [11], Sips [13], and dual site Langmuir–Freundlich [23]). Limitations of such models to reproduce perfectly the pure component equilibrium data, in addition to possible deviations from ideality in the adsorbed phase, can result in unrealistic estimates of selectivity [13]. In fact, Hao et al., [11] found that the IAST-predicted selectivities were notably lower than the estimates calculated according to the simplified approach given in Equation (1).

The apparent CO$_2$-over-N$_2$ selectivities reported in Table 4 are considerably higher than those reported by Hao et al., [11] (about 14 at 50 °C). It is interesting to note that apparent selectivity decreased with higher adsorption temperature, especially for wheat straw-derived carbons. This negative temperature dependence, which was also observed in previous studies [8,11,24,25], could be related to the higher isosteric heat of adsorption and/or lower diffusion coefficient of CO$_2$ with respect to N$_2$.

3.3. Breakthrough Experiments under Dry Conditions

The following variables were taken as key performance indicators for the dynamic adsorption experiments: (i) the total (i.e., at saturation) specific CO$_2$ uptake (q_{CO2}; in mmol g^{-1}), (ii) the percentage of total CO$_2$ uptake at breakthrough time (which represents the percentage of used bed and provides a measure of the efficiency of the adsorbent in terms of mass transfer), (iii) the regeneration efficiency (which is defined as the ratio between the total CO$_2$ uptake after the *n*th cycle and that obtained for the fresh adsorbent), and (iv) the selectivity towards CO$_2$ over N$_2$. Detailed information on how indicators were calculated is given in the Supplementary Materials.

Table 5 summarizes the results obtained from the dynamic adsorption experiments under dry conditions at 25 °C. The corresponding CO$_2$ breakthrough curves are displayed in Figure S1 (Supplementary Materials). From the results listed in Table 5, it should be highlighted that the best performance was obtained for the AC_VS_600 adsorbent, which exhibited a CO_2 uptake in the first cycle of 0.99 mmol g^{-1} (almost the same than that reported in Table 4 for static measurements at 25 °C and 14 kPa). This interesting finding could be explained by the relatively high volume of mesopores (0.027 cm^3 g^{-1}, as reported in Table 3), which can facilitate the diffusion of CO_2 molecules into ultra-micropores. Moreover, the AC_WS_500 adsorbent, for which we found the highest CO_2 uptake at equilibrium (1.3 mmol g^{-1} at 14 kPa), appears to be the worst material under dynamic conditions (just 0.90 mmol g^{-1} in cycle 1). This fact can also be related to the low volume of mesopores (0.011 cm^3 g^{-1}) measured for this material. Therefore, the results reported here seem to confirm the key role of mesopores to enhance adsorption kinetics.

Table 5. Summary of results from the breakthrough experiments (dry conditions) conducted at 25 °C.

		Activated Biochar			
		AC_VS_600	AC_VS_600_P	AC_WS_500	AC_WS_500_P
Mass of adsorbent (g)		15.1	16.8	26.9	27.8
Absolute pressure in adsorption steps (kPa)		111.8 ± 0.5	111.9 ± 0.7	111.4 ± 0.1	111.5 ± 0.2
Absolute pressure in desorption steps (kPa)		21.5 ± 0.1	31.3 ± 0.1	31.3 ± 0.1	31.1 ± 0.1
q_{CO2} (mmol g^{-1})	cycle 1	0.992	0.936	0.895	0.959
	cycle 2	0.891	0.799	0.763	0.869
	cycle 3	0.860	0.731	0.706	0.764
	cycle 4	0.852	0.696	0.676	0.727
	cycle 5	0.834	0.682	0.616	0.667
Percentage of total CO_2 uptake at breakthrough time	cycle 1	76.3	80.7	77.9	80.9
	cycle 2	76.7	77.5	71.0	78.8
	cycle 3	77.3	75.4	68.2	77.7
	cycle 4	76.6	76.3	65.8	74.8
	cycle 5	76.2	73.7	64.7	74.7
Regeneration efficiency (%)	cycle 1	–	–	–	–
	cycle 2	89.8	85.4	85.2	90.6
	cycle 3	86.7	78.1	78.9	79.7
	cycle 4	85.9	74.4	75.5	75.8
	cycle 5	84.1	72.9	68.8	69.6
Selectivity CO_2/N_2	cycle 1	56.7	58.6	49.4	48.6
	cycle 2	49.2	51.9	51.9	46.6
	cycle 3	44.9	49.4	45.2	45.9
	cycle 4	46.5	48.5	48.2	44.5
	cycle 5	46.6	50.8	44.2	45.7

Results shown in Table 5 (as well as breakthrough curves displayed in Figure S1) also indicate that the stability over the five adsorption/desorption cycles was dependent on the material tested. In this sense, the AC_VS_600 sample was the most stable with nearly constant percentages of used bed (76.2%–77.3%) and decent regeneration efficiencies (84.1%–89.8%). However, this fact could partly be attributed to the relatively low absolute pressure applied during regeneration for this adsorbent (21.5 kPa). In other words, the relatively poor results obtained in terms of regeneration efficiency for the rest of materials could be explained by an insufficient vacuum level during the regeneration step (absolute pressures of 31.3–31.5 kPa).

Regarding the experimental CO_2-over-N_2 selectivity at 25 °C, all the samples showed a very good behavior with values in the range of 44.5 to 58.6. These values, which were calculated according to Equation (A5), were considerably higher than those reported by González et al., (20–30) [13] and Shahkarami et al., (17.4–29.3) [14] from multicomponent adsorption measurements. Interestingly, the apparent CO_2-over-N_2 selectivities at 25 °C (see Table 4) were quite similar to those measured under dynamic conditions, with the exception of the AC_VS_600 sample. The fact that the selectivity significantly increased for this material under dynamic conditions could be explained by differences in pore size distribution. As can be deduced from the pore size distributions shown in Figure 4, the AC_VS_600 sample clearly exhibited a broader distribution, with a significant contribution in the ranges of relatively large micropores (0.8–2.0 nm) and relatively narrow mesopores (up to 15 nm). The higher availability of these pores can facilitate the adsorption of a relatively high amount of N_2 under static conditions. However, under dynamic conditions, the much shorter residence time of the gas phase (around 13 s) could lead to a decreased N_2 uptake, due to diffusional limitations.

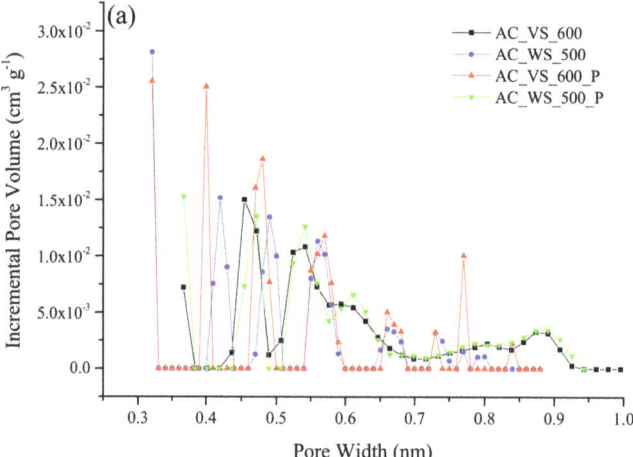

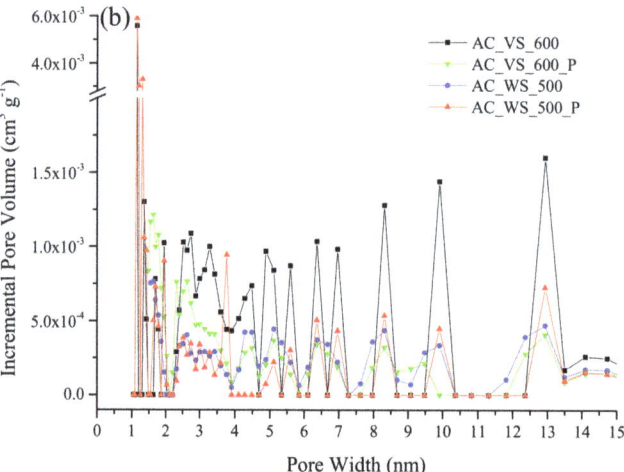

Figure 4. Pore size distribution of the adsorbents: (**a**) narrow micropores (deduced from the CO_2 adsorption isotherms at 0 °C), and (**b**) pores widths ranging from 1.0 to 15 nm (deduced from the N_2 adsorption isotherms at −196 °C).

On the basis of the results obtained at 25 °C, we decided to assess the performance at 50 °C of the AC_VS_600 and AC_WS_500_P materials. Table 6 summarizes the results obtained for this adsorption temperature, whereas Figure 5 shows the respective CO_2 breakthrough curves. The AC_VS_600 adsorbent exhibited a very good performance, especially in terms of CO_2 uptake (with values very close to that deduced from the adsorption isotherm) and stability over the adsorption/desorption cycles (see Figure 5a). The relatively low CO_2 adsorption capacity measured for the first cycle (0.54 mmol g^{-1})

and related regeneration efficiencies (higher than 100%) could be explained by slight differences in experimental conditions, especially the CO_2 partial pressure in the feeding gas stream, which was slightly lower than the average value used for the rest of cycles (15.2 kPa vs. 15.9 kPa). Regarding the performance of the AC_WS_500_P sample, a considerably decrease in the CO_2 uptake (18.2%) was observed after five cycles. As expected, the regeneration efficiency for the fifth cycle increased at 50 °C compared with that at 25 °C (81.6% and 69.6%, respectively). However, the pressure applied during desorption (24.9 ± 0.7 kPa) was still too low to ensure the complete regeneration of the AC_WS_500_P adsorbent, which showed a considerably lower mesopore volume compared to that of the AC_VS_600 material.

Table 6. Summary of results from the breakthrough experiments (dry conditions) conducted at 50 °C.

		Activated Biochar	
		AC_VS_600	AC_WS_500_P
Mass of adsorbent (g)		15.1	27.8
Absolute pressure in adsorption steps (kPa)		113.6 ± 0.6	113.3 ± 1.4
Absolute pressure in desorption steps (kPa)		23.5 ± 0.5	24.9 ± 0.7
q_{CO2} (mmol g^{-1})	cycle 1	0.543	0.624
	cycle 2	0.570	0.583
	cycle 3	0.565	0.555
	cycle 4	0.573	0.550
	cycle 5	0.571	0.511
Percentage of total CO_2 uptake at breakthrough time	cycle 1	75.3	68.0
	cycle 2	73.1	71.5
	cycle 3	73.2	71.4
	cycle 4	73.4	69.0
	cycle 5	76.5	69.7
Regeneration efficiency (%)	cycle 1	–	–
	cycle 2	105.0	93.3
	cycle 3	104.2	88.9
	cycle 4	105.6	88.2
	cycle 5	105.3	81.8
Selectivity CO_2/N_2	cycle 1	40.8	38.2
	cycle 2	39.1	43.1
	cycle 3	37.4	43.2
	cycle 4	34.6	35.0
	cycle 5	38.7	38.3

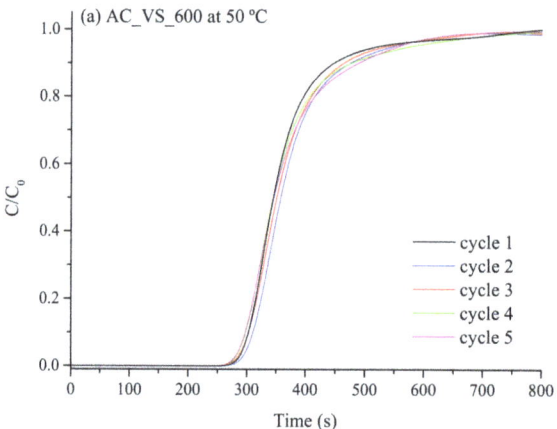

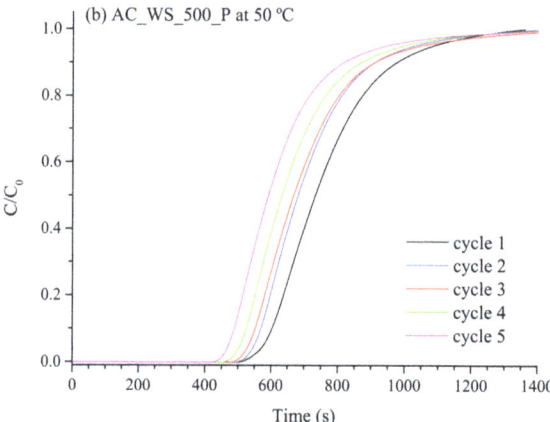

Figure 5. CO_2 breakthrough curves for five consecutive cycles (adsorption temperature = 50 °C) from a dry binary mixture of N_2 and CO_2 (with a CO_2 composition of 13.75–14.25 vol. %) using the AC_VS_600 (**a**) and AC_WS_500_P (**b**) materials.

3.4. Breakthrough Experiments under Humid Conditions

The summarized results given in Table 7 reveal some interesting findings with regard to the performance of the two tested materials. For the AC_VS_600 adsorbent, the total specific CO_2 uptake was found to be quite close to that obtained under dry conditions (in spite of the lower CO_2 partial pressure in the wet feeding gas), whereas a higher drop in this performance indicator was observed for the AC_WS_500_P adsorbent. However, a certain degree of variability in q_{CO2} among the 10 cycles was observed for both materials. This variability, which was also evident from the CO_2 breakthrough curves shown in Figure 6, could be explained by the extremely slower kinetics of water vapor adsorption on biomass-derived porous carbons in comparison with that of CO_2 [17].

Table 7. Summary of results from the breakthrough experiments (humid conditions) conducted at 50 °C.

		Activated Carbon	
		AC_VS_600	AC_WS_500_P
Mass of adsorbent (g)		15.1	27.8
Absolute pressure in adsorption steps (kPa)		118.8 ± 1.5	120.0 ± 0.9
Absolute pressure in desorption steps (kPa)		34.6 ± 3.0	36.9 ± 3.5
q_{CO2} (mmol g^{-1})	cycle 1	0.542	0.486
	cycle 2	0.447	0.393
	cycle 3	0.435	0.412
	cycle 4	0.487	0.391
	cycle 5	0.526	0.390
	cycle 6	0.526	0.385
	cycle 7	0.528	0.376
	cycle 8	0.498	0.363
	cycle 9	0.474	0.359
	cycle 10	0.533	0.371
Percentage of total CO_2 uptake at breakthrough time (excluding first cycle)		28.7 ± 1.6	42.8 ± 3.5
Regeneration efficiency (%)	cycle 1	–	–
	cycle 2	82.5	80.9
	cycle 3	80.2	84.8
	cycle 4	89.8	80.4
	cycle 5	97.0	80.2
	cycle 6	97.0	79.2
	cycle 7	97.4	77.4
	cycle 8	91.9	74.7
	cycle 9	87.4	73.9
	cycle 10	98.3	76.3
Selectivity CO_2/N_2 (excluding first cycle)		54.0 ± 4.1	40.9 ± 2.7

As already pointed out by Dasgupta et al., [26], the amount of water adsorbed during the first cycles could progressively increase, leading to a gradual decrease in the adsorption sites available for CO_2. This was especially evident during the two first cycles for both tested adsorbents, as reported in Table 7. Nevertheless, after a few number of cycles, the CO_2 uptake seemed to stabilize (for the AC_WS_500_P adsorbent) or even increase up to a closer value to that obtained for the first cycle (for the AC_VS_600 adsorbent). These differences observed in the performance of the two adsorbents could also be related to the effectiveness of the desorption step as well as the transient period to reach water equilibrium. In the case of the AC_VS_600 material, its more hierarchical pore size distribution could facilitate a relatively fast desorption of CO_2 at 50 °C and mild vacuum conditions, despite the fact that a certain volume of micro- and mesopores could be filled by water. The fact that the CO_2 uptake fluctuated over the ten cycles could also indicate that further cycles are required to saturate the bed with water. For the AC_WS_500_P adsorbent, however, the observed lower CO_2 uptake could be due to a higher impact of the water adsorption on its surface, which could result in a critical reduction in the volume of mesopores and wide micropores available for CO_2 diffusion during the desorption step. Nonetheless, the almost steady CO_2 adsorption capacity observed for the last cycles also indicated that the relatively short residence time used here during the adsorption steps was appropriate to avoid a gradual accumulation of water within the bed.

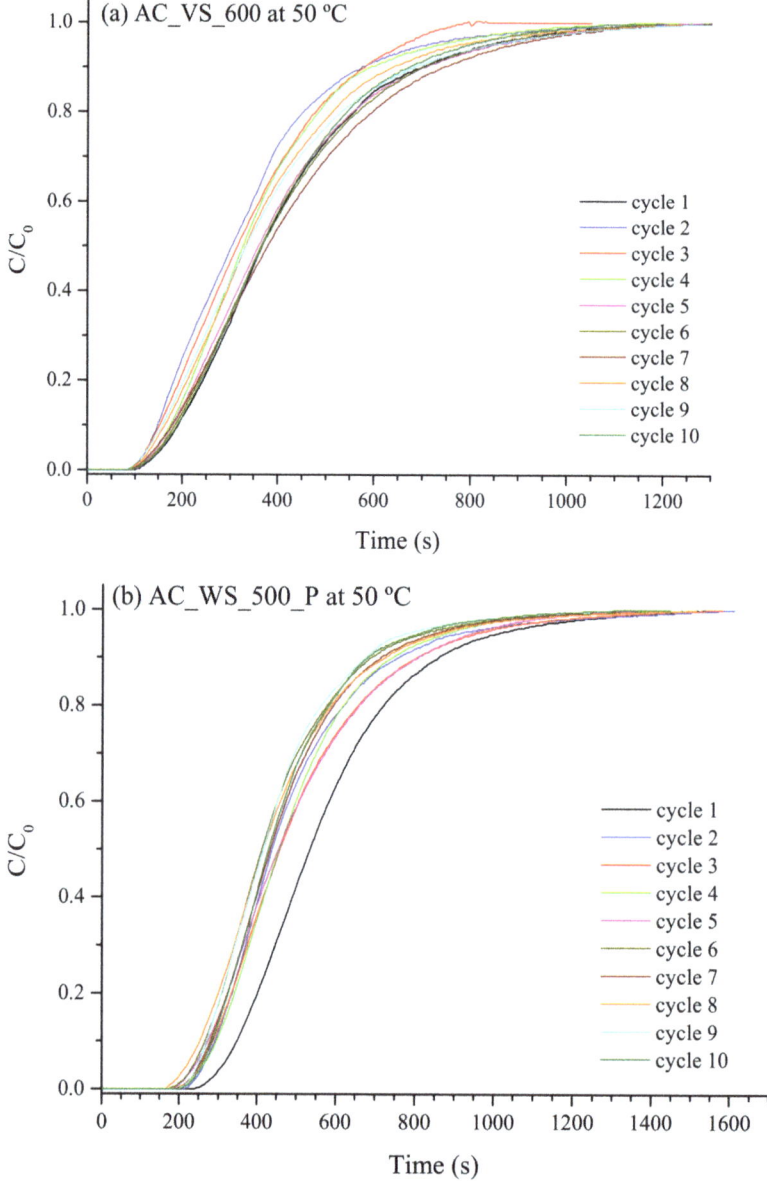

Figure 6. CO_2 breakthrough curves for ten consecutive cycles (adsorption temperature = 50 °C) from a wet mixture of N_2 and CO_2 (with a CO_2 composition of 13.75–14.25 vol. %; relative humidity (RH) = 100%) using the AC_VS_600 (**a**) and AC_WS_500_P (**b**) materials.

With regard to the CO_2-over-N_2 selectivity under humid conditions, similar (for AC_WS_500_P) or slightly higher (for AC_VS_600) values than those measured under dry conditions are reported in Table 7. A little impact of water vapor on the CO_2 purity was also observed by Xu et al., [16] for a coconut shell-derived activated carbon using a wet flue gas containing 4.6 vol. % H_2O. In a more recent study, focused on simulating a VSA process using activated carbon, You and Liu [27] also reported that the CO_2 purity was almost constant regardless of the relative humidity.

Although the promising results discussed above, the most negative impact of moisture was on the percentages of used bed, which were significantly lower than those measured under dry conditions, especially for the AC_VS_600 adsorbent (only 28.7% in average, as reported in Table 7). This loss of efficiency could be attributed to the fact that the presence of adsorbed water might substantially reduce the diffusion rate of CO_2 into ultra-micropores during the adsorption step. Better efficiencies were obtained for the AC_WS_500_P adsorbent, showing an average percentage of used bed of 42.8%. This finding suggests that the rate of water adsorption on this material could be lower and a relatively high number of adsorption sites could be available for CO_2 during the initial stage of the adsorption process.

As stated by Liu et al., in an excellent review [28], water adsorption on carbonaceous materials is a really complex process, which can be affected by the pore size distribution and the pore connectivity as well as the type of functional groups and their location and distribution on the surface. According to the mechanism firstly suggested by Pierce and Smith [29], water adsorption on activated carbon first takes place at the polar and oxygen-containing functional groups (OFGs) on the surface. Then, the water molecules can form clusters, which can grow and coalesce to fill the pores, in a greater or lower extent depending on the pore structure. Several previous studies revealed that the carboxyl group possessed the highest affinity to water molecules [30–32], due to both the accessibility of the carboxylic group to water molecules and the ability of this functional group to form hydrogen bonds. Bearing this in mind, one can hypothesize that differences in the concentration and accessibility of hydrophilic OFGs would explain the differences observed between the two adsorbents with respect to the water adsorption rate at the initial stage of the adsorption process. In this sense, advanced characterization studies (by using, e.g., in-situ NMR measurements) would be required in further studies to provide valuable insights into the interactions between water and OFGs at very low loadings.

In addition to the higher mass transfer efficiency observed for the AC_WS_500_P adsorbent, the fact that this material has a bulk density considerably higher than that of the AC_VS_600 adsorbent (368 vs. 200 kg m^{-3}, respectively) makes it more suitable for industrial-scale applications (i.e., reduced size of adsorption vessels).

4. Conclusions

From the results discussed above, we can conclude that, in general, physically activated biochars produced from both wheat straw and vine shoots are promising adsorbents for CO_2 capture under realistic postcombustion conditions. As expected, pore size distribution plays a key role in the CO_2 adsorption behavior. In this sense, biomass-derived activated carbons having a hierarchical structure (with a high volume of ultra-micropores but also a certain volume of mesopores) appear as the ideal candidates for an efficient and selective CO_2 adsorption from a dry flue gas under dynamic conditions. However, the presence of relatively high concentrations of water vapor in the feeding gas clearly interferes with the CO_2 adsorption mechanism, leading to significantly shorter breakthrough times (i.e., lower percentages of used bed). The differences observed here in the performance under humid conditions of two activated carbons (produced from different biochar precursors) suggest that surface chemistry, and especially the concentration and accessibility of hydrophilic OFGs, could exert considerable influence on the water adsorption rate. Therefore, future research should focus on synthetizing biomass-derived activated carbons with an appropriate hierarchical pore size distribution, relatively high bulk density, and low affinity to water molecules. For this purpose, the combined effects of the nature of the precursor, pyrolysis conditions, activation conditions, and degree of burn-off should properly be addressed.

Supplementary Materials: The following are available online at http://www.mdpi.com/2076-3417/10/1/376/s1. Figure S1: CO_2 breakthrough curves for five consecutive cycles (adsorption temperature = 25 °C) from a dry binary mixture of N_2 and CO_2 (with a CO_2 composition of 13.75–14.25 vol. %).

Author Contributions: Conceptualization, J.J.M.; methodology, D.G.-M., B.G. and J.J.M.; software, D.G.-M. and J.J.M.; validation and formal analysis, D.G.-M., B.G. and J.J.M.; writing—original draft preparation, J.J.M.; project administration, B.G.; funding acquisition, J.J.M. All authors have read and agreed to the published version of the manuscript.

Funding: This research received funding from the Spanish Ministry of Science, Innovation, and Universities (ERANET-MED Project MEDWASTE, ref. PCIN-2017-048).

Acknowledgments: The authors also acknowledge the funding from the Aragón Government (Ref. T22_17R), co-funded by FEDER 2014-2020 "Construyendo Europa desde Aragón".

Conflicts of Interest: The authors declare no conflict of interest.

References

1. Creamer, A.E.; Gao, B. Carbon-Based Adsorbents for Postcombustion CO_2 Capture: A Critical Review. *Environ. Sci. Technol.* **2016**, *50*, 7276–7289. [CrossRef]
2. Chen, J.; Yang, J.; Hu, G.; Hu, X.; Li, Z.; Shen, S.; Radosz, M.; Fan, M. Enhanced CO_2 Capture Capacity of Nitrogen-Doped Biomass-Derived Porous Carbons. *ACS Sustain. Chem. Eng.* **2016**, *4*, 1439–1445. [CrossRef]
3. Yang, J.; Yue, L.; Hu, X.; Wang, L.; Zhao, Y.; Lin, Y.; Sun, Y.; DaCosta, H.; Guo, L. Efficient CO_2 Capture by Porous Carbons Derived from Coconut Shell. *Energy Fuels* **2017**, *31*, 4287–4293. [CrossRef]
4. Coromina, H.M.; Walsh, D.A.; Mokaya, R. Biomass-derived activated carbon with simultaneously enhanced CO_2 uptake for both pre and post combustion capture applications. *J. Mater. Chem. A* **2016**, *4*, 280–289. [CrossRef]
5. Li, D.; Ma, T.; Zhang, R.; Tian, Y.; Qiao, Y. Preparation of porous carbons with high low-pressure CO_2 uptake by KOH activation of rice husk char. *Fuel* **2015**, *139*, 68–70. [CrossRef]
6. Deng, S.; Wei, H.; Chen, T.; Wang, B.; Huang, J.; Yu, G. Superior CO_2 adsorption on pine nut shell-derived activated carbons and the effective micropores at different temperatures. *Chem. Eng. J.* **2014**, *253*, 46–54. [CrossRef]
7. Hao, W.; Björkman, E.; Lliestråle, M.; Hedin, N. Activated carbons prepared from hydrothermally carbonized waste biomass used as adsorbents for CO_2. *Appl. Energy* **2013**, *112*, 526–532. [CrossRef]
8. Plaza, M.G.; González, A.S.; Pis, J.J.; Rubiera, F.; Pevida, C. Production of microporous biochars by single-step oxidation: Effect of activation conditions on CO_2 capture. *Appl. Energy* **2014**, *114*, 551–562. [CrossRef]
9. Manyà, J.J.; González, B.; Azuara, M.; Arner, G. Ultra-microporous adsorbents prepared from vine shoots-derived biochar with high CO_2 uptake and CO_2/N_2 selectivity. *Chem. Eng. J.* **2018**, *345*, 631–639. [CrossRef]
10. Yue, L.; Xia, Q.; Wang, L.L.; Wang, L.L.; DaCosta, H.; Yang, J.; Hu, X. CO_2 adsorption at nitrogen-doped carbons prepared by K_2CO_3 activation of urea-modified coconut shell. *J. Colloid Interface Sci.* **2018**, *511*, 259–267. [CrossRef]
11. Hao, W.; Björnebäck, F.; Trushkina, Y.; Oregui Bengoechea, M.; Salazar-Alvarez, G.; Barth, T.; Hedin, N. High-Performance Magnetic Activated Carbon from Solid Waste from Lignin Conversion Processes. 1. Their Use as Adsorbents for CO_2. *ACS Sustain. Chem. Eng.* **2017**, *5*, 3087–3095. [CrossRef]
12. Sevilla, M.; Falco, C.; Titirici, M.M.; Fuertes, A.B. High-performance CO_2 sorbents from algae. *RSC Adv.* **2012**, *2*, 12792–12797. [CrossRef]
13. González, A.S.; Plaza, M.G.; Rubiera, F.; Pevida, C. Sustainable biomass-based carbon adsorbents for post-combustion CO_2 capture. *Chem. Eng. J.* **2013**, *230*, 456–465. [CrossRef]
14. Shahkarami, S.; Dalai, A.K.; Soltan, J.; Hu, Y.; Wang, D. Selective CO_2 Capture by Activated Carbons: Evaluation of the Effects of Precursors and Pyrolysis Process. *Energy Fuels* **2015**, *29*, 7433–7440. [CrossRef]
15. Shafeeyan, M.S.; Daud, W.M.A.W.; Shamiri, A.; Aghamohammadi, N. Modeling of Carbon Dioxide Adsorption onto Ammonia-Modified Activated Carbon: Kinetic Analysis and Breakthrough Behavior. *Energy Fuels* **2015**, *29*, 6565–6577. [CrossRef]
16. Xu, D.; Xiao, P.; Zhang, J.; Li, G.; Xiao, G.; Webley, P.A.; Zhai, Y. Effects of water vapour on CO_2 capture with vacuum swing adsorption using activated carbon. *Chem. Eng. J.* **2013**, *230*, 64–72. [CrossRef]
17. Durán, I.; Álvarez-Gutiérrez, N.; Rubiera, F.; Pevida, C. Biogas purification by means of adsorption on pine sawdust-based activated carbon: Impact of water vapor. *Chem. Eng. J.* **2018**, *353*, 197–207. [CrossRef]

18. Manyà, J.J.; Azuara, M.; Manso, J.A. Biochar production through slow pyrolysis of different biomass materials: Seeking the best operating conditions. *Biomass Bioenergy* **2018**, *117*, 115–123. [CrossRef]
19. Greco, G.; Videgain, M.; Di Stasi, C.; González, B.; Manyà, J.J. Evolution of the mass-loss rate during atmospheric and pressurized slow pyrolysis of wheat straw in a bench-scale reactor. *J. Anal. Appl. Pyrolysis* **2018**, *136*, 18–26. [CrossRef]
20. Yin, G.; Liu, Z.; Liu, Q.; Wu, W. The role of different properties of activated carbon in CO_2 adsorption. *Chem. Eng. J.* **2013**, *230*, 133–140. [CrossRef]
21. Plaza, M.G.; Pevida, C.; Arias, B.; Fermoso, J.; Casal, M.D.; Martín, C.F.; Rubiera, F.; Pis, J.J. Development of low-cost biomass-based adsorbents for postcombustion CO_2 capture. *Fuel* **2009**, *88*, 2442–2447. [CrossRef]
22. Ahmad, F.; Daud, W.M.A.W.; Ahmad, M.A.; Radzi, R.; Azmi, A.A. The effects of CO_2 activation, on porosity and surface functional groups of cocoa (Theobroma cacao)—Shell based activated carbon. *J. Environ. Chem. Eng.* **2013**, *1*, 378–388. [CrossRef]
23. Chen, Y.; Lv, D.; Wu, J.; Xiao, J.; Xi, H.; Xia, Q.; Li, Z. A new MOF-505@GO composite with high selectivity for CO_2/CH_4 and CO_2/N_2 separation. *Chem. Eng. J.* **2017**, *308*, 1065–1072. [CrossRef]
24. Yang, J.; Zhang, P.; Zhang, Y.; Zeng, Z.; Liu, L.; Deng, S.; Wang, J. Controllable synthesis of bifunctional porous carbon for efficient gas-mixture separation and high-performance supercapacitor. *Chem. Eng. J.* **2018**, *348*, 57–66.
25. Yu, H.; Zhu, W.; Wang, X.; Krishna, R.; Chen, D.-L.; Xu, C. Utilizing transient breakthroughs for evaluating the potential of Kureha carbon for CO_2 capture. *Chem. Eng. J.* **2015**, *269*, 135–147. [CrossRef]
26. Dasgupta, S.; Divekar, S.; Aarti; Spjelkavik, A.I.; Didriksen, T.; Nanoti, A.; Blom, R. Adsorption properties and performance of CPO-27-Ni/alginate spheres during multicycle pressure-vacuum-swing adsorption (PVSA) CO_2 capture in the presence of moisture. *Chem. Eng. Sci.* **2015**, *137*, 525–531. [CrossRef]
27. You, Y.Y.; Liu, X.J. Modeling of CO_2 adsorption and recovery from wet flue gas by using activated carbon. *Chem. Eng. J.* **2019**, *369*, 672–685. [CrossRef]
28. Liu, L.; Tan, S.J.; Horikawa, T.; Do, D.D.; Nicholson, D.; Liu, J. Water adsorption on carbon—A review. *Adv. Colloid Interface Sci.* **2017**, *250*, 64–78. [CrossRef]
29. Pierce, C.; Smith, R.N. Adsorption—Desorption Hysteresis in Relation to Capillarity of Adsorbents. *J. Phys. Colloid Chem.* **1950**, *54*, 784–794. [CrossRef]
30. Fletcher, A.J.; Uygur, Y.; Mark Thomas, K. Role of surface functional groups in the adsorption kinetics of water vapor on microporous activated carbons. *J. Phys. Chem. C* **2007**, *111*, 8349–8359. [CrossRef]
31. Xiao, J.; Liu, Z.; Kim, K.; Chen, Y.; Yan, J.; Li, Z.; Wang, W. S/O-functionalities on modified carbon materials governing adsorption of water vapor. *J. Phys. Chem. C* **2013**, *117*, 23057–23065. [CrossRef]
32. Nguyen, V.T.; Horikawa, T.; Do, D.D.; Nicholson, D. Water as a potential molecular probe for functional groups on carbon surfaces. *Carbon* **2014**, *67*, 72–78. [CrossRef]

© 2020 by the authors. Licensee MDPI, Basel, Switzerland. This article is an open access article distributed under the terms and conditions of the Creative Commons Attribution (CC BY) license (http://creativecommons.org/licenses/by/4.0/).

Article

Potential for Developing Biocarbon Briquettes for Foundry Industry

Elsayed Mousa [1,3,*], Mania Kazemi [1], Mikael Larsson [1], Gert Karlsson [2] and Erik Persson [2]

[1] Swerim AB, Box 812, SE-97125 Luleå, Sweden; mania.kazemi@swerim.se (M.K.); mikael.larsson@swerim.se (M.L.)
[2] Volvo Group Truck Operations, SE-40508 Gothenburg, Sweden; gert.e.karlsson@volvo.com (G.K.); erik.persson.5@volvo.com (E.P.)
[3] Central Metallurgical Research and Development Institute, Cairo 12422, Egypt
* Correspondence: elsayed.mousa@swerim.se

Received: 16 October 2019; Accepted: 29 November 2019; Published: 4 December 2019

Abstract: The foundry industry is currently facing challenges to reduce the environmental impacts from application of fossil fuels. Replacing foundry coke with alternative renewable carbon sources can lead to significant decrease in fossil fuel consumption and fossil CO_2 emission. The low bulk density, low energy density, low mechanical strength and the high reactivity of biocarbon materials are the main factors limiting their efficient implementation in a cupola furnace. The current study aimed at designing, optimizing and developing briquettes containing biocarbon, namely, biocarbon briquettes for an efficient use in cupola furnace. Laboratory hydraulic press with compaction pressure of about 160 MPa and stainless-steel moulds (Ø = 40 mm and 70 mm) were used for compaction. The density, heating value, energy density, mechanical strength and reactivity of biocarbon briquettes were measured and evaluated. The compressive strength and splitting tensile strength of biocarbon briquettes were measured by a compression device. The reactivity of biocarbon briquettes was measured under controlled conditions of temperature and gas atmosphere using the thermogravimetric analysis technique (TGA). Different types of binders were tested for the compaction of commercial charcoal fines with/without contribution of coke breeze. The effect of charcoal ratio, particle size, binder type, binder ratio, moisture content and compaction pressure on the quality of the biocarbon briquettes was investigated. Molasses with hydrated lime and cement were superior in enhancing the biocarbon briquettes strength and energy density among other tested binders and additives. The briquettes' strength decreased as the biocarbon content increased. The optimum recipes consisted of 62% charcoal fines, 20% molasses, 10% hydrated lime and 8% cement. Cement is necessary to develop the tensile strength and hot mechanical strength of the briquettes. The charcoal with high ash content showed higher strength of briquettes but lower heating value compared to that with low ash content. Dispersion of silica suspension on charcoal particles during the mixing process was able to reduce the reactivity of biochar in the developed biocarbon briquettes. The biocarbon briquettes density and strength were increased by increasing the compaction pressure. Commercial powder hydrated lime was more effective in enhancing the briquettes' strength compared to slaked burnt lime. Upscaling of biocarbon briquettes (Ø = 70 mm) and testing of hot mechanical strength under load indicated development of cracks which significantly reduced the strength of briquettes. Further development of biocarbon briquettes is needed to fulfil the requirements of a cupola furnace.

Keywords: biocarbon; biomass; foundry industry; cupola furnace; CO_2 emission; briquetting

1. Introduction

The melting process in the manufacturing of ferrous metals is responsible for a large part of the energy consumption and carbon dioxide emissions from foundries. One of the most common melting technologies used for production of cast iron in large volumes is the cupola furnace, where foundry coke is the main source of energy. Technological developments and optimization of operation conditions have improved the energy efficiency and productivity of the cupola melting process. However, demands for decreasing the fossil greenhouse gas (GHG) emissions are new challenges for the foundries using cupola furnaces and for the steel industry. Using renewable carbon sources for partial or full replacement of the fossil carbon in the cupola furnace leads to lower CO_2 emissions and environmental effects from the foundries. The bio-carbon materials need to meet the requirements for density, strength and size to ensure smooth operation and product quality. Therefore, the compaction of bio-carbon materials has been studied for application in different types of processes.

Replacement of foundry coke in a cupola furnace with renewable carbon and biomass products is one option for securing the sustainability of this vital process; however, the relatively low mechanical strength, low energy density and high reactivity of biomass represent the main challenges for its efficient implementation [1,2]. Isnugroho et al. [3] tested biomass residues as a secondary energy source in a cupola furnace. Injection of biomass charcoal particles prepared from plant residues increased the temperature in the furnace and lowered the coke consumption. Echterhof et al. [4,5] investigated the potentials of using different biomass to partially replace fossil carbon sources in a cupola furnace. Injection tests were carried out in an industrial cupola furnace using biochar consisting of 78–80% fixed carbon and 18–19% volatiles. Injection of 100 kg biochar/h resulted in reduction of coke rate by 8.2–9.2%. Continuous injection of biochar was not possible due to increased temperature in the furnace chamber. The selected biomaterials were briquetted using different binders and the strength of the briquettes was compared to the strength of reference foundry cokes. Since the mechanical strength of the briquettes was low, they were not tested in the cupola furnace [4,5].

Several studies have been performed to develop anthracite briquettes to replace foundry coke in a cupola [6–9]. Briquettes made of anthracite grains, lignin, collagen and silicon were tested in two cupola furnaces and the replacement started from 12% up to 25% [6]. The process and melt conditions remained similar when coke was partially replaced by anthracite. The developed briquettes had a similar mechanical strength and efficiency as the foundry coke. However, the briquettes burnt faster than conventional coke when they reached the tuyeres [6]. In a recent publication by Noh et al. [7], it was shown that briquettes made of anthracite, plant by-products, collagen and silicon metal powder have suitable strength at high temperatures for replacing coke in foundries. Formation of silicon carbide nanowires in the briquettes after pyrolysis at 1400 °C for 2 h resulted in high mechanical strength for top charging in cupola furnace. Torielli et al. [8] discussed the effectiveness of different innovative technologies to reduce the consumption of foundry coke, such as replacing coke by anthracite briquettes, which resulted in 6% reduction of total carbon charge to the cupola. Gabra el al. [9] reported reducing of coke by ~38 kg/t of metal and increasing the energy efficiency from 43% to 62% by hybridizing the cupola furnace with a biomass wood gasifier. According to his assumptions, the biomass gasifier can be used to heat up the blast air going to cupola to about 475 °C which will save the foundry coke consumption in the cupola furnace.

Wang et al. [10] investigated the densification and gasification of biocarbon powders. Compaction was conducted using alkaline lignin and wheat starch as binders in roller press equipment. Addition of binders improved the strength of briquettes. Gasification of small samples by CO_2 was studied at 850 °C using thermogravimetric analysis (TGA). Higher gasification rates were observed in samples with alkaline lignin binder. This was contributed to by the presence of inorganic elements which catalyse the CO_2 gasification reaction. Their work showed that different binders affect the reactivity and strength of bio-carbon briquettes in different ways and, therefore, the final application of the briquettes is affected by the type of binders used. In a study by Rahman et al. [11] the variation of strength in biochar fuel briquettes with different sizes and shapes was evaluated. The briquettes were made of char fines (below 3 mm) and starch. They mixed char fines and starch solution using a surface dressing technique. During this process, the sharp and weak edges of the char particles were removed as a result of forces and friction between particles. This led to higher bulk density and reduced the surface defects of char particles. The mixtures were compacted using a hydraulic laboratory press and a pilot scale roller press. The authors observed that lower length to diameter ratio in the cylindrical briquettes lead to higher surface strength. The surface compression strength of cylindrical briquettes was 5 times higher than the point compression strength of pillow-shaped briquettes with the same recipes. Ifa el al. [12] investigated the briquetting of biochar derived from the pyrolysis of cashew nut waste. The biochar with different additives of starch (8–12%) was briquetted using a hydraulic press at a compaction pressure of 300 kg/cm^2. The biocarbon briquettes produced from biochar with particle size less than 74 μm and 12% addition of starch showed the highest compaction strength (7.6 kg/cm^2).

Despite the noticeable potentials of application of biomass in the foundries, limited numbers of studies have focused on testing renewable carbon briquettes on a large scale. Knowledge of the behaviour of briquettes at high process temperatures is scarce. The main challenge for developing the bio-carbon compacts is to fulfil the required characteristics suitable for utilization in cupola furnaces. This work has focused on evaluation of different additives and binders for compaction of bio-carbon and study of the behaviour of renewable materials at the laboratory-scale to build the foundation for upscaling and future developments for application of bio-carbon in a cupola furnace.

The studies in the present work started with the selection of raw materials and design of recipes. Different types and amounts of coke breeze, biochar, binders and additives were used. The briquettes, on a small-scale, were evaluated with respect to density, energy density, reactivity and mechanical strength. After comparison with reference samples, selected recipes were used and modified for developing larger briquettes. These briquettes were tested at high temperatures and compared to reference samples of foundry coke and coke breeze. This study provided useful information about effect of different parameters on compaction of biochar, selection of materials and performance of briquettes at high temperatures. The information is valuable for further developments of renewable carbon briquettes for industrial tests.

2. Materials and Methods

2.1. Coals and Particle Preparation

Two types of commercial charcoal with low and high ash contents, namely Charcoal 1 and Charcoal 2, and industrial coke breeze were used in the preparation of biocarbon briquettes. The particle size distributions of Charcoal 1 and 2 were measured and compared to coke breeze, as shown in Figure 1. Most particle sizes for all coals were in the range of 0.50–5.6 mm and the fines fractions were decreased in order of Charcoal 2 > Charcoal 1 > coke breeze. The chemical analysis of coke breeze and charcoals are given in Table 1.

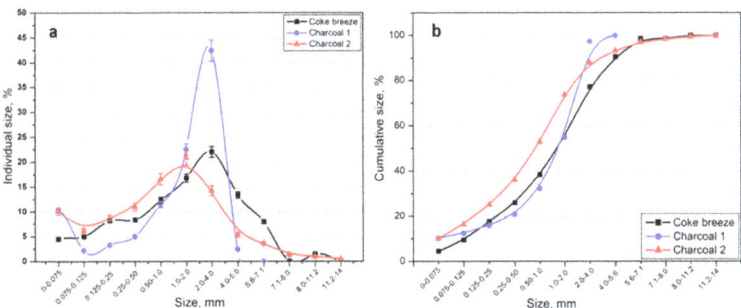

Figure 1. Particle size distribution of applied coals: (**a**) Individual size distribution. (**b**) Cumulative size distribution.

Table 1. Analysis of applied coals and binders.

Materials	Moisture, %	Ash, % (db)	VM, % (db)	C-Fix, % (db)	C-Total, % (db)	H, % (db)	N, % (db)	O, % (db)	S, % (db)	HHV, MJ/kg
Charcoal 1	3.5	0.7	18.6	80.7	87.0	3.4	0.2	8.3	0.0	31.1
Charcoal 2	2.4	39.2	11.6	49.2	54.2	1.6	0.1	4.8	0.0	18.4
Coke breeze	8.6	11.4	0.6	87.8	86.8	0.2	1.3	0.2	0.5	30.9
Molasses	19.0	5.6	88.4	6.1	42.7	3.8	0.9	52.6	0.78	15.0
Wood tar	12.6	4.1	67.9	28.0	40.0	6.3	0.1	53.7	0.0	21.6
Bitumen	0	1.7	22.0	70.0	85.7	5.3	1.9	3.6	1.8	41.4
Lignin	11.0	0.8	59.6	39.6	59.3	5.9	0.1	26.7	1.2	26.9
Nanocellulose	10.9	0.7	88.3	10.9	43.3	6.5	0.2	49.3	0.0	17.2

2.2. Binders Selected for Compaction

Six different organic binders; molasses/hydrated lime, wood tar, bitumen, lignin, nanocellulose and keracoal were selected for the compaction of charcoal fines with/without addition of Portland cement and coke breeze. The chemical compositions of binders used in the present study, except Keracoal which is a type of organic binder recently developed by Keramicalia [13], are given in Table 1.

2.3. Design of Recipes

The optimal design of recipe for biocarbon briquettes depending on binder type, binder ratio, and compaction pressure was derived from 27 tests. Reference recipe (R0) was produced from coke breeze and cement as binder, while recipes R1–R6, as given in Table 2, were produced from 45–50% of coke breeze, 25% of charcoal 1 and 10% of cement while different binders (molasses, wood tar, bitumen and lignin) were tested. The compaction pressure was fixed for all at 160 MPa.

Table 3 shows the composition of recipes R7–R15 that were designed for briquetting of Charcoal 2, which has lower carbon and higher ash content compared to Charcoal 1. The recipes were designed with/without coke breeze and different types of binders. Higher content (50–80%) of Charcoal 2 is applied in recipes R7–R9 compared to R0 (0% charcoal) or R4 (25% Charcoal 1). Recipes R10 and R11, with composition simulating R4, were used to produce briquettes under compaction pressure of 80 and 160 MPa, respectively, to investigate the effect of the compaction pressure and charcoal type on the briquettes' strength. The binding efficiency of Keracoal binder [13] was examined in recipes R11–R13. Recipe R14 was designed to simulate R4 (Charcoal 1), while R15 (10% cement) was designed to simulate R8 (without cement).

Recipes R16–R27 were designed with higher content of Charcoal 1, aiming at increasing the contribution of biocarbon in briquettes and optimizing the mechanical strength and reactivity by using different types of binders and other additives, as given in Table 4. R16–R21 were designed to investigate the effect of molasses content with/without cement and at different mixing ratios of hydrated lime. R16 and R17 were developed with same composition but different types of cement; Portland cement and calcium aluminate (CA) cement, respectively. The effect of silica suspension on charcoal reactivity was assessed in R22–R24. Recipes R25–R27 were designed to evaluate the effect of compaction pressure on briquette quality and the results were compared to R18, which was produced at 160 MPa.

After designing the recipes, the materials were thoroughly mixed to produce homogeneous mixtures and ensure that the binders are well-distributed in the mixtures. A laboratory hydraulic press with compaction pressures up to ~160 MPa and a stainless-steel mould (Ø = 40 mm) were used for briquetting.

Further recipes were designed for upscaling of the briquettes using a bigger-sized mould (Ø = 70 mm) and compaction pressure up to ~160 MPa, as given in Table 5. Several briquettes were produced for each recipe under the same compaction condition. Reproducibility of the results was checked for all measurements. The moisture contents of the mixtures and briquettes were measured using moisture analyzer HB43-S Halogen by Mettler Toledo.

The large-scale samples were tested in high-temperature experiments. Recipe M1 containing coke breeze and cement was used as reference for comparison. In recipes M2–M5, two types of hydrated lime were compared. The first type of hydrated lime was prepared in the laboratory by mixing burnt lime powder with water. After ensuring complete hydration, the hydrated lime with high water content (about 60%) was used in briquettes M2 and M3. The hydrated lime was first added to the molasses and the mixture was added to the dry charcoal and cement mixture. This method was also used in the development of smaller briquettes. The second type of hydrated lime (CL), used in M4 and M5, was a commercial hydrated lime in the form of fine powder with low moisture content (~10%). The commercial hydrated lime consisted of 91% calcium hydroxide ($Ca(OH)_2$) and 99% of particles were smaller than 0.074 mm. This material was selected considering its availability as a commercial product, suitability for large-scale production and improved preparation and mixing procedures. In each recipe, the commercial hydrated lime was mixed with the charcoal and cement before the addition of molasses. In recipes M6 to M14, commercial hydrated lime (CL) was used due to improved properties of briquettes.

Table 2. Design of reference and Charcoal 1–coke breeze recipes.

Recipe No.	Coke Breeze, wt.% (db *)	Charcoal 1, wt.% (db)	Cement, wt.% (db)	Organic Binder Type and Percent, wt.% (db)	Other Additives, wt.% (db)	Pressure, MPa
R0 (Ref.)	90	0	10	-	-	160
R1	45	25	10	15 Molasses	5 CaO	160
R2	50	25	10	15 Wood tar	-	160
R3	50	25	10	15 Bitumen	-	160
R4	45	25	10	20 Molasses	-	160
R5	55	25	10	9.5 Lignin	0.5 Collagen	160
R6	50	25	10	15 Nanocellulose	-	160

* db: dry basis.

Table 3. Design of Charcoal 2–coke breeze recipes.

Recipe No.	Coke Breeze, wt.%, dry basis (db)	Charcoal 2, wt.% (db)	Cement, wt.% (db)	Organic Binder Type and Percent, wt.% (db)	Other Additives, wt.% (db)	Pressure, MPa
R7	25	50	5	20 Molasses	-	160
R8	0	80	0	20 Molasses	-	160
R9	0	80	0	20 Bitumen	-	160
R10	45	25	10	20 Molasses	-	80
R11	50	34	10	5 Keracoal	1 Activator	16
R12	45	25	10	16 Keracoal	4 Activator	160
R13	45	25	10	16 Keracoal	4 Activator	80
R14	45	25	10	20 Molasses	-	160
R15	0	70	10	20 Molasses	-	160

Table 4. Design of Charcoal 1 recipes.

Recipe No.	Charcoal 1, wt.% (db)	Cement, wt.% (db)	Organic Binder Type and Percent, wt.% (db)	Other Additives, wt.% (db)	Pressure, MPa
R16	80	8	8 Molasses	4 Hydrated lime	160
R17	80	8 (CA)	8 Molasses	4 Hydrated lime	160
R18	74	8	12 Molasses	6 Hydrated lime	160
R19	62	8	20 Molasses	10 Hydrated lime	160
R20	69.5	8	22.5 Molasses	-	160
R21	82	0	12 Molasses	6 Hydrated lime	160
R22	77	0	12 Molasses	6 Hydrated lime 5 Silica suspension	160
R23	72	0	12 Molasses	6 Hydrated lime 10 Silica suspension	160

Table 4. Cont.

| R24 | 72 | 5 | 12 Molasses | 6 Hydrated lime
5 Silica suspension | 160 |
R25	74	8	12 Molasses	6 Hydrated lime	120
R26	74	8	12 Molasses	6 Hydrated lime	80
R27	74	8	12 Molasses	6 Hydrated lime	40

Table 5. Designed large-scale recipes.

Mixture Recipe	Coke Breeze	Charcoal 1	Charcoal 2	Molasses	Hydrated Lime	Portland Cement	SiO$_2$ Suspension	Charcoal Size, mm
				wt.%				
M1	90	0	0	0	0	10	0	-
M2	0	69.5	0	15	7.5	8	0	0–4
M3	0	62	0	20	10	8	0	0–4
M4	0	69.5	0	15	7.5 (CL)	8	0	0–4
M5	0	62	0	20	10 (CL)	8	0	0–4
M6	0	65	0	18	9 (CL)	8	0	0–4
M7	0	70	0	20	10 (CL)	0	0	0–4
M8	0	62	0	20	10 (CL)	4	4	0–4
M9	0	62	0	20	10 (CL)	0	8	0–4
M10	0	62	0	20	10 (CL)	8	0	0–0.5
M11	0	62	0	20	10 (CL)	8	0	0–2
M12	0	0	62	20	10 (CL)	8	0	0–2
M13	10	52	0	20	10 (CL)	8	0	0–2
M14	20	42	0	20	10 (CL)	8	0	0–2

2.4. Determination of Physical Properties and Reactivity

The cross-section cold compressive strength (CCS) and splitting tensile strength (STS) of biocarbon briquettes were measured by compression equipment. The compression device automatically records the peak point of strength after which the failure occurs in the briquette. The reactivity of selected biocarbon briquettes was measured using a Netzsch thermo-gravimetric analysis (TGA) technique STA 409, with sensitivity ±1 µg coupled with quadrupole mass spectroscopy (QMS). For all reactivity tests, regular cubic shapes (edge length ~15 mm) were prepared from the briquettes. The sample was placed on a shallow crucible and positioned in the centre of the reaction chamber of the furnace. The heating rate was 15 K/min from room temperature to 1100 °C under a continuous flow of CO_2 (200 mL/min) throughout the reaction and then fixed at this temperature for 60 min followed by fast cooling (20 K/min). The weight loss due to sample gasification and composition of the generated gases were recorded during the trials.

2.5. Evaluation of Behaviour of Briquettes at High Temperature

A test apparatus was designed to evaluate the behaviour of developed briquettes under well-controlled temperatures, mechanical load and gas atmospheres. The schematic illustration of the experimental setup is shown in Figure 2. The system includes a gas mixing station, an electrical gas heater, a pot furnace with load application system and an off-gas burner. The sample temperature was measured using thermocouples placed close to the briquettes. The gas composition for the experiments was selected considering the temperature reached in the sample and the correlated gas compositions in the cupola from a preceding study [14].

In each experiment, one of the upscaled briquettes (Ø = 70 mm) was tested. The following procedure was applied in the tests. The briquette was placed on the sample holder in the pot furnace and the system was sealed. In order to remove the remaining oxygen and protect the sample from oxidation, the system was flushed with high-purity nitrogen gas (99.5% purity) at room temperature. Meanwhile, N_2 was heated in an electrical heater and was bypassed. When the N_2 gas reached the target temperature and the temperature become stable, the furnace atmosphere was switched to the hot N_2 stream and the sample was heated up while the load was applied. The load was fixed in all tests and was equal to ~0.8 kg/cm^2. In a few experiments, hot N_2 was used and the rest of the tests were conducted using a reducing gas mixture consisting of 34 vol% CO, 7 vol% CO_2 and 59 vol% N_2. The sample was kept in this gas atmosphere under load for 60 min. Afterward, the load was removed and the furnace was flushed with N_2. The sample was cooled down using nitrogen gas and was taken out when it reached room temperature. The results were evaluated with regards to weight change and structural changes in the samples after the tests.

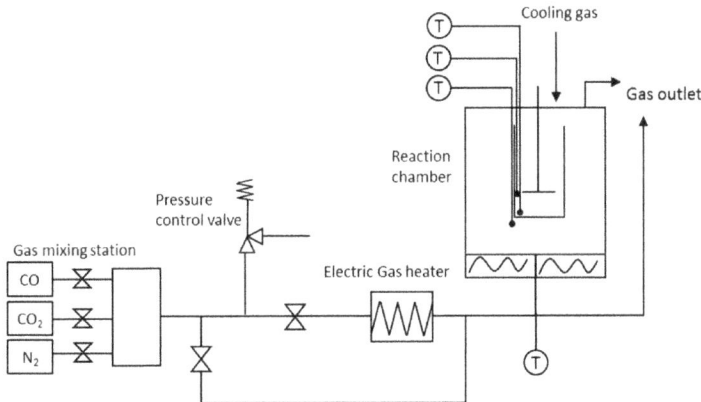

Figure 2. Schematic illustration of hot strength test equipment.

3. Results and Discussion

The physical, chemical and mechanical characteristics of biocarbon briquettes produced with different types and content of charcoals, binders and at different compaction pressures were evaluated in terms of density, cold mechanical strength, hot strength, reactivity and energy density, as described in the following subsections.

3.1. Densification of Biocarbon

Figure 3 shows the average density of biocarbon briquettes in comparison with commercial coke breeze briquettes (CBB) and the reference sample prepared with coke breeze and 10% cement (R0). Among all types of biocarbon-containing briquettes, only R10 was able to reach an equal density to that of CBB; however, it is still less than the reference briquettes (R0). The rest of the biocarbon briquettes showed densities lower than the target (CBB). In general, the density of the briquettes was decreased by increasing the content of charcoal. Recipes R7–R10 showed higher density compared to R15–R27 due to the higher ash content in Charcoal 2. Comparison of R4 and R10, with the same composition but different charcoal, indicated that the density increased with ash content. In general, the density of the briquettes was decreased by increasing the content of charcoal.

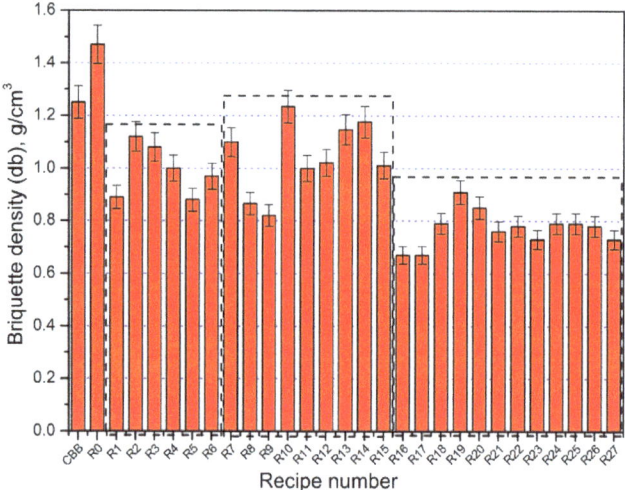

Figure 3. Average dry density of developed briquettes R0–R27.

3.2. Cold Mechanical Strength of Biocarbon Briquettes

The cold compression strength (CCS) was measured for all recipes, as shown in Figure 4. Most of the biocarbon recipes exhibited higher mechanical strength compared to CBB. R1 was spontaneously disintegrated, while the highest values of CCS were demonstrated by R8 and R10, which was higher than that of R0. This indicated that molasses with lime can provide an adequate mechanical strength for biocarbon briquettes. R11 and R13–R15 showed similar values of CCS to that of R0, which indicated that 25% of charcoal can be added to coke breeze without deteriorating the mechanical strength of the briquettes.

The splitting tensile strength (STS) was measured for all the recipes, as shown in Figure 5. The STS for CBB was not measured as it has a different geometry (hexagonal prism), while the reference and biocarbon briquettes have a cylindrical shape, which allowed for the load to be applied on the peripheral surface of the briquettes. The highest STS were demonstrated by R15, R13 and R10. This indicated that both molasses and Keracoal were superior in terms of the enhancement of the STS of biocarbon briquettes. The comparison between the mechanical strength of R4 (CCS = 66.25 kg/cm^2 and STS = 5.18 kg/cm^2) and R14 (CCS = 88.37 kg/cm^2 and STS = 6.38 kg/cm^2) indicated that the briquettes produced from Charcoal 2 exhibited higher mechanical strength compared to that of Charcoal 1. This can be attributed to the higher ash content in Charcoal 2 compared to that of Charcoal 1. The comparison between R8 and R15 indicated that addition of cement deteriorated the CCS but improved STS. No significant difference was found by changing the type of cement from normal Portland cement in R16 to calcium aluminate cement (CA) in R17.

In general, most of the briquettes that exhibited high CCS, demonstrated relatively higher STS. The significant difference between the CCS and the STS for all recipes can be attributed to the effect of the applied pressure on the area under load. In case of CCS, the load is distributed on the whole surface area of the briquettes, while in the case of splitting tensile strength, the pressure is applied on a narrow plane in the briquettes. This effect must be considered if the briquette geometry (shape, size and dimension) is changed and the measured strength can be used qualitatively rather than quantitatively to evaluate the large-scale biocarbon briquettes. Similar differences in strength of cylindrical briquettes were observed in the work by Rahman et al. [11].

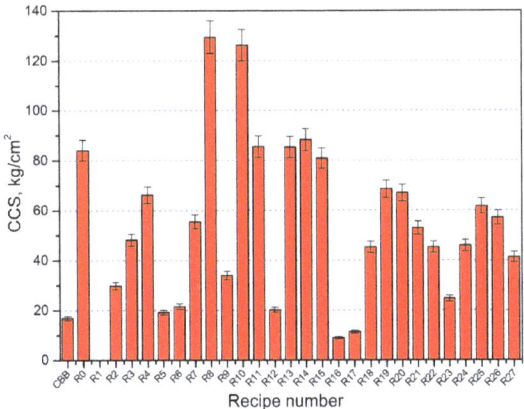

Figure 4. Average cold compression strength of developed biocarbon briquettes R0–R27.

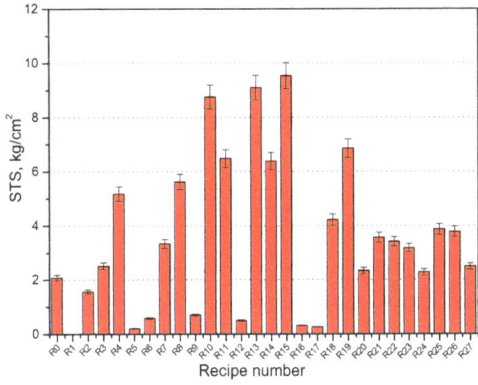

Figure 5. Average splitting tensile strength (STS) of developed biocarbon briquettes R0–R27.

3.3. Reactivity of Biocarbon Briquettes

The reactivity was measured for biocarbon briquettes from selected recipes (R10, R18 and R24) and the results are compared to fossil carbon briquettes (R0, CBB and FC), as shown in Figure 6. FC exhibited the lowest weight loss (~10%) and consequently, the lowest reactivity after gasification with CO_2 at 1100 °C for 1 h, while R18 demonstrated the highest weight loss (~54%) among the tested briquettes, which can be attributed to the higher content of biocarbon (74% charcoal). The gasification trend of CBB, FC and R0 were similar and the weight loss was in the order of CBB > R0 > FC. The lowest weight loss (~31%) of biocarbon briquettes was revealed by R10, which can be attributed to the lower content of biocarbon (25% Charcoal 2). It can be concluded that the biocarbon content in the briquettes is the predominant factor affecting the reactivity of the briquettes. All biocarbon samples showed a weight loss of ~8–12% at 250 °C and ~15–30% at ~800 °C, which can be attributed to the decomposition of molasses, dehydration of lime and cement and the devolatilization of charcoal. At temperatures between 800 and 1100 °C, the gasification took place in all types of briquettes with different rates according to the reactivity of embedded carbon. The effect of volatiles and dehydration can be neglected at 1100 °C and the change in weight can be only attributed to the gasification of fixed carbon with surrounding CO_2 gas.

To ensure that the gasification rate is not related to the differences in starting carbon content in the sample, normalization was conducted to neutralize the effect of carbon content. Equations (1) and (2) are used to calculate the apparent gasification rate ($R_{Apparent}$) and the normalized gasification rate ($R_{Normalized}$), respectively. Figure 7 shows the normalized gasification rate in fossil and biocarbon briquettes. In general, the reactivity trend can be classified into three sets; low-reactive briquettes (FC, R0, CBB), middle-reactive briquettes (R10) and high-reactive briquettes (R18 and R24). The relatively lower apparent gasification rate of R24 compared to R18 can be attributed to the effect of silica suspension, which reduces the reactivity of carbon, as was reported elsewhere [15].

$$R_{Apparent} = \frac{\Delta W_t}{W_0 \times \Delta t} \quad (1)$$

$$R_{Normalized} = \frac{R_{Apparent}}{X_{Fixed\ C}} \quad (2)$$

where $R_{Apparent}$: apparent gasification rate (%/min); ΔW_t: weight change of sample at time t (mg); W_0: initial weight of sample (mg); $R_{Normalized}$: normalized gasification rate (1/min); $X_{Fixed\ C}$: total content of fixed carbon in the recipe (%).

3.4. Heating Value and Energy Density of Biocarbon Briquettes

Figure 8 shows the composition of developed briquettes for all recipes classified as fossil fixed carbon, fixed biocarbon, volatile matter (VM) and ash content. The bitumen's carbon in R3 is accounted as fossil carbon, while Keracoal binder is assumed to have a similar composition to molasses. Cement and lime are included in the ash content. The total fixed biocarbon can reach up to ~67% in briquettes prepared from 94% biomaterials (R21). The recipes (R7–R15) prepared from Charcoal 2 has lower carbon content and higher ash content compared to that prepared from Charcoal 1 (R1–R6 and R16–R27) due to the effect of charcoal grade.

The higher heating value (HHV) has been calculated for all types of briquettes by applying Equation (3). Biocarbon briquettes prepared from Charcoal 1 showed higher HHV compared to biocarbon briquettes made from Charcoal 2 or CBB or as shown in Figure 9. The relatively high calorific value of these briquettes can be attributed to the higher carbon content and lower ash content for Charcoal 1 and used binders. Figure 10 shows the calculated values of energy density for different recipes. The energy density for the briquettes containing coke breeze or Charcoal 2 was relatively higher than that prepared from Charcoal 1. This indicated that the higher compaction density of biocarbon briquettes prepared form Charcoal 2 was able to compensate its lower heating value.

$$HHV\ (MJ/kg\ briq.) = (34.834 * Fixed\ C + 115.12 * H + 10.467 * S + 6.28 * N - 10.802 * O)/100 \quad (3)$$

where C: fixed carbon; H: hydrogen content; S: sulphur content; N: nitrogen content; O: oxygen content.

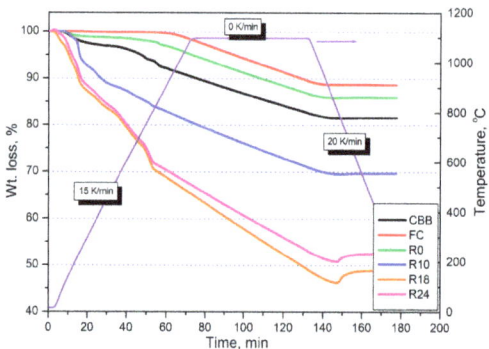

Figure 6. Changes in sample weight and experimental temperature vs. time.

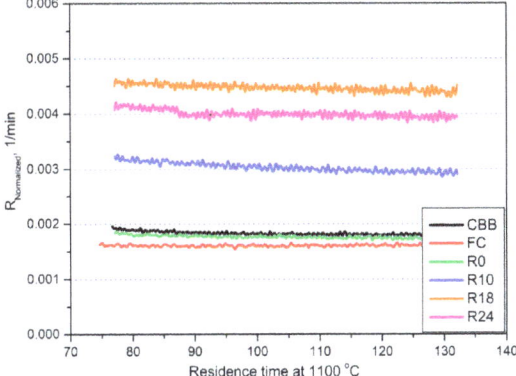

Figure 7. Comparison of the normalized gasification rate for the coke samples.

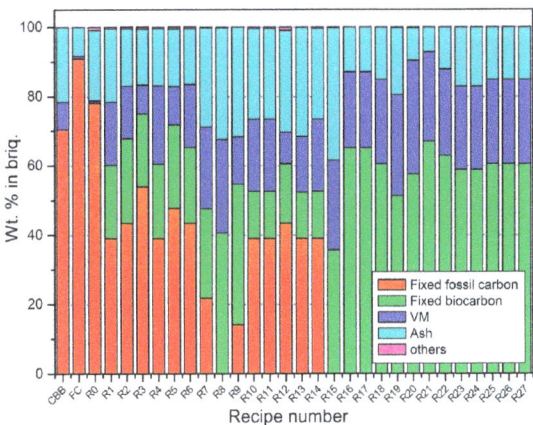

Figure 8. Carbon, volatile matter and ash content of briquettes in all recipes.

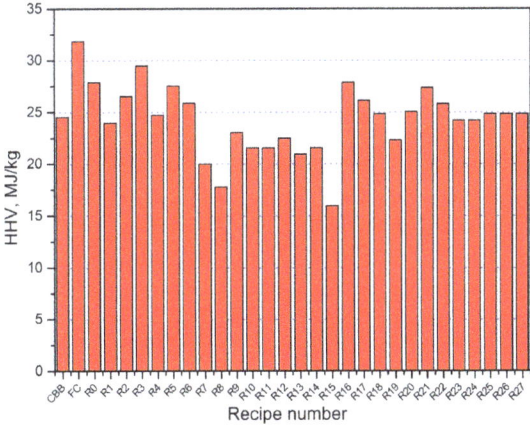

Figure 9. Higher heating values of briquettes in all recipes.

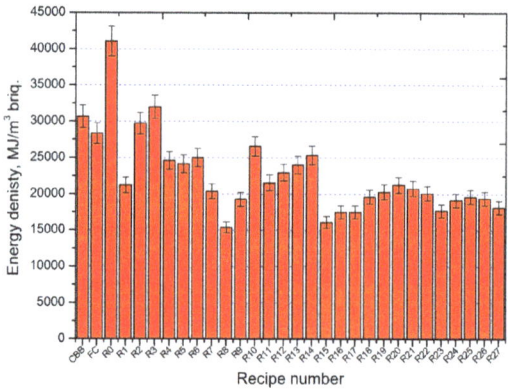

Figure 10. Energy density of briquettes in all recipes.

3.5. Behaviour of Upscaled Biocarbon Briquettes at High Temperature

The experimental conditions are summarized in Table 6. Five trials were conducted for biocarbon briquettes and compared to CBB and industrial foundry coke (FC). Figure 11 shows the procedure for changing the gas flow rates and variation of temperature in the experiments. In the first three experiments, the biocarbon briquettes were tested in a pure N_2 atmosphere and the samples disintegrated; therefore, the weight change is not presented in Table 6. Examples of a foundry coke reference sample and a biocarbon briquette before and after the experiment are shown in Figure 12. Development of cracks in the biocarbon briquette and oxidation on the surface of the sample is clearly seen in the picture, while the changes in the foundry coke were minor. Slight weight changes were observed in the reference samples, while in all biocarbon briquettes, the average weight loss was around 35% due to the higher content of moisture, volatiles and binders.

Table 6. Test conditions for high temperature performance of briquettes.

Test No.	Sample	Gas Composition (vol%)			Total Load (t/m^2)	Total Weight Loss (%)
		CO	CO$_2$	N$_2$		
1, 2, 3	M5	-	-	100	0	-
4	M5	-	-	100	8	35.3
5, 6	M5	34	7	59	8	36
7	CBB	34	7	59	8	5.3
8, 9	Foundry coke (FC)	34	7	59	8	0.6
10	M11	34	7	59	8	36.5
11	M13	34	7	59	8	34.7

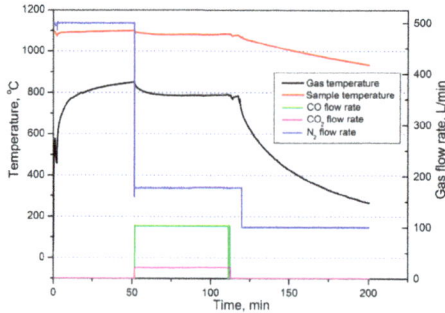

Figure 11. Changes in temperatures and gas flow rates during experiments.

Figure 12. Foundry coke and biocarbon briquettes (**a**) foundry coke before test, (**b**) foundry coke after test, (**c**) biochar briquette before test, (**d**) biochar briquette after test.

During the high-temperature tests, opening of cracks and disintegration of biocarbon briquettes took place even in nitrogen atmosphere and without application of load. The cracks followed a similar pattern to the initial fine cracks seen in the briquettes after drying. In general, three types of cracks were identified: (i) surface cracks caused by friction between the material and the die wall during compaction and during ejection of briquettes from the die, (ii) vertical cracks developed during drying of briquettes and moisture evaporation, and (iii) cracks developed from the pressure gradient during compaction, resulting in the formation and separation of a cone cap. It is expected that changes in mechanical stresses during the compaction cycle such as compression, decompression and ejection of briquettes from the mould play an important role in the formation of the cone shape cap in briquettes. Similar behaviour has been observed in the work reported by Wu et al., on the compaction of pharmaceutical powders [16]. Using modelling and calculation of the mechanical stresses in the compacts, they showed a clear correlation between the separation of cone and stress gradients during removal of sample from the die in the cylindrical compacts [16].

Efforts were made to decrease the formation of cracks by improving die surface quality, testing finer charcoal particles, and curing the briquettes in closed containers with high moisture content to slow down the drying process. Major improvement was obtained by slowing down the drying process. Although the briquette qualities were improved, it was clear that the cracks are caused by a combination of different parameters. Therefore, further investigation of the bonding mechanisms, compaction method and microstructure of the compacted material is necessary. The occurrence of a density gradient in the briquettes is inevitable in one punch die pressing method; therefore, using this compaction technique for producing materials with suitable dimensions for cupola demands are still needed for further improvements in material selection and compaction conditions. To guarantee efficient operation and permeability in the cupola furnace, the minimum size for foundry coke has to be 90 × 90 mm. In 2017, the annual capacity of the world's casting production was reached; about 110 million tons (Mt) of cast iron [17] with average foundry coke consumption of about 110 kg/t of cast iron [18]. The partial replacement of foundry coke with developed biocarbon briquettes can be a good start to mitigate the fossil CO_2 emission in a cupola furnace. Replacing 25% of fossil foundry coke with biochar can contribute in cutting of more than 1.1 Mt of fossil CO_2 emissions annually from the cupola furnaces worldwide. Beside developing biocarbon briquettes to meet the requirements of a cupola furnace, our future work will focus on conducting a system analysis of a bio-based cupola furnace to evaluate the potential of pre-treated biomass in replacing the foundry coke and reducing the fossil CO_2 emission.

4. Concluding Remarks

Replacement of foundry coke with renewable and neutral carbon biomass products is one critical to secure the sustainability of this process; however, the relatively low mechanical strength, low energy density and high reactivity of biomass represent the main challenges for its efficient implementation. The present study focused on developing biocarbon briquettes with adequate energy density, mechanical strength, and reactivity for a cupola furnace. The trials were performed on a lab-scale briquetting press to investigate and optimize the recipe's composition, biochar particle size, binder type and ratio, compaction pressure and drying conditions. The main finding can be summarized in the following points:

Compaction can significantly enhance the density and energy density of biocarbon-containing briquettes. The compaction pressure and the amount of bio-carbon in the recipe significantly affect the strength of the briquettes. The mechanical strength of the briquettes increased with increasing compaction pressure, while it decreased with increasing the biocarbon content in the briquettes.

1. In this study, combination of 20% molasses with 10% lime and 8% cement is the most suitable binder compared to other types of binders tested for enhancing the mechanical strength of biocarbon briquettes. Cement is required to improve the tensile strength and hot strength of biocarbon briquettes. The results illustrated that the optimum amount of binders varies with the compaction method and size of produced briquettes.
2. The normalized gasification rate of the fossil and biocarbon samples increased by increasing the total biocarbon content in the recipes due to higher reactivity and the more porous nature of biocarbon compared to fossil carbon. Addition of a silica suspension to the mixtures can be a potential method to reduce the surface area for reaction and, therefore, lower the reactivity of biocarbon-containing briquettes.
3. Upscaling of biocarbon briquettes and testing of hot strength under a mechanical load resulted in generation of cracks and disintegration of briquettes, which can be mainly attributed to the presence of mechanical stresses during the compaction cycle from compression, decompression, ejection of briquettes to curing and drying of briquettes.

Further investigation of different compaction technologies, in-depth understanding of the bonding mechanisms, development of high-temperature strength and control of biocarbon reactivity in the briquettes are necessary for developing biocarbon briquettes to meet the requirements of a cupola furnace and replace foundry coke.

Author Contributions: E.M. and M.K. designed and performed the experiments, carried out the data analysis and results evaluation and drafted the manuscript. M.L. supervised the work, contributed to results evaluation and conducted final revision of the manuscript. All authors discussed the design of tests, the results and contributed to the final manuscript.

Funding: This research was funded by the Swedish Energy Agency and Volvo Powertrain AB and the work carried out at Swerim AB.

Acknowledgments: The support provided for this work from the Swedish Energy Agency for financing Project number 45365-1 "Investigation of behaviour of bio-carbon briquettes at elevated temperatures" and Volvo Powertrain AB is greatly acknowledged.

Conflicts of Interest: The authors declare no conflict of interest.

References

1. Mousa, E.A.; Wang, C.; Riesbeck, J.; Larsson, M. Biomass Applications in Iron and Steel Industry: An Overview of Challenges and Opportunities. *Renew. Sustain. Energy Rev.* **2016**, *65*, 1247–1266. [CrossRef]

2. Suopajärvia, H.; Umekib, K.; Mousa, E.A.; Hedayati, A.; Romar, H.; Kemppainen, A.; Wang, C.; Phounglamcheik, A.; Tuomikoski, S.; Norberg, N.; et al. Use of biomass in integrated steelmaking- Status quo, future needs and comparison to other low-CO_2 steel production technologies. *Appl. Energy* **2018**, *213*, 384–407. [CrossRef]
3. Isnugroho, K.; Birawidha, D.C.; Hendronursito, Y. The biomass waste use as a secondary energy source for metal foundry process. In Proceedings of the 2016 Conference on Fundamental and Applied Science for Advanced Technology (ConFAST 2016), Yogyakarta, Indonesia, 25–26 January 2016; Volume 1746, pp. 020001-1–020001-5.
4. Schulten, M.; Pena Chipatecua, G.; Quicker, P.; Seabra, S. Investigations on the application of biochar as an alternative for foundry coke. In Proceedings of the 21st European Biomass Conference, Copenhagen, Denmark, 3–7 June 2013.
5. Echterhof, T.; Demus, T.; Schulten, M.; Noel, Y.; Pfeifer, H. Substituting fossil carbon sources in the electric arc and cupola furnace with biochar. In *European Steel Environment & Energy Congress (ESEC)*; Teesside University: Middlesbrough, UK, 2014.
6. Nieto-Delgado, C.; Cannon, F.S.; Paulsen, P.D.; Furness, J.C.; Voigt, R.C.; Pagnotti, J.R. Bindered anthracite briquettes as fuel alternative to metallurgical coke: Full scale performance in cupola furnaces. *Fuel* **2014**, *121*, 39–47. [CrossRef]
7. Noh, Y.D.; Komarneni, S.; Cannon, F.S.; Brown, N.R.; Katsuki, H. Anthracite briquettes with plant byproducts as an ecofriendly fuel for foundries. *Fuel* **2016**, *175*, 210–216. [CrossRef]
8. Torielli, R.M.; Cannon, F.S.; Voigt, R.C.; Considine, T.J.; Furness, J.C.; Fox, J.T.; Goudzwaard, J.E.; Huang, H. The environmental performance and cost of innovative technologies for ductile iron foundry production. *Int. J. Met.* **2014**, *8*, 37–48. [CrossRef]
9. Gabra, M.H.; Jain, R.K.; Tiwari, A. Energy efficient cupola furnace via hybridization with a biomass gasifier. *Int. J. Emerg. Technol. Eng. Res.* **2017**, *5*, 54–62.
10. Wang, L.; Buvarp, F.; Skreiberg, Ø.; Bartocci, P.; Fantozzi, F. A study on densification and CO_2 gasification of biocarbon. *Chem. Eng. Trans.* **2018**, *65*, 145–150.
11. Rahman, A.N.E.; Masood, M.A.; Prasad, C.S.N.; Venkatesham, M. Influence of size and shape on the strength of briquettes. *Fuel Process. Technol.* **1989**, *23*, 185–195. [CrossRef]
12. Ifa, L.; Kusuma, H.; Sabara, Z.; Mahfud, M. Production of bio-briquette from biochar derived from pyrolysis of cashew nut waste. *Ecol. Environ. Conserv.* **2019**, *25*, 125–131.
13. Keracoal Binder. Available online: https://www.keramicalia.com/binders/6-keracoal-binder (accessed on 10 November 2019).
14. Aristizabal, R.E.; Silva, C.M.; Perez, P.A. Studies of a quenched cupola Part I: Overview of experimental studies. *AFS Trans.* **2009**, *117*, 681–691.
15. Iwai, Y.; Ishiwata, N.; Murai, R.; Matsuno, H. Control technique of coke rate in shaft furnace by controlling coke reactivity. *ISIJ Int.* **2016**, *56*, 1723–1727. [CrossRef]
16. Wu, C.Y.; Hancock, B.C.; Mills, A.; Bentham, A.C.; Best, S.M.; Elliott, J.A. Numerical and experimental investigation of capping mechanisms during pharmaceutical table compaction. *Powder Technol.* **2008**, *181*, 121–129. [CrossRef]
17. The European Foundry Industry at a Glance. Available online: https://www.caef.eu/statistics/ (accessed on 10 November 2019).
18. Rao, N.M. Iron Foundry and Pig Iron Industries in India. Available online: http://www.isrinfomedia.com/main/archive/1 (accessed on 10 November 2019).

© 2019 by the authors. Licensee MDPI, Basel, Switzerland. This article is an open access article distributed under the terms and conditions of the Creative Commons Attribution (CC BY) license (http://creativecommons.org/licenses/by/4.0/).

Article

Low-Cost Activated Grape Seed-Derived Hydrochar through Hydrothermal Carbonization and Chemical Activation for Sulfamethoxazole Adsorption

Elena Diaz *, Francisco Javier Manzano, John Villamil, Juan Jose Rodriguez and Angel F. Mohedano

Chemical Engineering Department, Universidad Autonoma de Madrid, C/Francisco Tomás y Valiente 7, 28049 Madrid, Spain; francisco.manzano@uam.es (F.J.M.); john.villamil@uam.es (J.V.); juanjo.rodriguez@uam.es (J.J.R.); angelf.mohedano@uam.es (A.F.M.)
* Correspondence: elena.diaz@uam.es; Tel.: +34-914978035; Fax: +34-914973516

Received: 23 August 2019; Accepted: 23 November 2019; Published: 27 November 2019

Featured Application: Hydrothermal carbonization is presented as an alternative management way to valorize biomass wastes, and transform them in high-value solid products.

Abstract: Activated carbons were prepared by chemical activation with KOH, $FeCl_3$ and H_3PO_4 of the chars obtained via hydrothermal carbonization of grape seeds. The hydrochars prepared at temperatures higher than 200 °C yielded quite similar proximate and ultimate analyses. However, heating value (24.5–31.4 $MJ·kg^{-1}$) and energy density (1.04–1.33) significantly increased with carbonization temperatures between 180 and 300 °C. All the hydrochars showed negligible BET surface areas, while values between 100 and 845 $m^2·g^{-1}$ were measured by CO_2 adsorption at 273 K. Activation of the hydrochars with KOH (activating agent to hydrochar ratio of 3:1 and 750 °C) led to highly porous carbons with around 2200 $m^2·g^{-1}$ BET surface area. Significantly lower values were obtained with $FeCl_3$ (321–417 $m^2·g^{-1}$) and H_3PO_4 (590–654 $m^2·g^{-1}$), showing these last activated carbons important contributors to mesopores. The resulting materials were tested in the adsorption of sulfamethoxazole from aqueous solution. The adsorption capacity was determined by the porous texture rather than by the surface composition, and analyzed by FTIR and TPD. The adsorption equilibrium data (20 °C) fitted the Langmuir equation well. The KOH-activated carbons yielded fairly high saturation capacity reaching up to 650 $mg·g^{-1}$.

Keywords: grape seeds; hydrothermal carbonization; hydrochar; activated carbon; adsorption; sulfamethoxazole

1. Introduction

Biomass is a widely available source of energy, particularly important in developing countries, but cannot be considered a technically ideal fuel due to physical and chemical properties, each often having fibrous nature, high moisture content, volatile components, alkali and alkaline earth metallic content and a relatively low bulk density and heating value [1–3]. Thus, pre-treatment of biomass is in many cases convenient for efficient use as energy source. A broad range of biological (mainly anaerobic digestion and fermentation) and thermochemical (torrefaction and pyrolysis) treatments are typically used to improve the fuel properties of raw biomass [2,4,5]. From those processes, liquid or gaseous biofuels and even some valuable products are derived.

As an alternative to the classical thermochemical methods, hydrothermal carbonization (HTC), also referred to as wet torrefaction, is becoming an increasingly attractive way of biomass conversion. It operates in presence of water, at comparatively mild temperatures (180–300 °C) and a corresponding saturation pressure [6–11]. The resulting solid product, usually called hydrochar (HC), is more stable and has higher carbon content than the starting substrate and improved (higher) heating value with respect to the char resulting from slow-pyrolysis or conventional carbonization at the same temperature [12–14]. In addition to the hydrochar, HTC gives rise to a high organic load aqueous stream [15,16], and a gas consisting mainly of CO_2.

Hydrochars have several industrial and environmental applications, such as for soil improvers [17] and solid fuel, whether upon direct combustion [18] or through gasification [12]. HTC chars may have also potential uses as sorbents for CO_2 sequestration, methane and hydrogen storage, in energy storage devices (Li/Na ion batteries, supercapacitors, fuel cells) and as precursors for activated carbon preparation [7,18–23].

In general, hydrochars show relatively low BET surface areas and pore volumes, which hinders their application as adsorbents [24]. Nevertheless, Titirici [23] reported the presence of ultramicropores (narrow micropores) in several sugar-derived hydrochars, as determined by CO_2 adsorption. Low-cost adsorbents can be obtained from HC upon further activation with different agents (CO_2, steam, acids, bases, or salts), allowing substantial pore volume and surface area increases, and even changes of the morphological structures of HTC chars [23–25]. A number of works have reported on the application of activated hydrochars in aqueous-phase adsorption for the removal of heavy metals [24,26–28], dyes [24,29], phenolic compounds [30] and emerging pollutants [18,31]. Gas-phase applications have been also studied, including, VOCs [32] and CO_2 [31,33] adsorption. Hydrochars treated by H_2O_2 oxidation have been used for lead removal from water [34].

Agricultural wastes are recognized as interesting feedstocks for inexpensive carbon materials [35]. Among them, grape seeds represent up to 15% of the solid wastes from the wine industry, where they are mostly burnt as fuel. Some papers have reported on the preparation of activated carbons by both physical and chemical activation from this precursor [36–40]. Activated carbon is widely available at low cost in many regions and its structural characteristics (granular morphology, size and preferential distribution of the lignocellulosic material in the periphery of the seed) make it particularly attractive for that purpose [36]. Essentially microporous activated carbons have been obtained by physical (CO_2) and chemical (KOH and K_2CO_3) activation [36,40]. Using phosphoric acid as the activating agent allowed obtaining also predominantly microporous carbons but with a broader porous texture with some significant contribution of mesoporosity [36]. Recently, Purnomo et al. [41] reported a two-stage process consisting of HTC followed by KOH-activation of the resulting hydrochar for the preparation of microporous carbons with high BET surface area under milder operating conditions than those used for direct single-step activation with the same agent.

The aim of this work was to study the valorization of grape seeds into carbon materials with the double purpose of obtaining high energy-density solid fuels upon hydrothermal carbonization and low-cost activated carbons by chemical activation of the resulting hydrochar. Temperature and biomass:water ratio have been analysed as operating conditions for HTC and KOH. $FeCl_3$ and H_3PO_4 have been tested as activating agents under different conditions. Hydrochars and activated carbons have been characterized by several techniques covering proximate and ultimate analyses, 77 K N_2 adsorption-desorption and CO_2 adsorption isotherms for porous texture, SEM for morphological examination and FTIR and TPD for surface functional groups assessment. Sulfamethoxazole, a sulfonamide antibiotic used to treat urinary infections was selected as the model compound to test the potential of the activated carbons as adsorbents for the removal of water pollutants. That compound is one of the most frequently emerging contaminants found in municipal wastewater, and in many cases is still present in the effluents from sewage treatment plants.

2. Materials and Methods

2.1. Preparation of Hydrochars

Grape seeds (GSs) provided by a local winery ("Tinta de Toro", Zamora) were used as the hydrochar precursor. Table 1 shows the proximate and ultimate analyses. The HTC process was carried out in a Teflon-lined stainless steel vessel (100 mL), using 20 g of dried grape seeds and different water to GS ratios (10–60 wt.% of GSs). The reactor was sealed, inserted in a muffle furnace (Hobersal serie 8B Mod 12 PR/400, Hobersal, Barcelona, Spain) and heated up to 180–300 °C for 16 h. The hydrochar was recovered by filtration, washed with distilled water and oven-dried at 105 °C for 24 h (Nabertherm R 60/750/12-C6, Nabertherm, Bremen, Germany). The hydrochars were denoted as GS followed by the carbonization temperature and the dry GS percentage in the mixture. For instance, GS-260-40 represents the hydrochar obtained from grape seeds at 260 °C with 40% of dry GS. Table 1 includes also the hydrochar yield values calculated as the mass of the hydrochar per unit mass of GS, both on a dry basis.

Table 1. Proximate and ultimate analyses of grape seeds (GSs) and hydrochars (wt.% dry basis).

Sample	Yield (%)	Proximate Analysis (%)			Ultimate Analysis (%)				
		Volatile Matter	Fixed Carbon	Ash	C	H	N	S	O [1]
GS	-	73.3	24.3	2.4	56.5	6.6	1.8	0.13	32.6
GS-180-40	75	66.0	31.0	3.0	61.7	6.6	1.6	0.02	27.1
GS-200-40	77	59.2	37.6	3.2	63.8	6.4	1.7	0.01	24.9
GS-220-40	59	55.9	40.7	3.4	69.0	6.4	1.7	0.02	19.5
GS-240-40	59	55.2	41.6	3.2	69.2	6.2	2.0	0.02	19.4
GS-260-10	53	50.2	46.8	3.0	70.5	6.1	1.7	0.08	18.6
GS-260-20	62	49.8	46.6	3.6	70.9	6.4	1.8	0.04	17.3
GS-260-30	64	50.4	46.3	3.3	71.2	6.2	1.9	0.06	17.4
GS-260-40	63	50.3	46.4	3.3	70.7	6.3	1.9	0.05	17.8
GS-260-50	62	49.6	46.8	3.6	69.8	6.1	2.0	0.03	18.5
GS-260-60	64	50.6	45.9	3.5	70.0	6.2	2.1	0.02	18.1
GS-280-40	62	47.0	49.8	3.2	72.6	6.1	2.2	0.02	15.9
GS-300-40	48	43.8	51.8	4.4	74.0	5.8	2.3	0.01	13.5

[1] Calculated by difference O = 100 − (C + H + N + S + Ash).

The energy yield or energy recovery efficiency is calculated by [42] (also known as the energetic biomass utilization efficiency (BUE_E) [43]):

$$\text{Energy recovery efficiency} = \text{Hydrochar yield} \cdot \text{Energy density},$$

where the energy density is the ratio of the higher heating values of the char relative to that of the grape seeds.

2.2. Preparation of Activated Carbons

The activated carbons (ACs) were prepared using either potassium hydroxide (KOH), iron chloride ($FeCl_3$) or phosphoric acid (H_3PO_4), which were physically mixed with the ground hydrochar at different activating agent to HC ratios of 4:1, 3:1 and 2:1 (w:w). In the case of H_3PO_4, the mixture was left overnight at 60 °C and then heated at 500 °C for 2 h under continuous N_2 flow (100 NmL·min^{-1}) in a stainless steel tube placed in an electrical furnace. The activation temperature was reached at 10 °C·min^{-1} heating rate. Activation with KOH and $FeCl_3$ was performed similarly but at 750 °C for 1 h [44–47]. After cooling under nitrogen flow, the ACs were washed with HCl or NaOH 0.1 M until neutral pH; then were washed with distilled water; and finally, were dried at 105 °C overnight. The ACs were denoted as GSHC, followed by the activating agent and the activating agent:hydrochar (w:w) ratio. For instance, GSHC-KOH-3 represents the activated carbon obtained with KOH using 3:1 activating agent to hydrochar ratio.

2.3. Characterization of Hydrochar and Activated Carbons

The proximate analyses were performed by the ASTM methods D3173-11 (moisture), D3174-11 (ash) and D3175-11 (volatile matter (VM)) using a Mettler Toledo apparatus (TGA/SDTA851e). The elemental analyses were carried out in a LECO Model CHNS-932 elemental analyzer. The higher heating value (HHV) of hydrochar samples was determined using an IKA Calorimeter System C2000 according to the ASTM D5865 procedure.

The porous texture of the AC samples was characterized by 77 K N_2 adsorption-desorption in a Micromeritics apparatus (Tristar 3020, Micromeritics, Narcross, GA, USA). The samples were previously degassed (150 °C for 6 h) in a Micromeritics VacPrep 061 device (Micromeritics, Narcross, GA, USA). The surface area was calculated by the BET equation within the 0.05–0.30 relative pressure range and the micropore volume (V_{micro}) was obtained by the t-method. The difference between the volume of N_2 adsorbed at 0.95 relative pressure (as liquid) and the micropore volume was taken as the mesopore volume (V_{meso}). The external or non-micropore surface area (A_{ext}) was also obtained from the t method. Surface area and micropore volume of the samples were also determined by CO_2 adsorption at 273 K in the same equipment. The corresponding surface area (S_{DA}) and the micropore volume ($V_{micro-DA}$) values were calculated by the Dubinin–Astakhov equation [48].

The surface composition of the hydrochars and activated carbons was analyzed by FTIR spectroscopy using a FTIR Bruker IFS66v spectrophotometer (Bruker Corporation, Billerica, MA, USA), with a KBr disc, having a resolution of 4 cm^{-1} from 4000 to 550 cm^{-1} and 250 scans. Temperature programmed desorption (TPD) was also used to assess the amount of oxygen surface groups. The samples (0.1 g) were heated up to 900 °C at 10 °C·min^{-1} in a vertical quartz tube under continuous N_2 flow of 1 NL·min^{-1}. The amounts of CO_2 and CO evolved were analyzed by non-dispersive infrared absorption in a Siemens model Ultramat 22 equipment (Siemens Aktiengesellschoff, Munich, Germany).

The morphologies of hydrochars and activated carbon samples were studied by scanning electron microscopy (SEM) with a Hitachi S-3000N apparatus (Hitachi Ltd, Tokyo, Japan). The samples were metalized with gold using a Sputter Coater SC502 (Quorum, East Sussex, UK). Images were obtained in the high vacuum mode under an accelerating voltage of 20 kV, using secondary and backscattered electrons.

The pH slurry of the carbons was determined measuring the pH (pH-meter, Crison) of an aqueous suspension of the sample (1 g) in distilled water (10 mL) [49].

2.4. Adsorption Tests

The potential application of the activated carbons as adsorbents in aqueous phase was evaluated using sulfamethoxazole (SMX) as model impurity, which has a solubility in water of 610 mg·L^{-1} at 298 K. Samples of AC (12.5 mg, ≈100 µm) were contacted in stoppered glass bottles with 50 mL of SMX aqueous solutions (25 to 150 mg·L^{-1}). A commercial activated carbon (C: 89.5 wt.%, A_{BET}: 800 $m^2·g^{-1}$; Vmicro: 0.67 $cm^3·g^{-1}$; Vmeso: 0.53 $cm^3·g^{-1}$), supplied by Merck, was also used as reference for comparison purposes. Experiments were carried out at 20 °C and the natural pH of the SMX solution (4.6) in a thermostatized shaker (Optic Ivymen System, Biotech, Madrid, Spain) at equivalent 200 rpm. A kinetic experiment showed that 5 days was time enough to reach the adsorption equilibrium in the operating conditions. SMX concentration was determined by UV–vis spectrophotometry (Cary 60 UV-Vis, Agilent Technologies, Santa Clara, CA, USA) at 265 nm wavelength. The Langmuir equation was used to fit the equilibrium data. The reported results are the average values from triplicate runs, being the standard errors always below 5%.

3. Results and Discussion

3.1. Characterization of the Hydrochars

Table 1 shows the hydrochar yields and the proximate and ultimate analyses of the raw GSs and hydrochars. Hydrochar yields were almost constant (≅60–65 wt.%) in the HTC runs performed with a GS relative amount higher than 20% in the GS + water mixture within the temperature range of 220–280 °C (aprox.). At a lower GS percentage the HC yield decreased since a higher relative amount of water favors the extraction and transfer of solid components to the liquid phase [12,50]. The effect of the temperature in the HTC process can be seen at fixed GS percentage in the starting mixture (40 wt.%) and temperatures below and above the aforementioned range. At temperatures ≤200 °C, high HC yields were obtained associated to lower carbonization as can be seen from the C content of the samples. At temperature above 280 °C gasification reactions gain significance, giving rise to lower yield but a more carbonized solid [51]. It is well known that dehydration and decarboxylation reactions are the main responsible of the mass loss upon HTC [52,53].

A van Krevelen diagram was plotted from the ultimate analyses (Figure 1) where the points corresponding to the HCs have been placed. As the HTC temperature increases, the resulting HCs evolve from a typical peat composition towards the lignite region, approaching even sub-bituminous coal at the highest HTC temperature tested (300 °C).

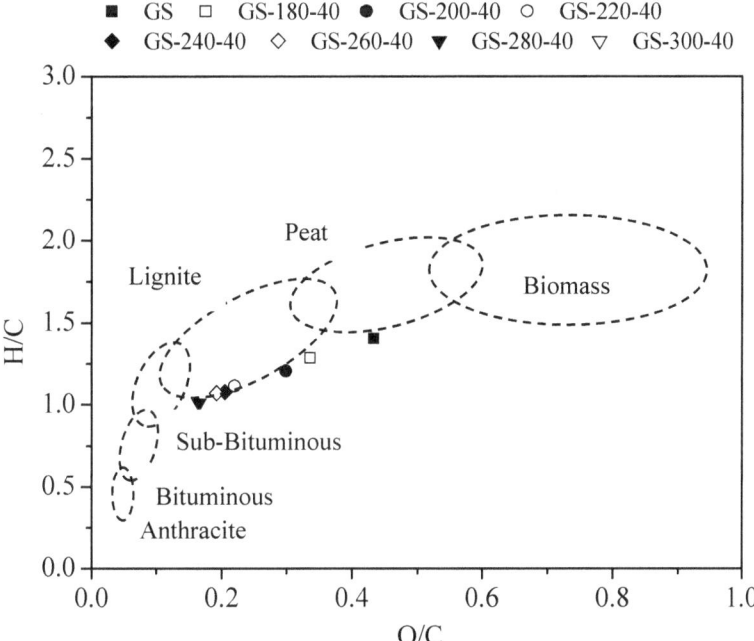

Figure 1. Van Krevelen diagram for grape seeds and hydrochars obtained at different temperatures (180–300 °C) with an initial solid content of 40 wt.%.

Table 2 summarizes the energy properties of hydrochars, where, clearly, there was a significant increase of the higher heating value as the HTC temperature increased. The energy density of the most carbonized solid was improved by one-third with respect to the starting precursor. Taking into account the GS to HC yield, the highest energy recovery efficiency does not correspond to the highest HTC temperature tested but to a significantly lower value somewhere around 200 °C. However, since the energy density of the resulting fuel is also an important issue, the information provided by Table 2 points to an optimum HTC temperature in the range of 260–280 °C. Moreover, this determination, also known as a BUE_E, provides an extra information because it allows a comparison between different fuels and options for biomass utilization [43].

Table 2. Energy properties of raw grape seeds and hydrochars.

Sample	HHV (MJ·kg^{-1})	Energy Density	Energy Recovery Efficiency [1]
GS	23.6	-	-
GS-180-40	24.5	1.04	0.78
GS-200-40	26.8	1.14	0.87
GS-220-40	27.6	1.17	0.69
GS-240-40	29.1	1.23	0.73
GS-260-40	30.3	1.29	0.81
GS-280-40	30.9	1.31	0.81
GS-300-40	31.4	1.33	0.64

[1] Also known as BUE_E.

The 77 K N_2 adsorption isotherms of the hydrochars (supporting material) corresponded to Type II of the IUPAC classification, associated to non-porous solids. The highest BET surface area was only 8 m^2·g^{-1}, corresponding to the GS-220-40 sample. To learn more on the porous texture of these hydrochars the CO_2 adsorption isotherms at 273 K were obtained (Supplementary Material). The surface area values from those isotherms (S_{DA}) varied from 100 to 845 m^2·g^{-1}, the highest corresponding to the above mentioned sample, with a micropore volume of 0.513 cm^3·g^{-1}. These huge differences between the BET and S_{DA} values are indicative of microporous texture consisting essentially of very narrow micropores, the so-called ultramicropores (≤0.5 nm width), corresponding most probably to slit-shaped pores. This type of microporosity has been also observed in chars from HTC of different sugars at 180 °C calcined at high temperatures in an inert atmosphere [23].

3.2. Activated Carbons by Chemical Activation of Hydrochars

Previous works have proven that hydrochars can be more effective adsorbents than thermal (pyrolytic) biochars due to their diverse structures and surface functional groups [54]. Looking at the characteristics of the hydrochars, the GS-220-40 was selected for further activation to achieve a well-developed porosity typical of a carbon-based adsorbent. In previous works [23] HTCs obtained at 180–240 °C were more prone to yield a better developed porosity upon KOH-activation than those obtained above 260 °C, although from fairly different precursors than in the current work.

Table 3 summarizes the porous textures of the activated carbons. Representative N_2 adsorption-desorption isotherms are depicted in Figure 2. All of them correspond to essentially microporous solids; there was a higher contribution of mesoporosity in the case of H_3PO_4-activation. The rest of isotherms are given as Supplementary Material (Figures S1–S3). KOH was by far the activating agent allowing the highest porosity development, with BET surface area values up to ca. 2200 m^2·g^{-1}. These carbons were essentially microporous with increased mesoporosity at increasing KOH to hydrochar mass ratio. The activated carbon prepared at 4:1 mass ratio had almost 1800 m^2·g^{-1} BET surface area in spite of its high ash content (38.2%), mostly due to the remaining inorganic matter derived from the activating agent. In fact, its BET surface area represented as much as ca. 2900 m^2·g^{-1} on an ash-free basis.

The lowest surface area development occurred with $FeCl_3$ as activating agent, due in part to the very high ash content of the resulting activated carbons. The highest BET surface area of these carbons on an ash-free basis was 770 m^2·g^{-1}, only slightly more than one-quarter that of the abovementioned KOH-activated carbon. $FeCl_3$-activation can be addressed to the preparation of carbon-supported Fe catalysts. Bedia et al. [44] and Mena et al. [55] prepared this type of catalyst, and tested theirs in catalytic wet peroxide oxidation (CWPO), from $FeCl_3$-activation of sewage sludge.

Activation with H_3PO_4 gave rise to microporous carbons but with much higher relative contribution of mesoporosity, showing Type IV isotherms of the IUPAC classification. The BET surface area values are relatively low compared to the obtained by the same method with other precursors. The ash content was significantly lower than by the two other activation procedures. Increasing the H_3PO_4 to HC ratio decreased the BET surface area, affecting mostly, the mesopore contribution.

Table 3. Textural characteristics of the activated carbons.

Sample	C (wt.% Dry Basis)	Ash (wt.% Dry Basis)	A_{BET} ($m^2 \cdot g^{-1}$)	A_{ext} ($m^2 \cdot g^{-1}$)	V_{micro} ($cm^3 \cdot g^{-1}$)	V_{meso} ($cm^3 \cdot g^{-1}$)
GSHC-KOH-2	75.1	9.8	1215	35	0.57	0.02
GSHC-KOH-3	73.3	16.3	2194	121	0.98	0.05
GSHC-KOH-4	45.6	38.2	1780	260	0.74	0.18
GSHC-FeCl$_3$-2	51.1	44.4	394	44	0.17	0.02
GSHC-FeCl$_3$-3	49.7	46.0	417	54	0.17	0.02
GSHC-FeCl$_3$-4	35.7	49.2	312	50	0.12	0.02
GSHC-H$_3$PO$_4$-2	63.3	4.6	654	255	0.19	0.23
GSHC-H$_3$PO$_4$-3	61.4	7.6	596	242	0.17	0.21
GSHC-H$_3$PO$_4$-4	69.7	10.8	590	172	0.20	0.16

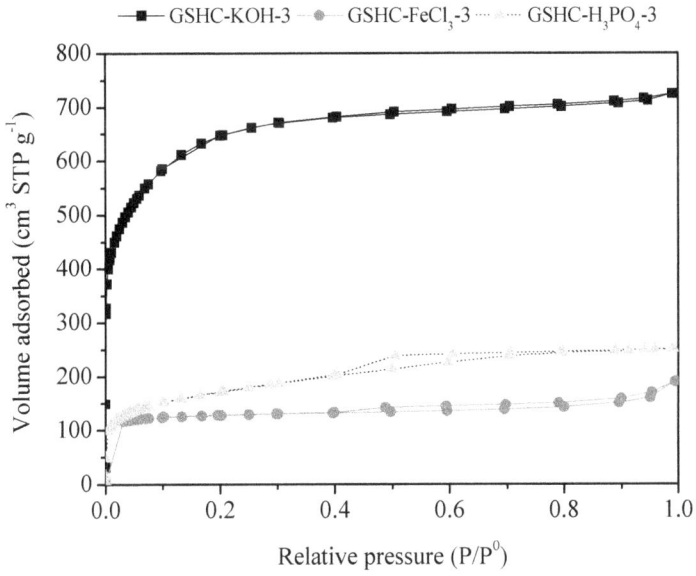

Figure 2. N_2 adsorption-desorption isotherms at 77 K of the activated carbons obtained at 3:1 activating agent to hydrochar ratio.

Figure 3 shows SEM images of the grape seeds, hydrochars and activated carbons. The raw GSs exhibited a well-defined morphological structure in three different layers, while in the hydrochar image shows some partial degradation of the structure occurred, although the morphology was still preserved in great part, which indicates good thermomechanical stability for the GSs. This allows obtaining granular activating carbons from this starting material. KOH activation produced a spongy-like structure with many holes. In the case of FeCl$_3$-activation, the formation of microspheres was visible; after the activation with H$_3$PO$_4$, prismatic particles can be seen.

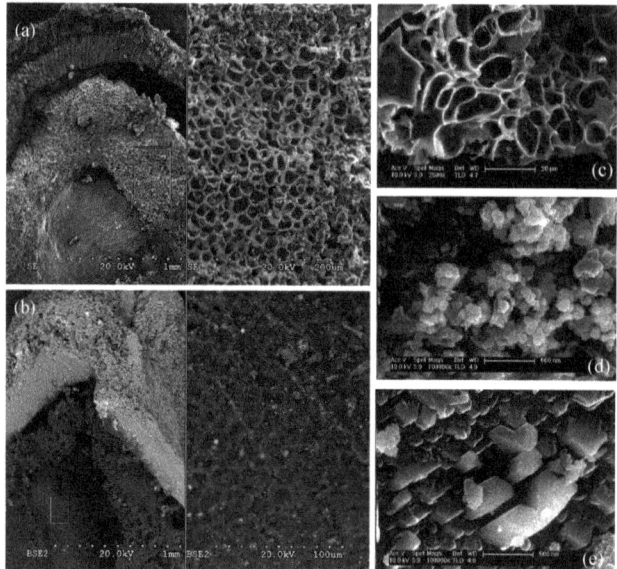

Figure 3. SEM images of cross-section of GS (**a**), GS-220-40 hydrochar (**b**), and GSAC-KOH-3 (**c**), GSAC-FeCl$_3$-3 (**d**) and GSAC-H$_3$PO$_4$-3 (**e**) activated carbons.

Figure 4 shows the FTIR spectra of the raw GS, GS-220-40 hydrochar and the activated carbons obtained with an activating agent to hydrochar ratio of 3. A decreasing intensity of the O–H stretching of the hydroxyl groups at 3400 cm^{-1} was observed for the hydrochar, because the dehydration occurred upon HTC [56,57]. The bands appearing in the 2950 to 2750 cm^{-1} range, associated to aliphatic carbon –CH$_x$ stretching vibration, tended to be less intense for the GS-220-40 hydrochar, probably due to the evolution of nonpolar alkyl carbon structure [58]. In addition, transmission intensities at 1750–750 cm^{-1} arising from C=O stretching of ketones and other carbonyl structures were reduced. Furthermore, the decline of the peak at 1030 cm^{-1} (the C–O and C–O–C stretching) indicates that both decarboxylation and dehydration occur, consistently with the van Krevelen representation of Figure 1.

The FTIR spectra of activated carbons showed a dramatic decrease of functionalities in agreement with the reported in general in the literature relative to activation of carbon precursors. A broad band (3600–3100 cm^{-1}) appears in the three activated carbons, related to the existence of hydroxyl functional groups. Those related to C=C and aromatic rings vibration (1600–1500 cm^{-1}) and the associated to C–O bonds at 1157–1090 cm^{-1}, related to the presence of alcohol groups were more intense in the GSHC-FeCl$_3$-3, and, mainly, in the GS-H$_3$PO$_4$-3 activated carbons [59,60].

The amount of surface oxygen groups in the activated carbons was assessed in terms of CO$_2$ and CO evolved upon TPD. The results are summarized in the inside Table of Figure 4. In all the cases, the ratio CO$_2$/CO was within 27–30%, suggesting that the oxygen groups must be ascribed majorly to CO-evolving functionalities (phenol, carbonyl, anhydride ether, quinone), mainly associated to weak acid and neutral/basic surfaces. However, there are also some significant amounts of evolved CO$_2$, such that carboxylic, lactone and anhydride groups may be also present in some extent [61,62]. H$_3$PO$_4$, and in particular, FeCl$_3$, appear to be more prone than KOH to create oxygen groups on the carbon surface upon activation. Rey et al. [62] arrived to similar conclusion working with Fe(NO$_3$)$_3$ as a precursor.

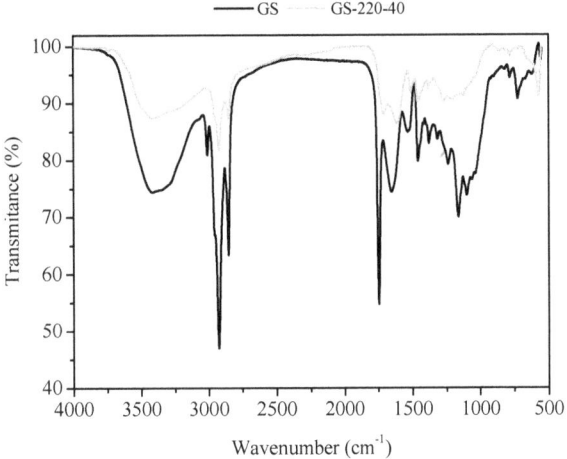

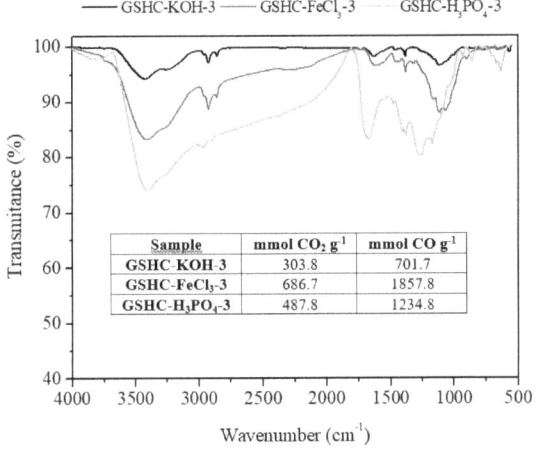

Figure 4. FTIR spectra for GS, GS-220-40 hydrochar and activated carbons. Results of TPD of the activated carbons in the inside Table.

3.3. Adsorption of Sulfamethoxazole

Figure 5 shows the adsorption isotherms (20 °C) of SMX with the GSAC-KOH-3, GSAC-FeCl$_3$-3 and GSAC-H$_3$PO$_4$-3 activated carbons and a commercial active carbon. The adsorption capacity seems to be mainly determined by the BET surface area of the carbons rather than the amount and character of the surface groups (Figure 4). Consistently with its much higher surface area, GSHC-KOH-3 yielded the highest adsorption capacity. The only deviation of that behavior occurred with the GSHC-H$_3$PO$_4$-3 activated carbon, which showed lower adsorption than the expected from its BET surface area. This can be explained in terms of SMX ionization and the pH slurry of this carbon. At solution pH 4.6, SMX (pKa1: 1.7, pKa2: 5.6 [63]) must be majorly made of neutral and anion species; since the pH slurry of GSHC-H$_3$PO$_4$-3 is 2.2, its surface will be negatively charged, which induces repulsion forces between the molecules of SMX and the carbon surface [64,65].

The experimental data were fitted to the Langmuir equation:

$$q_e = \frac{q_L \cdot K_L \cdot C_e}{1 + K_L \cdot C_e},$$

where q_e is the equilibrium adsorbate loading onto the adsorbent (mg·g^{-1}), C_e the equilibrium liquid-phase concentration of the adsorbate (mg·L^{-1}), q_L the monolayer saturation capacity (mg·g^{-1}) and K_L the Langmuir constant (L·mg^{-1}). This equation described the equilibrium data well, as can be seen in Figure 5. The values of the fitting parameters and the correlation coefficients are given in Table 4. The calculated monolayer Langmuir capacity of GSHC-KOH-3 reached 650 mg·g^{-1} (notice that in Figure 5 the represented experimental SMX values did not reach the saturation level). The SMX adsorption capacity of this material is higher than the exhibited by other adsorbents, such as activated carbon from pine tree or coconut shell, coal, carbon nanotubes, graphene and organo-montmorillonites at similar operating conditions, which are characterized by maximum adsorption capacities within the range of 122–242 mg·g^{-1} [65–67].

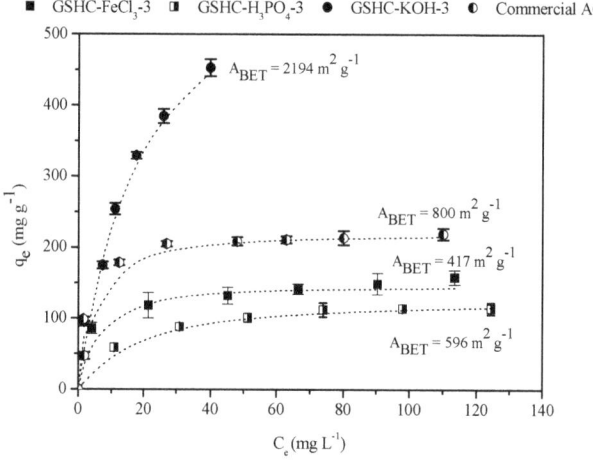

Figure 5. Adsorption isotherms of SMX on GSAC-KOH-3, GSAC-FeCl$_3$-3, GSAC-H$_3$PO$_4$-3 and commercial activated carbon at 20 °C (symbols: experimental values; short dot lines: fitting to the Langmuir equation). Error bars are included.

Table 4. Langmuir parameters for SMX adsorption on the activated carbons of Figure 5.

Sample	pH Slurry	q_L (mg·g^{-1})	K_L (L·mg^{-1})	R^2
GSHC-KOH-3	7.6	650.8	0.055	0.983
GSHC-FeCl$_3$-3	8.1	147.4	0.295	0.963
GSHC-H$_3$PO$_4$-3	2.2	128.6	0.069	0.988
Commercial AC	7.7	221.8	0.296	0.951

4. Conclusions

Hydrothermal carbonization of grape seeds is a promising way of valorization of that biomass waste from the wine industry. It allows obtaining chars of fairly good properties to be used as improved solid fuel. Further activation of those chars leads to higher value-added products, activated carbons. Activation with FeCl$_3$, H$_3$PO$_4$ and KOH led to high BET surface area carbons of predominantly microporous texture, which can be useful adsorbents for liquid-phase applications, as in water pollutant removal. In fact, a fairly high adsorption capacity was experimentally found with sulfamethoxazole, an emerging contaminant, as the target compound.

Supplementary Materials: The following are available online at http://www.mdpi.com/2076-3417/9/23/5127/s1: Figure S1: N_2 adsorption-desorption isotherms at 77 K of hydrochars, Figure S2: CO_2 adsorption isotherms at 273 K of hydrochars, Figure S3: N_2 adsorption-desorption isotherms at 77 K of activated carbons not included in Figure 2.

Author Contributions: Conceptualization, E.D., J.J.R. and A.F.M.; methodology, E.D., F.J.M. and J.V.; investigation, E.D., F.J.M. and J.V.; writing—original draft preparation, E.D. and F.J.M.; writing—review and editing, E.D., J.J.R. and A.F.M.; supervision, J.V., J.J.R. and A.F.M.; funding acquisition, E.D., J.J.R. and A.F.M.

Funding: The authors greatly appreciate financial support from the Spanish MINECO (CTM2016-76564-R), Comunidad de Madrid (S2018/EMT-4344) and UAM-Santander (2017/EEUU/07). F.J. Manzano wishes to thank the Comunidad de Madrid (PEJ16/AMB/AI-1327) for a research grant.

Conflicts of Interest: The authors declare no conflict of interest.

References

1. Demirbas, A. Combustion characteristics of different biomass fuels. *Prog. Energy Combust. Sci.* **2004**, *30*, 219–230. [CrossRef]
2. Goyal, H.B.; Seal, D.; Saxena, R.C. Biofuels from thermochemical conversion of renewable resources: A review. *Renew. Sustain. Energy Rev.* **2008**, *12*, 504–517. [CrossRef]
3. Yip, K.; Tian, F.; Hayashi, J.-I.; Wu, H. Effect of alkali and alkaline earth metallic species on biochar reactivity and syngas compositions during steam gasification. *Energy Fuel.* **2009**, *24*, 173–181. [CrossRef]
4. Saxena, R.; Adhikari, D.; Goyal, H. Biomass-based energy fuel through biochemical routes: A review. *Renew. Sustain. Energy Rev.* **2009**, *13*, 167–178. [CrossRef]
5. Kambo, H.S.; Dutta, A. A comparative review of biochar and hydrochar in terms of production, physico-chemical properties and applications. *Renew. Sustain. Energy Rev.* **2015**, *45*, 359–378. [CrossRef]
6. Berge, N.D.; Ro, K.S.; Mao, J.; Flora, J.R.V.; Chappell, M.A.; Bae, S. Hydrothermal carbonization of municipal waste streams. *Environ. Sci. Technol.* **2011**, *45*, 5696–5703. [CrossRef]
7. Falco, C.; Marco-Lozar, J.P.; Salinas-Torres, D.; Morallón, E.; Cazorla-Amorós, D.; Titirici, M.M.; Lozano-Castelló, D. Tailoring the porosity of chemically activated hydrothermal carbons: Influence of the precursor and hydrothermal carbonization temperature. *Carbon* **2013**, *62*, 346–355. [CrossRef]
8. Funke, A.; Ziegler, F. Hydrothermal carbonization of biomass: A summary and discussion of chemical mechanisms for process engineering. *Biofuels Bioprod. Biorefin.* **2010**, *4*, 160–177. [CrossRef]
9. Hitzl, M.; Corma, A.; Pomares, F.; Renz, M. The hydrothermal carbonization (HTC) plant as a decentral biorefinery for wet biomass. *Catal. Today* **2015**, *257*, 154–159. [CrossRef]
10. Hrnčič, M.K.; Kravanja, G.; Knez, Ž. Hydrothermal treatment of biomass for energy and chemicals. *Energy* **2016**, *116*, 1312–1322. [CrossRef]
11. Sermyagina, E.; Saari, J.; Kaikko, J.; Vakkilainen, E. Hydrothermal carbonization of coniferous biomass: Effect of process parameters on mass and energy yields. *J. Anal. Appl. Pyrolysis* **2015**, *113*, 551–556. [CrossRef]
12. Álvarez-Murillo, A.; Román, S.; Ledesma, B.; Sabio, E. Study of variables in energy densification of olive stone by hydrothermal carbonization. *J. Anal. Appl. Pyrolysis* **2015**, *113*, 307–314. [CrossRef]
13. Román, S.; Nabais, J.M.V.; Laginhas, C.; Ledesma, B.; González, J.F. Hydrothermal carbonization as an effective way of densifying the energy content of biomass. *Fuel Process. Technol.* **2012**, *103*, 78–83. [CrossRef]
14. Wüst, D.; Rodriguez Correa, C.; Suwelack, K.U.; Köhler, H.; Kruse, A. Hydrothermal carbonization of dry toilet residues as an added-value strategy—Investigation of process parameters. *J. Environ. Manag.* **2019**, *234*, 537–545. [CrossRef]
15. De la Rubia, M.A.; Villamil, J.A.; Rodriguez, J.J.; Mohedano, A.F. Effect of inoculum source and initial concentration on the anaerobic digestion of the liquid fraction from hydrothermal carbonisation of sewage sludge. *Renew. Energy* **2018**, *127*, 697–704. [CrossRef]
16. Villamil, J.A.; Mohedano, A.F.; Rodriguez, J.J.; De la Rubia, M.A. Anaerobic co-digestion of the aqueous phase from hydrothermally treated waste activated sludge with primary sewage sludge. *J. Environ. Manag.* **2019**, *231*, 726–733. [CrossRef]

17. Bargmann, I.; Rillig, M.C.; Kruse, A.; Greef, J.-M.; Kücke, M. Effects of hydrochar application on the dynamics of soluble nitrogen in soils and on plant availability. *J. Plant Nutr. Soil Sci.* **2014**, *177*, 48–58. [CrossRef]
18. Fernandez, M.E.; Ledesma, B.; Román, S.; Bonelli, P.R.; Cukierman, A.L. Development and characterization of activated hydrochars from orange peels as potential adsorbents for emerging organic contaminants. *Bioresour. Technol.* **2015**, *183*, 221–228. [CrossRef]
19. Guo, S.; Dong, X.; Wu, T.; Zhu, C. Influence of reaction conditions and feedstock on hydrochar properties. *Energy Convers. Manag.* **2016**, *123*, 95–103. [CrossRef]
20. Fang, J.; Zhan, L.; Ok, Y.S.; Gao, B. Minireview of potential applications of hydrochar derived from hydrothermal carbonization of biomass. *J. Ind. Eng. Chem.* **2018**, *57*, 15–21. [CrossRef]
21. Heidari, M.; Dutta, A.; Acharya, B.; Mahmud, S. A review of the current knowledge and challenges of hydrothermal carbonization for biomass conversion. *J. Energy Inst.* **2019**, *92*, 1779–1799. [CrossRef]
22. Nizamuddin, S.; Baloch, H.A.; Griffin, G.J.; Mubarak, N.M.; Bhutto, A.W.; Abro, R.; Mazari, S.A.; Ali, B.S. An overview of effect of process parameters on hydrothermal carbonization of biomass. *Renew. Sustain. Energy Rev.* **2017**, *73*, 1289–1299. [CrossRef]
23. Titirici, M.M. Hydrothermal Carbons: Synthesis, Characterization, and Applications. In *Novel Carbon Adsorbents*, 1st ed.; Tascón, J.M.D., Ed.; Elsevier Ltd.: Oxford, UK, 2012; pp. 351–399.
24. Fang, J.; Gao, B.; Zimmerman, A.R.; Ro, K.S.; Chen, J.J. Physically (CO_2) activated hydrochars from hickory and peanut hull: Preparation, characterization, and sorption of methylene blue, lead, copper, and cadmium. *RSC Adv.* **2016**, *6*, 24906–24911. [CrossRef]
25. Wang, B.; Gao, B.; Fang, J. Recent advances in engineered biochar productions and applications. *Crit. Rev. Environ. Sci. Technol.* **2017**, *47*, 2158–2207. [CrossRef]
26. Aliakbari, Z.; Younesi, H.; Ghoreyshi, A.A.; Bahramifar, N.; Heidari, A. Sewage sludge-based activated carbon: Its application for hexavalent chromium from synthetic and electroplating wastewater in batch and fixed-bed column adsorption. *Desalin. Water Treat.* **2017**, *93*, 61–73. [CrossRef]
27. Regmi, P.; Moscoso, J.L.G.; Kumar, S.; Cao, X.; Maob, J.; Schafran, G. Removal of copper and cadmium from aqueous solution using switchgrass biochar produced via hydrothermal carbonization process. *J. Environ. Manag.* **2012**, *109*, 61–69. [CrossRef]
28. Zhang, X.; Zhang, L.; Li, A. Eucalyptus sawdust derived biochar generated by combining the hydrothermal carbonization and low concentration KOH modification for hexavalent chromium removal. *J. Environ. Manag.* **2018**, *206*, 989–998. [CrossRef]
29. Islam, A.; Ahmed, M.J.; Khanday, W.A.; Asif, M.; Hameed, B.H. Mesoporous activated coconut shell-derived hydrochar prepared via hydrothermal carbonization-NaOH activation for methylene blue adsorption. *J. Environ. Manag.* **2017**, *203*, 237–244. [CrossRef]
30. Jain, A.; Balasubramanian, R.; Srinivasan, M.P. Tuning hydrochar properties for enhanced mesopore development in activated carbon by hydrothermal carbonization. *Microporous Mesoporous Mater.* **2015**, *203*, 178–185. [CrossRef]
31. Puccini, M.; Stefanelli, E.; Tasca, A.L.; Vitolo, S. Pollutant Removal from gaseous and aqueous phases using hydrochar-based activated carbon. *Chem. Eng. Trans.* **2018**, *67*, 637–642.
32. Zhang, X.; Gao, B.; Fang, J.; Zou, W.; Dong, L.; Cao, C.; Zhang, J.; Li, Y.; Wang, H. Chemically activated hydrochar as an effective adsorbent for volatile organic compounds (VOCs). *Chemosphere* **2019**, *218*, 680–686. [CrossRef]
33. Rodríguez-Correa, C.; Bernardo, M.; Ribeiro, R.P.P.L.; Esteves, I.A.A.C.; Kruse, A. Evaluation of hydrothermal carbonization as a preliminary step for the production of functional materials from biogas digestate. *J. Anal. Appl. Pyrolysis* **2017**, *124*, 461–474. [CrossRef]
34. Xue, Y.; Gao, B.; Yao, Y.; Inyang, M.; Zhang, M.; Zimmerman, A.R.; Ro, K.S. Hydrogen peroxide modification enhances the ability of biochar (hydrochar) produced from hydrothermal carbonization of peanut hull to remove aqueous heavy metals: Batch and column tests. *Chem. Eng. J.* **2012**, *200*, 673–680. [CrossRef]
35. Zhang, T.; Wu, X.; Fan, X.; Tsang, D.C.W.; Li, G.; Shen, Y. Corn waste valorization to generate activated hydrochar to recover ammonium nitrogen from compost leachate by hydrothermal assisted pretreatment. *J. Environ. Manag.* **2019**, *236*, 108–117. [CrossRef]
36. Al Bahri, M.; Calvo, L.; Gilarranz, M.A.; Rodriguez, J.J. Activated carbon from grape seeds upon chemical activation with phosphoric acid: Application to the adsorption of diuron from water. *Chem. Eng. J.* **2012**, *203*, 348–356. [CrossRef]

37. Al Bahri, M.; Calvo, L.; Gilarranz, M.A.; Rodriguez, J.J. Diuron multilayer adsorption on activated carbon from CO_2 activation of grape seeds. *Chem. Eng. Commun.* **2016**, *203*, 103–113. [CrossRef]
38. Jimenez-Cordero, D.; Heras, F.; Alonso-Morales, N.; Gilarranz, M.A.; Rodriguez, J.J. Porous structure and morphology of granular chars from flash and conventional pyrolysis of grape seeds. *Biomass Bioenergy* **2013**, *54*, 123–132. [CrossRef]
39. Jimenez-Cordero, D.; Heras, F.; Gilarranz, M.A.; Raymundo-Piñero, E. Grape seed carbons for studying the influence of texture on supercapacitor behaviour in aqueous electrolytes. *Carbon* **2014**, *71*, 27–138. [CrossRef]
40. Okman, I.; Karagöz, S.; Tay, T.; Erdem, M. Activated carbons from grape seeds by chemical activation with potassium carbonate and potassium hydroxide. *Appl. Surf. Sci.* **2014**, *293*, 138–142. [CrossRef]
41. Purnomo, C.; Castello, D.; Fiori, L. Granular activated carbon from grape seeds hydrothermal char. *Appl. Sci.* **2018**, *8*, 331. [CrossRef]
42. Park, K.Y.; Lee, K.; Kim, D. Characterized hydrochar of algal biomass for producing solid fuel through hydrothermal carbonization. *Bioresour. Technol.* **2018**, *258*, 119–124. [CrossRef]
43. Iffland, K.; Sherwood, J.; Carus, M.; Raschka, A.; Farmer, T.; Clark, J. Calculation and Comparison of the "Biomass Utilization Efficiency (BUE)" of Various Bio-based Chemicals, Polymers and Fuels. Available online: http://bio-based.eu/?did=32321&vp_edd_act=show_download (accessed on 5 September 2019).
44. Bedia, J.; Monsalvo, V.M.; Rodriguez, J.J.; Mohedano, A.F. Iron catalysts by chemical activation of sewage sludge with $FeCl_3$ for CWPO. *Chem. Eng. J.* **2017**, *318*, 224–230. [CrossRef]
45. Bedia, J.; Belver, C.; Ponce, S.; Rodriguez, J.; Rodriguez, J.J. Adsorption of antipyrine by activated carbons from $FeCl_3$-activation of Tara gum. *Chem. Eng. J.* **2018**, *333*, 58–65. [CrossRef]
46. Rosas, J.M.; Bedia, J.; Rodríguez-Mirasol, J.; Cordero, T. HEMP-derived activated carbon fibers by chemical activation with phosphoric acid. *Fuel* **2009**, *88*, 19–26. [CrossRef]
47. Sevilla, M.; Ferrero, G.A.; Fuertes, A.B. Beyond KOH activation for the synthesis of superactivated carbons from hydrochar. *Carbon N. Y.* **2017**, *114*, 50–58. [CrossRef]
48. Dubinin, M.M. Fundamentals of the theory of adsorption in micropores of carbon adsorbents: Characteristics of their adsorption properties and microporous structures. *Carbon* **1989**, *27*, 457–467. [CrossRef]
49. Rey, A.; Zazo, J.A.; Casas, J.A.; Bahamonde, A.; Rodriguez, J.J. Influence of the structural and surface characteristics of activated carbon on the catalytic decomposition of hydrogen peroxide. *Appl. Catal. A: Gen.* **2011**, *402*, 146–155. [CrossRef]
50. Sabio, E.; Álvarez-Murillo, A.; Román, S.; Ledesma, B. Conversion of tomato-peel waste into solid fuel by hydrothermal carbonization: Influence of the processing variables. *Waste Manag.* **2016**, *47*, 122–132. [CrossRef]
51. Falco, C.; Baccile, N.; Titirici, M.M. Morphological and structural differences between glucose, cellulose and lignocellulosic biomass derived hydrothermal carbons. *Green Chem.* **2011**, *13*, 3273–3281. [CrossRef]
52. Benstoem, F.; Becker, G.; Firk, J.; Kaless, M.; Wuest, D.; Pinnekamp, J.; Kruse, A. Elimination of micropollutants by activated carbon produced from fibers taken from wastewater screenings using hydrothermal carbonization. *J. Environ. Manag.* **2018**, *211*, 278–286. [CrossRef]
53. Sevilla, M.; Fuertes, A.B. The production of carbon materials by hydrothermal carbonization of cellulose. *Carbon N. Y.* **2009**, *47*, 2281–2289. [CrossRef]
54. Jian, X.; Zhuang, X.; Li, B.; Xu, X.; Wei, Z.; Song, Y.; Jiang, E. Comparison of characterization and adsorption of biochars produced from hydrothermal carbonization and pyrolysis. *Environ. Technol. Innov.* **2018**, *10*, 27–35. [CrossRef]
55. Mena, I.F.; Diaz, E.; Moreno-Andrade, I.; Rodriguez, J.J.; Mohedano, A.F. Stability of carbon-supported iron catalysts for catalytic wet peroxide oxidation of ionic liquids. *J. Environ. Chem. Eng.* **2018**, *6*, 6444–6450. [CrossRef]
56. Sevilla, M.; Maciá-Agulló, J.A.; Fuertes, A.B. Hydrothermal carbonization of biomass as a route for the sequestration of CO_2: Chemical and structural properties of the carbonized products. *Biomass Bioenerg.* **2011**, *35*, 3152–3159. [CrossRef]
57. Wang, T.; Zhai, Y.; Zhu, Y.; Li, C.; Zeng, G. A review of the hydrothermal carbonization of biomass waste for hydrochar formation: Process conditions, fundamentals, and physicochemical properties. *Renew. Sustain. Energy Rev.* **2018**, *90*, 223–247. [CrossRef]

58. He, C.; Zhao, J.; Yang, Y.; Wang, J.-Y. Multiscale characteristics dynamics of hydrochar from hydrothermal conversion of sewage sludge under sub- and near-critical water. *Bioresour. Technol.* **2016**, *211*, 486–493. [CrossRef]
59. Laginhas, C.; Nabais, J.M.V.; Titirici, M.M. Activated carbons with high nitrogen content by a combination of hydrothermal carbonization with activation. *Microporous Mesoporous Mater.* **2016**, *226*, 125–132. [CrossRef]
60. Rodríguez Correa, C.; Stollovsky, M.; Hehr, T.; Rauscher, Y.; Rolli, B.; Kruse, A. Influence of the carbonization process on activated carbon properties from lignin and lignin-rich biomasses. *ACS Sustain. Chem. Eng.* **2017**, *5*, 8222–8233. [CrossRef]
61. Figueiredo, J.L.; Pereira, M.F.R.; Freitas, M.M.A.; Órfão, J.J.M. Modification of the surface chemistry of activated carbons. *Carbon* **1999**, *37*, 1379–1389. [CrossRef]
62. Rey, A.; Hungria, A.B.; Duran-Valle, C.J.; Faraldos, M.; Bahamonde, A.; Casas, J.A.; Rodriguez, J.J. On the optimization of activated carbon-supported iron catalysts in catalytic wer peroxide oxidation process. *Appl. Catal. Environ. B.* **2016**, *181*, 249–259. [CrossRef]
63. Çalışkan, E.; Göktürk, S. Adsorption characteristics of sulfamethoxazole and metronidazole on activated carbon. *Sep. Sci, Technol.* **2010**, *45*, 244–255. [CrossRef]
64. Bajpai, A.K.; Rajpoot, M.; Mishra, D.D. Studies on the correlation between structure and adsorption of sulfonamide compounds. *Colloids Surf. A Physicochem. Eng. Asp.* **2000**, *168*, 193–205. [CrossRef]
65. Tonucci, M.C.; Gurgel, L.V.A.; de Aquino, S.F. Activated carbons from agricultural byproducts (pine tree and coconut shell), coal, and carbon nanotubes as adsorbents for removal of sulfamethoxazole from spiked aqueous solutions: Kinetic and thermodynamic studies. *Ind. Crops Prod.* **2015**, *74*, 111–121. [CrossRef]
66. Lu, X.; Berge, N.D. Influence of feedstock chemical composition on product formation and characteristics derived from the hydrothermal carbonization of mixed feedstocks. *Bioresour. Technol.* **2014**, *166*, 120–131. [CrossRef]
67. Rostamian, R.; Behnejad, H. A comparative adsorption study of sulfamethoxazole onto graphene and graphene oxide nanosheets through equilibrium, kinetic and thermodynamic modeling. *Process. Saf. Environ. Prot.* **2016**, *102*, 20–29. [CrossRef]

© 2019 by the authors. Licensee MDPI, Basel, Switzerland. This article is an open access article distributed under the terms and conditions of the Creative Commons Attribution (CC BY) license (http://creativecommons.org/licenses/by/4.0/).

Article

Mixed Hardwood and Sugarcane Bagasse Biochar as Potting Mix Components for Container Tomato and Basil Seedling Production

Ping Yu [1], Qiansheng Li [2], Lan Huang [3], Genhua Niu [4] and Mengmeng Gu [2,*]

1. Department of Horticultural Sciences, Texas A&M University, 2133 TAMU, College Station, TX 77843, USA; yuping520@tamu.edu
2. Department of Horticultural Sciences, Texas A&M AgriLife Extension Service, 2134 TAMU, College Station, TX 77843, USA; qianshengli@tamu.edu
3. Institute of Urban Agriculture, Chinese Academy of Agricultural Sciences, Chengdu 610000, China; huanglan_92@163.com
4. Texas A&M AgriLife Research and Extension Center at Dallas, 17360 Coit Road, Dallas, TX 75252, USA; gniu@ag.tamu.edu
* Correspondence: mgu@tamu.edu; Tel.: +1-979-845-8567

Received: 30 September 2019; Accepted: 3 November 2019; Published: 5 November 2019

Abstract: To investigate the potential of biochar as a propagation mix component, three experiments were conducted. A phytotoxicity test was conducted with water extract of sugarcane bagasse biochar (SBB), SBB mixes (10%, 30%, 50%, and 70% SBB with 30% perlite (P) and the rest being peat moss (PM); by vol.), mixed hardwood biochar (HB) mixes (10%, 30%, 50%, 70% and 100% HB with PM; by vol.), PM, P, 70%PM:30%P, and a commercial propagation mix (exp. 1). None of the mixes caused phytotoxicity. The same biochar mixes (except 100% HB) were used for the seedling growth test (exp. 2). Both tomato and basil seedlings grown in all of the biochar mixes (except 50% HB) had significantly lower fresh weight, dry weight and growth index (GI) compared to a commercial propagation mix. Six seedlings from each biochar mix were transplanted into a commercial growing mix and grown for four weeks (exp. 3). Tomato seedlings from all biochar mixes (except 30% SBB) had similar SPAD (Soil-Plant Analyses Development) and GI to the control. Basil seedlings from all HB mixes, 70% and 100% SBB mixes had similar GI to the control. In conclusion, 70% HB could be amended with PM for tomato and basil seedling production without negative effects on plant biomass.

Keywords: biochar; greenhouse; production; seedlings

1. Introduction

Peat moss (PM) has been widely used as a horticultural substrate due to its ideal physical and chemical properties, such as low bulk density (BD), high water holding capacity, high aeration ratio, and high cation exchange capacity [1–3]. Domestic PM sales in the US were 0.25 M m^3 in 2016 and almost 91% PM was sold to the horticultural industry [4]. The marketable PM estimated value in the US was $13.0 million in 2018 [4]. Peat moss mining, however, has been questioned due to the peatland ecosystem disturbance and/or loss, and its environmental consequences. Hence, alternative materials such as pretreated manure composts and processed timber by-products have been introduced as PM replacements [5].

Biochar, a carbon-rich by-product from biomass pyrolysis, has potential for substituting PM as greenhouse growing media [6]. Pyrolysis biochar is generated from biomass thermo-chemical decomposition in oxygen-depleted or oxygen-limited atmosphere [7–9]. Biochar has been considered as a sustainable material because it can be derived from various sources, such as pinewood [3,10,11], green waste [12], wood, sugarcane bagasse [13], straw [14–18], bark [19], rice hull [20], and wheat straw [16,21]. For the same reason, biochar properties can vary widely [22]. Most greenhouse trials have used biochar derived from lignin-based materials, which has appropriate properties for plant growth [12]. Graber [23] reported that citrus wood biochar has potential to improve pepper and tomato plant growth in a systematic way, increasing the leaf area, canopy and yield. Guo [6] found that incorporating pinewood biochar with PM-based commercial substrate increased poinsettia growth. Huang's [24] study showed that mixing hardwood biochar with two different composts could lead to similar or better plant growth in basil and tomato plants in comparison to those in PM-based commercial substrates. Tian [12] confirmed that the total biomass could be significantly increased (by 22%) by mixing green waste biochar with a PM-based substrate. When adding biochar in composted green waste, the shoot fresh weight, shoot dry weight, root fresh weight, and root dry weight of *Calathea insignis* were increased by 57.3%, 79.7%, 64.5%, and 82.0%, respectively [25]. Similar works had also been reported on Easter lily [6,26,27]. The biochar from red oak feedstock mixed with vermiculite also increased hybrid poplar total biomass and shoot biomass [28].

Biochar that affects greenhouse seedling production or subsequent seedling growth has seldom been reported. As biochar from different resources has varied properties, some may have adverse effects on plant growth due to possible phytotoxicity [29]. Phytotoxicity assessment is critical for successful soil/soilless amendment with bioenergy by-products such as biochar [30], and the germination test is a reliable procedure for different types of biochar phytotoxicity examinations [30]. We conducted this study to test the phytotoxicity of two biochars from different raw materials and to explore the use of the two biochars in subsequent container seedling production.

2. Materials and Methods

2.1. Experiment 1: Media Phytotoxicity and Property Test

Sugarcane bagasse biochar (SBB, American Biocarbon LLC White Castle, Louisiana, USA) was mixed with P (30%, by vol., Kinney Bonded Warehouse, Tyler, TX, USA) at rates of 10%, 30%, 50%, 70% and 100% (by vol.), with the rest being PM (Voluntary purchasing Group Inc., Bonham, Texas, USA) when SBB and P did not add up to 100%. No P or PM were added to 100% SBB mix. Mixed hardwood biochar (HB, Proton Power Inc. Lenouir City, Tennessee, USA) was mixed with PM at rates of 10%, 30%, 50%, 70% and 100% (by vol.), and no P was incorporated. Another mix was formulated by mixing PM and P at a 7:3 ratio (70%PM:30%P; by vol.). Peat moss, P, and a commercial propagation substrate (CS, BM2, Berger, Saint-Modest, Quebec, Canada) were also included in this study. The commercial propagation mix contained 70–80% of fine sphagnum moss with the rest being fine P and fine vermiculite. The United States Department of Agriculture-Agricultural Research Service, Sugarcane Research Unit (Houma, Louisiana, USA) provided the SBB, which was produced with proprietary methods, and the Proton Power Inc. (Lenouir City, Tennessee, USA) provided HB, which was a by-product from fast pyrolysis of mixed hardwood. Sugarcane bagasse biochar had a pH of 5.9 and HB had a pH of 10.1. The electrical conductivity (EC) of the two biochars were 753 μS/cm (SBB) and 1,058 μS/cm (HB), respectively [31]. Because SBB had a similar pH to PM (SBB 5.9, PM 5.0, Table 1) and the SBB particle size was smaller (mean 0.17mm, resulting in low air space (AS)) [31], when formulating mixes with SBB, 30% P was incorporated to increase the pH of the AS and the mix. As HB had a higher pH (10.1) than PM (5.0), and the HB particle size was larger (67.3% > 2.0 mm, resulting in high AS) [25], no P was incorporated when formulating mixes with HB. The properties of all the components used in this study are shown in Table 1.

Table 1. The pH, electrical conductbity (EC), total porosity (TP), container capacity (CC), air space (AS) and bulk density (BD) of substrate components used in this study.

Substrate Component [z]	pH	EC (μS/cm)	TP (%)	CC (%)	AS (%)	BD (g/cm^3)
SBB	5.9	753	74 ± 2	71 ± 1	3 ± 1	0.11 ± 0.00
HB	10.1	1,058	87 ± 1	66 ± 1	20 ± 1	0.13 ± 0.00
PM:P (70:30)	5.6	162	79 ± 1	62 ± 1	16 ± 1	0.09 ± 0.00
CS	6.8	745	75 ± 2	66 ± 1	9 ± 1	0.09 ± 0.00
P	7.3	57	92 ± 1	59 ± 1	34 ± 0	0.05 ± 0.00
PM	5.0	179	69 ± 1	58 ± 1	11 ± 0	0.11 ± 0.00

Note: [z] SBB=Sugarcane bagasse biochar; HB=Mixed hardwood biochar; CS=Commercial propagation substrate; P=Perlite; PM=Peat moss. Numbers in parens indicated the ratio of different components, by vol.

All of the mixes were subjected to a phytotoxicity test with Gravel's method [32]. Briefly, water extract was obtained by soaking the mixes with 100 mL deionized (DI) water and shaking for 24 hours. The mixtures were filtered through 11cm-diameter VWR Grade 415 filter paper (quantitative) (VWR International, LLC, Randor, Pennsylvania, USA) and 3 mL extract was used to saturate another filter paper placed in a petri dish. Deionized water was used as the control in this experiment. Twenty-five basil (*Ocimum basilicum*) (Johnny's Selected Seeds, Winslow, Maine, USA) seeds were placed in each petri dish. The emergence percentage (EP) of basil seeds was calculated after incubating the petri dishes at 25 °C in the dark for 7 days by using the following formula: EP = (no. of emerged seedlings/total no. of seeds) × 100%. This experimental design was a complete randomized design with six replicates.

All of the physical properties of the media, including bulk density (BD), total porosity (TP), air space (AS) and container capacity (CC), were determined using the North Carolina State University Horticultural Substrates Laboratory Porometers [33]. The substrate pH and EC were measured by using a handheld pH-EC meter (Hanna Instrument, Woonsocket, Rhode Island, USA) according to the pour-through extraction method [34]. Three replications of each substrate were measured.

2.2. Experiment 2: Biochar as Amendments for Greenhouse Media for Seedling Production

Tomato (*Solanum lycopersicum* 'Red Robin'™') (Fred C. Gloeckner & Company Inc., Harrison, NY, USA) and basil (*Ocimum basilicum*) (Johnny's Selected Seeds, Winslow, ME, USA) seeds were soaked in DI water for 24 h before sowing in 72-cell (cell depth: 5 cm; cell top length and width: 4 cm; volume: 55 ml) plug trays with one seed per cell on 16 February, 2019.

Five SBB:P substrates were formulated by mixing SBB at 10%, 30%, 50%, 70%, and 100% (by vol.) with 30% P (Kinney Bonded Warehouse, Tyler, TX, USA, except for the 100% SBB) and the rest being Peat moss (PM) (Voluntary purchasing Group Inc., Bonham, TX, USA) when SBB and P did not add up to 100%. Four HB:PM substrates were formulated by mixing HB at 10%, 30%, 50%, and 70% (by vol.) with PM and a commercial propagation mix (CS, BM2, Berger, Saint-Modest, QC, Canada) was used as the control.

All of the mixes had four replications (10 cells per replication), which were arranged in completely randomized blocks in the greenhouse located on Texas A&M University campus, College Station, TX, USA. The average greenhouse temperature, relative humidity and dew point were 22.8 °C, 79.7% and 18.3 °C, respectively. Before the true leaves (tomato or basil) emerged, the trays were irrigated with DI water. After the true leaves emerged, trays were irrigated with 50 mg N· L^{-1} (20N-4.3P-16.6K) Peters®Professional (Everris NA Inc, Dublin, Ohio, USA) nutrient solution.

At the end of this experiment (27 March, 2019), the height of four randomly-selected seedlings from each mix was measured from the medium surface to the highest point of the plants, and the widest width (width 1) and its perpendicular width (width 2) were measured. The growth index (GI) was calculated as: GI= Height/2 + (width 1+ width 2)/4 [6]. Leave greenness was indicated by Soil-Plant Analyses Development (SPAD) readings. (SPAD 502 Plus Chlorophyll Meter, Spectrum Technologies, Inc., Plainfield, Illinois USA). For each mix, shoots (stalk and leaf) and roots of four seedlings were harvested separately. The total fresh weight (TFW) was determined at harvest by adding up the fresh weights of the stalk and leaf. Shoot dry weight (SDW) and root dry weight (RDW, after being washed) were determined after drying at 80 °C in an oven until a constant weight was reached. Roots were washed and root length, surface area, root average diameter, and the number of tips were measured by using a root scanner (WinRHIZO, Regent Instruments Inc., Canada). The total dry weight (TDW) was calculated by adding up the SDW and RDW.

2.3. Experiment 3: The Subsequent Growth Evaluation of Seedlings Produced in Biochar-Amended Media

At the end of the second experiment, six seedlings in each mix with similar GI were selected and transplanted into 6-inch azalea pots (depth: 10.8 cm; top diameter: 15.5 cm; bottom diameter: 11.3 cm; volume: 1330 ml) with a commercial growing substrate (CS1, Jolly Gardener, Oldcastle Lawn & Garden Inc. Atlanta, Georgia, USA). The commercial growing mix contained 55% (by vol.) aged pine bark, with the other ingredients being Canadian sphagnum PM, P and vermiculite. The growth index was measured biweekly and the SPAD was measured on 2 April, 2019. After four weeks of growing, plants' leaves and stems were harvested separately, and the stem DW (SDW), leaf DW (LDW) and flower or fruit DW (FDW) were determined after drying in an oven at 80 °C until a constant weight was reached. The total dry weight (TDW) was calculated by adding up the SDW, LDW and FDW.

2.4. Statistical Analysis

The experiments were set up in a completely randomized block design. Data were analyzed with one-way analysis of variance using JMP Statistical Software (version Pro 12.2.0; SAS Institute, Cary, North Carolina, USA) and means were separated using Dunnett's test when treatments were significantly different from control at $p < 0.05$.

3. Results

3.1. Media Phytotoxicity and Properties

The water extract of the commercial propagation mix and the P had significantly higher EP than DI water (the control). All other biochar-amended mixes, the PM:P mix, and PM had a similar EP compared to the control (Figure 1).

The pH of HB-amended mixes had positive linear correlations with the biochar incorporation rate, while SBB-amended mixes showed quadratic correlations. The electrical conductivity (EC) of all biochar-amended mixes increased with an increasing biochar incorporation rate, and had quadratic correlations (Figure 2). All of the mixes' TPs were within the recommended range (50% to 85%). The TPs of the HB-amended mixes had positive linear correlations with the biochar incorporation rate; however, the SBB-amended mixes showed quadratic correlations. All of the SBB-amended mixes' CCs were also within the recommended range (45% to 65%), except for 100% SBB mixes (71%). The CC of 10% HB-amended mixes was within the recommended range (62%), while all the other HB-amended mixes' CCs were slightly beyond the range (68% the highest). The air space of all the biochar-amended mixes was within the recommended range (10% to 30%), except for the 50%, 70%, and 100% SBB-amended mixes. The air space of all SBB-amended mixes decreased as the biochar incorporation rate increased; however, the AS of all HB-amended mixes increased with the biochar rate. The bulk density (BD) of both SBB- and HB-amended mixes were similar (Figure 3).

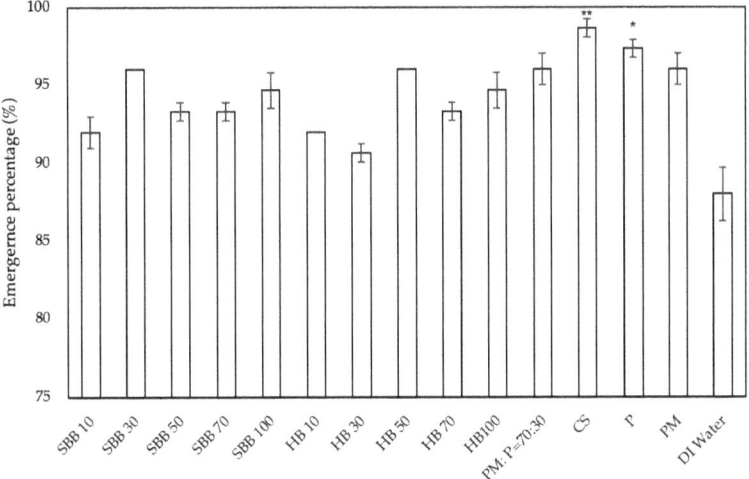

Figure 1. The emergence percentage of basil seedlings in the water extract of different mixes. *, ** indicate a significant difference from the control (DI water) using Dunnett's test at $p \leq 0.05$ and $p \leq 0.01$, respectively. SBB = sugarcane bagasse biochar; HB = mixed hardwood biochar; CS = commercial propagation substrate; SBB (10%, 30%, 50%, 70% and 100%; by vol.) incorporated with 30% perlite with the rest being peat moss; HB (10%, 30%, 50% and 70%; by vol.) incorporated with peat moss.

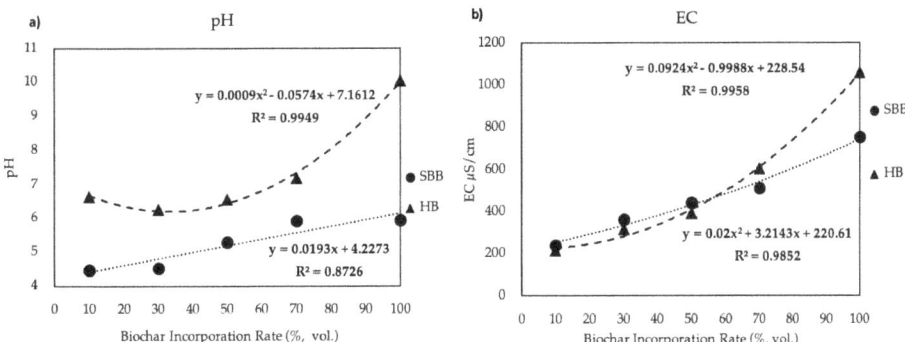

Figure 2. The correlation between pH (**a**) and electrical conductivity (EC) (**b**) and biochar incorporation rate. SBB = sugarcane bagasse biochar; HB = mixed hardwood biochar. SBB (10%, 30%, 50%, 70% and 100%; by vol.) incorporated with 30% perlite with the rest being peat moss; HB (10%, 30%, 50% and 70%; by vol.) incorporated with peat moss.

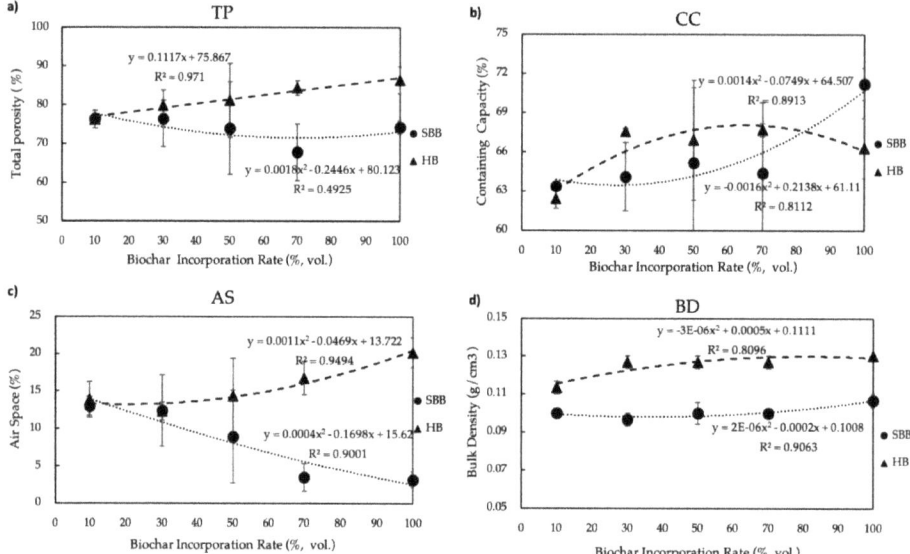

Figure 3. The correlation between substrate total porosity (**TP, a**), containing capacity (**CC, b**), air space (**AS, c**) and bulk density (**BD, d**) and biochar incorporation rate. SBB = sugarcane bagasse biochar; HB = mixed hardwood biochar. SBB (10%, 30%, 50%, 70% and 100%; by vol.) incorporated with 30% perlite with the rest being peat moss; HB (10%, 30%, 50% and 70%; by vol.) incorporated with peat moss.

3.2. Biochars as Amendment for Greenhouse Media for Seedling Production

3.2.1. Tomato Seedling Growth

The total fresh weight (TFW), total dry weight (TDW) and GI of SBB-amended mixes had positive linear correlations with the biochar incorporation rate, while HB-amended mixes showed quadratic correlations (Figure 4a–c). All TFWs, TDWs and GIs in biochar-amended mixes were significantly lower than the control, except for those in 50% HB-amended mixes. Tomato seedlings grown in all SBB-amended mixes had similar SPAD to the control (except 100% SBB, Figure 4d); however, seedlings grown in all HB-amended mixes had significantly lower SPAD (except 10% HB).

All tomato seedlings grown in biochar-amended mixes had significantly shorter root lengths than the control (except 30% HB, Table 2). Except for seedlings grown in 50% SBB, 30% HB and 50% HB, all tomato seedlings grown in biochar-amended mixes had significantly smaller root surface areas than the control. Seedlings grown in all biochar-amended mixes had similar or wider root diameters compared to the control; however, they all had fewer root tips (except 30% HB and 50% HB).

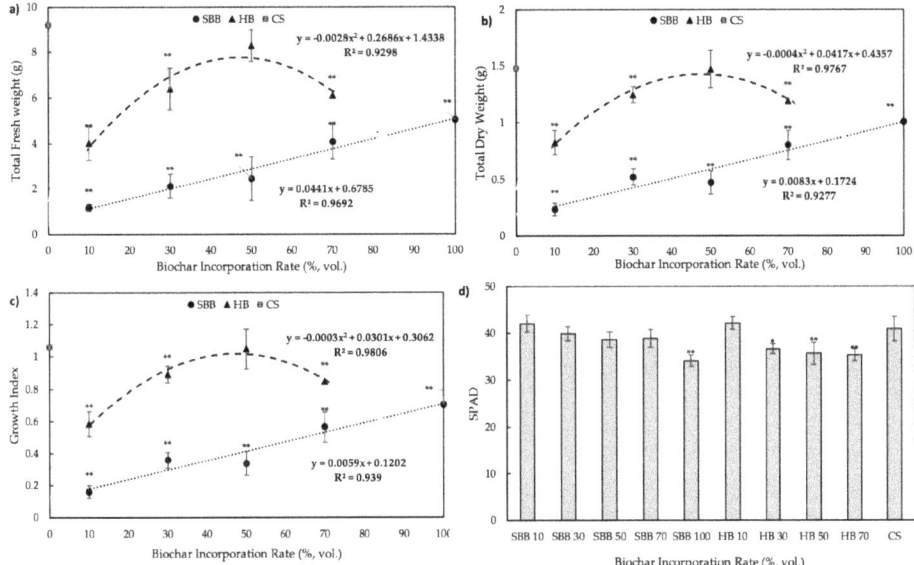

Figure 4. The correlations between total fresh weight (**TFW, a**), total dry weight (**TDW, b**), growth index (**GI, c**) and biochar incorporation rate and the soil-Plant Analyses development (SPAD) (**d**) of tomato seedlings grown in biochar-amended mixes. SBB = sugarcane bagasse biochar; HB = mixed hardwood biochar; CS = commercial propagation substrate. SBB (10%, 30%, 50%, 70% and 100%; by vol.) incorporated with 30% perlite with the rest being peat moss; HB (10%, 30%, 50% and 70%; by vol.) incorporated with peat moss. (*, ** indicated significant difference from the control using Dunnett's test at $p \leq 0.05$ and $p \leq 0.01$, respectively).

Table 2. Root growth of tomato seedlings grown in different mixes (*, ** indicate a significant difference from the control using Dunnett's test at $p \leq 0.05$ and $p \leq 0.01$, respectively).

Mixes [z]	Root Length (cm)	Root Surface Area (cm^2)	Average Diameter (mm)	Number of Tips
SBB:PM:P(10:60:30)	125 ± 10**	27 ± 3**	0.69 ± 0.05**	410 ± 45**
SBB:PM:P(30:40:30)	209 ± 8**	37 ± 1**	0.57 ± 0.01	625 ± 60**
SBB:PM:P(50:20:30)	277 ± 27**	55 ± 4	0.64 ± 0.03*	789 ± 120*
SBB:PM:P(70:0:30)	259 ± 26**	49 ± 4*	0.60 ± 0.02	657 ± 26**
SBB:PM:P(100:0:0)	281 ± 50*	52 ± 7*	0.60 ± 0.03	718 ± 91*
HB:PM(10:90)	243 ± 36**	46 ± 5**	0.62 ± 0.04	648 ± 30*
HB:PM(30:70)	350 ± 26	64 ± 4	0.58 ± 0.04	817 ± 64
HB:PM(50:50)	278 ± 31**	56 ± 3	0.66 ± 0.05*	1055 ± 148
HB:PM(70:30)	271 ± 21**	50 ± 2*	0.60 ± 0.02	746 ± 47**
Control	432 ± 35	68 ± 4	0.50 ± 0.01	1147 ± 141

Note: [z] SBB = sugarcane bagasse biochar; HB = mixed hardwood biochar; P = perlite; PM = peat moss; Control = commercial propagation substrate. Numbers in parentheses indicate the ratio of different components, by vol.

3.2.2. Basil Seedling Growth

The total fresh weight (TFW), total dry weight (TDW) and GI of seedlings in SBB-amended mixes had positive linear correlations with the biochar incorporation rate, while seedlings in HB-amended mixes showed quadratic correlations (Figure 5a–c). All TFWs (except 30% and 50% HB), TDWs and GIs (except 50% HB) in biochar-amended mixes were significantly lower than the control. Basil seedlings grown in all biochar-amended mixes had similar or higher SPAD than the control (except 100% SBB, Figure 5d).

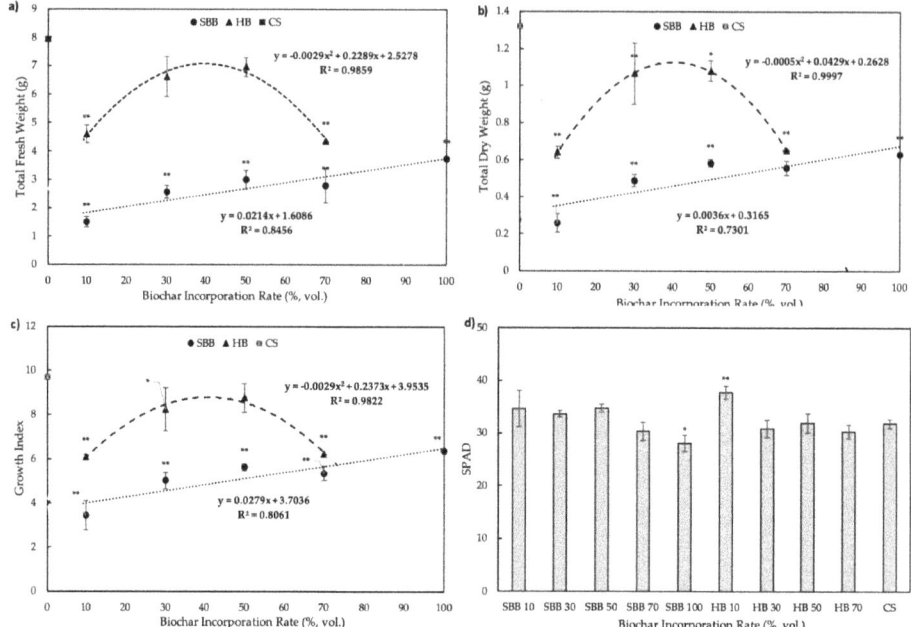

Figure 5. The correlations between total fresh weight (**TFW, a**), total dry weight (**TDW, b**), growth index (**GI, c**) and biochar incorporation rate and the SPAD (**d**) of basil seedlings grown in biochar-amended mixes. SBB = sugarcane bagasse biochar; HB = mixed hardwood biochar; CS = commercial propagation substrate. SBB (10%, 30%, 50%, 70% and 100%; by vol.) incorporated with 30% perlite with the rest being peat moss; HB (10%, 30%, 50% and 70%; by vol.) incorporated with peat moss. (*, ** indicate a significant difference from the control using Dunnett's test at $p \leq 0.05$ and $p \leq 0.01$, respectively).

All basil seedlings grown in biochar-amended mixes had significantly shorter root lengths, smaller root surface areas and fewer root tips than the control (Table 3); however, they all had similar root diameters.

Table 3. Root growth of basil seedlings grown in different mixes (*, ** indicate a significant difference from the control using Dunnett's test at $p \leq 0.05$ and $p \leq 0.01$, respectively).

Mixes [z]	Root Length (cm)	Root Surface Area (cm^2)	Average Diameter (mm)	Number of Tips
SBB:PM:P(10:60:30)	121 ± 13**	16 ± 1**	0.43 ± 0.03	480 ± 42**
SBB:PM:P(30:40:30)	295 ± 523**	34 ± 11**	0.37 ± 0.01	819 ± 88**
SBB:PM:P(50:20:30)	433 ± 23**	51 ± 6**	0.37 ± 0.01	1408 ± 235**
SBB:PM:P(70:0:30)	617 ± 92**	72 ± 22**	0.37 ± 0.01	1204 ± 118**
SBB:PM:P(100:0:0)	841 ± 95*	97 ± 15**	0.37 ± 0.02	1584 ± 163**
HB:PM(10:90)	331 ± 29**	40 ± 7**	0.39 ± 0.01	873 ± 45**
HB:PM(30:70)	757 ± 67**	88 ± 19**	0.37 ± 0.01	1758 ± 177**
HB:PM(50:50)	793 ± 145**	96 ± 35**	0.39 ± 0.01	1761 ± 167**
HB:PM(70:30)	690 ± 44**	85 ± 6**	0.39 ± 0.01	1446 ± 194**
Control	1181 ± 67	145± 21	0.39 ± 0.02	3001 ± 214

Note: [z] SBB = sugarcane bagasse biochar; HB = mixed hardwood biochar; P = perlite; PM = peat moss; Control = commercial propagation substrate. Numbers in parentheses indicate the ratio of different components, by vol.

3.3. The After-Growth Evaluation of Seedlings Produced in Biochar-Amended Media

3.3.1. Tomato Plant Growth

Tomato seedlings from different biochar-amended mixes (except 50% HB) all had significantly lower GI at transplanting compared to those from the commercial propagation mix (Figure 6a). However, after growing in CS1 for four weeks, all plants from SBB- and HB-amended mixes had similar GI (except 30% SBB) and SPAD to the control (Figure 6a,b). In addition, tomato plants from all biochar-amended mixes had similar SDW (except 10% SBB, 30% SBB and 10% HB) and leaf dry weight (LDW) (except 10% SBB) in comparison to the control (Table 4); however, they had significantly lower FDW and TDW than the control.

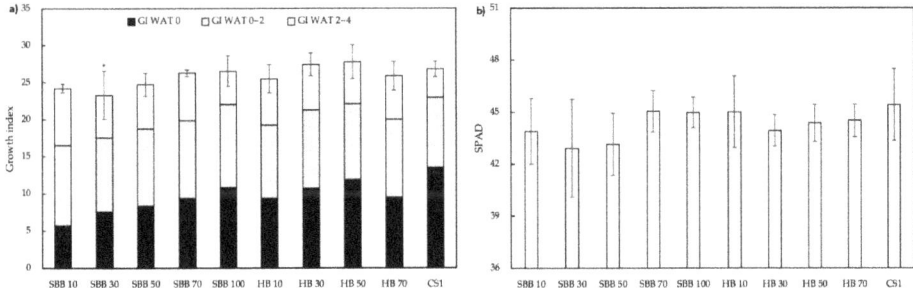

Figure 6. The growth index (**a**) and SPAD (**b**) of tomato seedlings from biochar-amended mixes after four weeks in commercial substrate. SBB = sugarcane bagasse biochar; HB = mixed hardwood biochar; CS1 = commercial growing substrate. SBB (10%, 30%, 50%, 70% and 100%; by vol.) incorporated with 30% perlite with the rest being peat moss; HB (10%, 30%, 50% and 70%; by vol.) incorporated with peat moss. (*, ** indicate a significant difference from the control using Dunnett's test at $p \leq 0.05$ and $p \leq 0.01$, respectively).

Table 4. Stalk, leaf, and fruit dry weight (g) of tomato seedlings from biochar-amended mixes after four weeks in the commercial substrate. (*, ** indicate a significant difference from the control using Dunnett's test at $p \leq 0.05$ and $p \leq 0.01$, respectively).

Mixes [z]	Stalk DW(g)	Leaf DW(g)	Fruit DW(g)	Total DW (g)
SBB:PM:P(10:60:30)	1.7 ± 0.1**	6.5 ± 0.1 **	0.3 ± 0. 0**	8.5 ± 0.2***
SBB:PM:P(30:40:30)	2.0 ± 0.2**	7.6 ± 0.4	1.0 ± 0.3**	10.6 ± 0.9***
SBB:PM:P(50:20:30)	2.7 ± 0.2	8.8 ± 0.5	0.9 ± 0.1**	12.5 ± 0.6***
SBB:PM:P(70:0:30)	3.2 ± 0.3	8.7 ± 0.4	1.3 ± 0.2**	13.2 ± 0.9**
SBB:PM:P(100:0:0)	3.5± 0.2	8.6 ± 0.2	2.4 ± 0.2**	14.5 ± 0.4**
HB:PM(10:90)	2.4 ± 0.3*	8.9 ± 0.5	1.3 ± 0.2**	12.6 ± 0.8***
HB:PM(30:70)	2.9 ± 0.3	7.8 ± 0.4	3.2 ± 0.7**	13.9 ± 1.0**
HB:PM(50:50)	3.2 ± 0.2	8.1 ± 0.2	4.0 ± 0.5*	15.4 ± 0.5*
HB:PM(70:30)	2.6 ± 0.2	8.4 ± 0.2	2.0 ± 0.3**	12.9 ± 0.6***
Control	3.4 ± 0.1	8.1 ± 0.5	7.7 ± 2.3	19.2 ± 2.1

Note: [z] SBB = sugarcane bagasse biochar; HB = mixed hardwood biochar; P = perlite; PM = peat moss; Control = commercial growing substrate. Numbers in parentheses indicate the ratio of different components, by vol.

3.3.2. Basil Plant Growth

Basil seedlings from different biochar-amended mixes (except 50% HB) all had significantly lower GI at transplanting compared to those from the commercial propagation mix (Figure 7a). However, after growing in CS1 for four weeks, all plants from SBB- and HB-amended mixes (except 10% SBB, 30% SBB and 50% SBB) had similar GI and SPAD (except 30% SBB, 70% SBB, 100% SBB, 30% HB, 50% HB and 70% HB) to the control (Figure 7a,b). In addition, basil plants from all biochar-amended mixes had similar LDW (except 10% SBB, 30% SBB and 50% SBB) and FDW (except 10% SBB, 30% SBB, 50%

SBB and 10% HB) in comparison to the control (Table 5). Plants from SBB-amended mixes all had significantly lower SDW and TDW compared to the control; however, plants from HB-amended mixes all had similar SDW (except 10% HB) and TDW (except 10% HB and 70% HB) to the control.

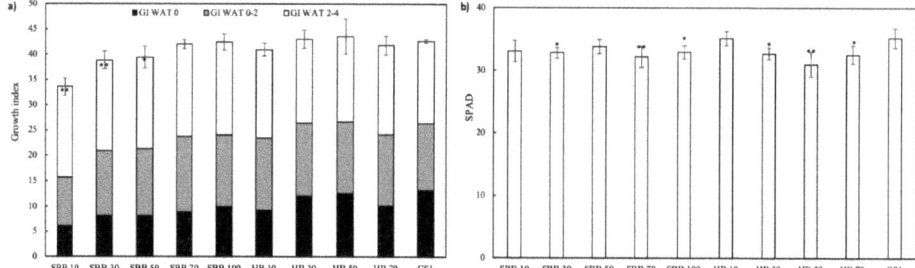

Figure 7. The growth index (**a**) and SPAD (**b**) of basil plants transplanted from biochar-amended mixes after four weeks. SBB = sugarcane bagasse biochar; HB = mixed hardwood bioTA CS1 = commercial growing substrate. SBB (10%, 30%, 50%, 70% and 100%; by vol.) incorporated with 30% perlite with the rest being peat moss; HB (10%, 30%, 50% and 70%; by vol.) incorporated with peat moss. (*, ** indicate a significant difference from the control using Dunnett's test at $p \leq 0.05$ and $p \leq 0.01$, respectively).

Table 5. Biomass of basil plants transplanted from biochar-amended mixes after four weeks. (*, ** indicate a significant difference from the control using Dunnett's test at $p \leq 0.05$ and $p \leq 0.01$, respectively).

Mixes [z]	Stalk DW(g)	Leaf DW(g)	Flower DW (g)	Total DW (g)
SBB:PM:P(10:60:30)	1.8 ± 0.2**	4.3 ± 0.4**	0.4 ± 0.1**	6.4 ± 0.7***
SBB:PM:P(30:40:30)	3.3 ± 0.2**	6.2 ± 0.2*	1.1 ± 0.2**	10.5 ± 0.5***
SBB:PM:P(50:20:30)	3.1 ± 0.2**	6.0 ± 0.4**	0.8 ± 0.2**	9.9 ± 0.5***
SBB:PM:P(70:0:30)	3.3 ± 0.1**	5.9 ± 0.2	1.0 ± 0.1**	10.2 ± 0.2***
SBB:PM:P(100:0:0)	3.9 ± 0.08**	6.9 ± 0.3	1.7 ± 0.3	12.5 ± 0.3*
HB:PM(10:90)	3.6 ± 0.1**	6.4 ± 0.1	1.2 ± 0.2*	11.3 ± 0.3***
HB:PM(30:70)	4.3 ± 0.1	7.3 ± 0.3	2.0 ± 0.2	13.5 ± 0.4
HB:PM(50:50)	4.4 ± 0.1	7.3 ± 0.2	1.9 ± 0.4	13.6 ± 0.6
HB:PM(70:30)	4.1 ± 0.3	6.4 ± 0.6	1.5 ± 0.5	11.9 ± 0.7**
Control	4.8 ± 0.2	7.7 ± 0.3	2.5 ± 0.2	14.9 ± 0.6

Note: [z] SBB = sugarcane bagasse biochar; HB = mixed hardwood biochar; P = perlite; PM = peat moss; Control = commercial growing substrate. Numbers in parentheses indicate the ratio of different components, by vol.

4. Discussion

4.1. Media Phytotoxicity and Substrate Properties

Prior to incorporating biochar into any soilless substrate, simple germination tests could be used to test the phytotoxicity of biochar, and a phytotoxicity assessment is essential for successful soilless amendment [30]. The soilless petri dish bioassay (also known as the germination test) is a rapid and simple preliminary test recommended by Solaiman to test potential biochar toxicity [35]. In this study, the extracts of SBB, HB, their mixes with PM, PM, P, 70%PM:30%P and the commercial propagation mix all showed no phytotoxicity, which is consistent with Taek–Keun's findings [36].

Biochar may or may not have phytotoxic effects on plants depending on the original feedstock and process conditions [30]. For instance, the biochar from hardwood, corn and switchgrass under different process conditions had no effect on germination rate [29], while biochar from olive mill waste was phytotoxic [37]. In this study, the germination rates of all basil seeds in the aqueous extracts of biochar-amended mixes were higher than those in DI water, which indicated no phytotoxicity for the biochar used in this study. This is similar to what had been found in Rogovska's work [29]. However, biochar-amended mixes having no effects on seedling germination rates does not necessarily mean they had no inhibition on shoot growth because plant's shoot growth can be inhibited by polycyclic aromatic hydrocarbons present in aqueous extracts [29], or by the poor physical–chemical properties of the mixes [2]. The seedlings grown in SBB-amended mixes had significantly lower DW than the control, which may be due to their low AS [38].

Even though the effects of biochar on substrate properties also varies depending on original feedstock and process conditions [22,30], some biochar types have been proven to improve the physical properties of the growing media [39]. For instance, pinewood biochar from fast pyrolysis of pinewood at 450 °C can make the substrate better for poinsettia and Easter lily to grow [6,27]. Mixed hardwood biochar from fast pyrolysis can also improve the substrate properties for tomato and basil plant growth [24]. Sugarcane bagasse biochar and pinewood biochar improved the growing mix properties for bean and cucurbit seedlings production [31]. The pruning residue biochar produced from pyrolysis at 500°C can improve growing media properties for soilless vegetable production [13,39]. The biochar could also replace perlite and has a liming effect when incorporated into a soilless substrate [40,41]. These improvements were also observed in this study, especially for HB.

4.2. Biochar Effects on Plant Growth

The effects of biochar on plant growth could be positive, null and negative [6,42,43], depending on the types of biochar and the incorporation rates. Incorporating biochar made from woodchips of *Pinus densiflora* and *Quercus acutissima* and rice husk at 20% (by vol.) with growing media, *Zelkova serrata* seedlings showed better performance than the control in a containerized production system; however, biochar made from crab shell had negative effects on seedling growth [44]. In this study, seedling biomass increased with the SBB (10–100%) and HB (10–50%) incorporation rate, which is slightly different from Webber's results [31]. Tomato plants from all biochar-amended mixes had significantly lower FDW (yield), basil plants from biochar-amended mixes (except for 10% SBB, 30% SBB and 50% SBB) had similar LDW (yield) to the control.

Even though the effects of biochar on plant biomass can be variable [6,42,43], the effects of biochar on plant GI is more often positively reported [6,27,45]. The GI of plants can be an important parameter for landscape plants such as *Magnolia, Ilex, Lagerstroemia* and other species [46]. Biochar has also been reported to have positive effects on some ornamental plant GIs, such as poinsettia, Easter lily and "Firework" *Gomphrena* [6,27,45]. In this study, even though seedlings grown in biochar-amended mixes (SBB-, 10%, 30% and 70% HB-amended) had significantly lower TDW than the control, after growing in CS1 for four weeks, plants from biochar-amended mixes (except 30% SBB for tomato, 10%, 30% SBB for basil) all had a similar GI to the control. As landscape plants need more time to grow from seedling to a marketable size, the biochar-amended mixes used in this study might be used more successfully for landscape plants seedling production. More biochar studies on landscape plants should be conducted in the future.

Detailed studies on biochar–root interactions are few [47], but plant roots are the first contacting points to biochar particles. Plants with longer root length, larger surface area and more root tips may be able to obtain more nutrients and grow better [47,48]. In this study, root length, surface area and the number of tips of seedlings grown in biochar-amended mixes (except for 50%HB-amended) were all shorter, smaller or less than those grown in the control, which can explain why seedlings grown in most biochar-amended mixes did not perform as well as the control.

5. Conclusions

The biochar-amended mixes used in this experiment had acceptable BD, CC, AS, and TP (except 50%, 70% and 100% SBB). The results from this experiment indicated PM mixed with up to 50% HB could be used for tomato and basil seedling production in a greenhouse. Both tomato and basil seedlings grown in 50% HB-amended mixes exhibited greater or similar growth compared to those in a commercial propagation mix, as reflected by similar seedling FW, DW, GI, SPAD and root development.

Seedlings grown in 70% HB-amended mixes had significantly lower DW than the control, however, after growing in commercial growing media for four weeks, their DWs were similar to the control. Up to 70% of HB could be amended with PM for tomato and basil seedling production without negative effects on plant biomass.

Author Contributions: This work was a product of the combined effort of all the authors. All authors conceptualized and designed the study. P.Y. performed the experiments, collected and analyzed the data, and wrote the manuscript with assistance from all other authors, mainly M.G. L.H., Q.L. and G.N. provided technical advice and assistance when the study was conducted, and revised and improved the manuscript.

Funding: This research received no external funding.

Acknowledgments: The authors would like to thank American Biocarbon LLC White Castle, Louisiana, Proton Power Inc. Lenouir City, Tennessee, for supplying the biochar for the experiment. The work would have been impossible without the support from Elizabeth Pierson and Kevin Crosby, for supplying experimental instruments. The authors would like to thank the Texas A&M University Open Access to Knowledge Fund (OAKFund, supported by the University Libraries and the Office of the Vice President for Research) for partially covering the open access publishing fees. The authors would also like to thank the Agriculture Women Excited to Share Opinions, Mentoring and Experiences (AWESOME) faculty group of the College of Agriculture and Life Sciences at Texas A&M University for assistance with editing the manuscript.

Conflicts of Interest: The authors declare no conflict of interest.

References

1. Chalk, P.; Souza, R.D.F.; Urquiaga, S.; Alves, B.; Boddey, R. The role of arbuscular mycorrhiza in legume symbiotic performance. *Soil Biol. Biochem.* **2006**, *38*, 2944–2951. [CrossRef]
2. Nelson, P. Root substrate. In *Greenhouse Operation and Management*, 7th ed.; Prentice Hall: Upper Saddle River, NJ, USA, 2012; pp. 161–194.
3. Peng, D.H.; Gu, M.M.; Zhao, Y.; Yu, F.; Choi, H.S. Effects of biochar mixes with peat-moss based substrates on growth and development of horticultural crops. *Hortic. Sci. Technol.* **2018**, *36*, 501–512. [CrossRef]
4. United States Geological Survey (USGS). PEAT. In *Mineral Commodity Summaries*; Center, N.M.I., Ed.; USGS: Reston, VA, USA, 2019; pp. 118–119.
5. Alexander, P.; Bragg, N.; Meade, R.; Padelopoulos, G.; Watts, O. Peat in horticulture and conservation: The UK response to a changing world. *Mires Peat* **2008**, *3*, 1–10.
6. Guo, Y.; Niu, G.; Starman, T.; Volder, A.; Gu, M. Poinsettia growth and development response to container root substrate with biochar. *Horticulturae* **2018**, *4*, 1. [CrossRef]
7. Demirbas, A.; Arin, G. An overview of biomass pyrolysis. *Energy Sources* **2002**, *24*, 471–482. [CrossRef]
8. Lehmann, J. A handful of carbon. *Nature* **2007**, *447*, 143–144. [CrossRef]
9. Nartey, O.D.; Zhao, B. Biochar preparation, characterization, and adsorptive capacity and its effect on bioavailability of contaminants: An overview. *Adv. Mater. Sci. Eng.* **2014**, *2014*, 715398. [CrossRef]
10. Liu, R.; Gu, M.; Huang, L.; Yu, F.; Jung, S.-K.; Choi, H.-S. Effect of pine wood biochar mixed with two types of compost on growth of bell pepper (*Capsicum annuum* L.). *Hortic. Environ. Biotechnol.* **2019**, *60*, 313–319. [CrossRef]
11. Choi, H.-S.; Zhao, Y.; Dou, H.; Cai, X.; Gu, M.; Yu, F. Effects of biochar mixtures with pine-bark based substrates on growth and development of horticultural crops. *Hortic. Environ. Biotechnol.* **2018**, *59*, 345–354. [CrossRef]
12. Tian, Y.; Sun, X.; Li, S.; Wang, H.; Wang, L.; Cao, J.; Zhang, L. Biochar made from green waste as peat substitute in growth media for *Calathea rotundifola* cv. *Fasciata*. *Sci. Hortic.* **2012**, *143*, 15–18. [CrossRef]
13. Webber, C.L., III; White, P.M., Jr.; Spaunhorst, D.J.; Lima, I.M.; Petrie, E.C. Sugarcane biochar as an amendment for greenhouse growing media for the production of cucurbit seedlings. *J. Agric. Sci.* **2018**, *10*, 104–115. [CrossRef]
14. Hansen, V.; Hauggaard-Nielsen, H.; Petersen, C.T.; Mikkelsen, T.N.; Müller-Stöver, D. Effects of gasification biochar on plant-available water capacity and plant growth in two contrasting soil types. *Soil Tillage Res.* **2016**, *161*, 1–9. [CrossRef]
15. Spokas, K.; Koskinen, W.; Baker, J.; Reicosky, D. Impacts of woodchip biochar additions on greenhouse gas production and sorption/degradation of two herbicides in a Minnesota soil. *Chemosphere* **2009**, *77*, 574–581. [CrossRef] [PubMed]

16. Vaughn, S.F.; Kenar, J.A.; Thompson, A.R.; Peterson, S.C. Comparison of biochars derived from wood pellets and pelletized wheat straw as replacements for peat in potting substrates. *Ind. Crop. Prod.* **2013**, *51*, 437–443. [CrossRef]
17. Hansen, V.; Müller-Stöver, D.; Ahrenfeldt, J.; Holm, J.K.; Henriksen, U.B.; Hauggaard-Nielsen, H. Gasification biochar as a valuable by-product for carbon sequestration and soil amendment. *Biomass Bioenergy* **2015**, *72*, 300–308. [CrossRef]
18. Spokas, K.A.; Baker, J.M.; Reicosky, D.C. Ethylene: Potential key for biochar amendment impacts. *Plant Soil* **2010**, *333*, 443–452. [CrossRef]
19. Hina, K.; Bishop, P.; Arbestain, M.C.; Calvelo-Pereira, R.; Maciá-Agulló, J.A.; Hindmarsh, J.; Hanly, J.; Macìas, F.; Hedley, M. Producing biochars with enhanced surface activity through alkaline pretreatment of feedstocks. *Soil Res.* **2010**, *48*, 606–617. [CrossRef]
20. Locke, J.C.; Altland, J.E.; Ford, C.W. Gasified rice hull biochar affects nutrition and growth of horticultural crops in container substrates. *J. Environ. Hortic.* **2013**, *31*, 195–202.
21. Xu, G.; Zhang, Y.; Sun, J.; Shao, H. Negative interactive effects between biochar and phosphorus fertilization on phosphorus availability and plant yield in saline sodic soil. *Sci. Total Environ.* **2016**, *568*, 910–915. [CrossRef]
22. Huang, L.; Gu, M. Effects of biochar on container substrate properties and growth of plants—A Review. *Horticulturae* **2019**, *5*, 14. [CrossRef]
23. Graber, E.R.; Harel, Y.M.; Kolton, M.; Cytryn, E.; Silber, A.; David, D.R.; Tsechansky, L.; Borenshtein, M.; Elad, Y. Biochar impact on development and productivity of pepper and tomato grown in fertigated soilless media. *Plant Soil* **2010**, *337*, 481–496. [CrossRef]
24. Huang, L.; Niu, G.; Feagley, S.E.; Gu, M. Evaluation of a hardwood biochar and two composts mixes as replacements for a peat-based commercial substrate. *Ind. Crop. Prod.* **2019**, *129*, 549–560. [CrossRef]
25. Zhang, L.; Sun, X.-Y.; Tian, Y.; Gong, X.-Q. Biochar and humic acid amendments improve the quality of composted green waste as a growth medium for the ornamental plant *Calathea insignis*. *Sci. Hortic.* **2014**, *176*, 70–78. [CrossRef]
26. Guo, M.; He, Z.; Uchimiya, S.M. Introduction to biochar as an agricultural and environmental amendment. *Agric. Environ. Appl. Biochar Adv. Barriers* **2016**, *63*, 1–14.
27. Guo, Y.; Niu, G.; Starman, T.; Gu, M. Growth and development of Easter lily in response to container substrate with biochar. *J. Hortic. Sci. Biotechnol.* **2018**, *94*, 80–86. [CrossRef]
28. Headlee, W.L.; Brewer, C.E.; Hall, R.B. Biochar as a substitute for vermiculite in potting mix for hybrid poplar. *Bioenergy Res.* **2013**, *7*, 120–131. [CrossRef]
29. Rogovska, N.; Laird, D.; Cruse, R.; Trabue, S.; Heaton, E. Germination tests for assessing biochar quality. *J. Environ. Qual.* **2012**, *41*, 1014–1022. [CrossRef] [PubMed]
30. Gell, K.; van Groenigen, J.; Cayuela, M.L. Residues of bioenergy production chains as soil amendments: Immediate and temporal phytotoxicity. *J. Hazard. Mater.* **2011**, *186*, 2017–2025. [CrossRef]
31. Webber, C.L., III; White, P.M., Jr.; Gu, M.; Spaunhorst, D.J.; Lima, I.M.; Petrie, E.C. Sugarcane and pine biochar as amendments for greenhouse growing media for the production of bean (*Phaseolus vulgaris* L.) seedlings. *J. Agric. Sci.* **2018**, *10*, 58–68. [CrossRef]
32. Gravel, V.; Dorais, M.; Ménard, C. Organic potted plants amended with biochar: Its effect on growth and Pythium colonization. *Can. J. Plant Sci.* **2013**, *93*, 1217–1227. [CrossRef]
33. Fonteno, W.; Hardin, C.; Brewster, J. *Procedures for Determining Physical Properties of Horticultural Substrates Using the NCSU Porometer*; Horticultural Substrates Laboratory, North Carolina State University: Raleigh, NC, USA, 1995.
34. LeBude, A.; Bilderback, T. Pour-through extraction procedure: A nutrient management tool for nursery crops. *North Carol. Coop. Ext.* **2009**. AG-717-W.
35. Solaiman, Z.M.; Murphy, D.V.; Abbott, L.K. Biochars influence seed germination and early growth of seedlings. *Plant Soil* **2012**, *353*, 273–287. [CrossRef]
36. Taek–Keun, O.; Shinogi, Y.; Chikushi, J.; Yong–Hwan, L.; Choi, B. Effect of aqueous extract of biochar on germination and seedling growth of lettuce (*Lactuca sativa* L.). *J. Fac. Agric. Kyushu Univ.* **2012**, *57*, 55–60.
37. Fornes, F.; Belda, R.M. Biochar versus hydrochar as growth media constituents for ornamental plant cultivation. *Sci. Agric.* **2018**, *75*, 304–312. [CrossRef]

38. Yeager, T.; Fare, D.; Lea-Cox, J.; Ruter, J.; Bilderback, T.; Gilliam, C.; Niemiera, A.; Warren, S.; Whitewell, T.; White, R. *Best Management Practices: Guide for Producing Container-Grown Plants*; Southern Nursery Association: Atlanta, GA, USA, 2007.
39. Nieto, A.; Gascó, G.; Paz-Ferreiro, J.; Fernández, J.; Plaza, C.; Méndez, A. The effect of pruning waste and biochar addition on brown peat based growing media properties. *Sci. Hortic.* **2016**, *199*, 142–148. [CrossRef]
40. Northup, J. Biochar as a Replacement for Perlite in Greenhouse Soilless Substrates. Master's Thesis, Iowa State University, Ames, IA, USA, 2013.
41. Berek, A.K.; Hue, N.; Ahmad, A. Beneficial use of biochar to correct soil acidity. *Food Provid. Hanai Ai* **2011**, *9*, 1–3.
42. Vaughn, S.F.; Eller, F.J.; Evangelista, R.L.; Moser, B.R.; Lee, E.; Wagner, R.E.; Peterson, S.C. Evaluation of biochar-anaerobic potato digestate mixtures as renewable components of horticultural potting media. *Ind. Crop. Prod.* **2015**, *65*, 467–471. [CrossRef]
43. Dunlop, S.J.; Arbestain, M.C.; Bishop, P.A.; Wargent, J.J. Closing the loop: Use of biochar produced from tomato crop green waste as a substrate for soilless, hydroponic tomato production. *HortScience* **2015**, *50*, 1572–1581. [CrossRef]
44. Cho, M.S.; Meng, L.; Song, J.-H.; Han, S.H.; Bae, K.; Park, B.B. The effects of biochars on the growth of *Zelkova serrata* seedlings in a containerized seedling production system. *For. Sci. Technol.* **2017**, *13*, 25–30. [CrossRef]
45. Gu, M.; Li, Q.; Steele, P.H.; Niu, G.; Yu, F. Growth of 'Fireworks' gomphrena grown in substrates amended with biochar. *J. Food Agric. Environ.* **2013**, *11*, 819–821.
46. Ruter, J.M. Growth and landscape performance of three landscape plants produced in conventional and pot-in-pot production systems. *J. Environ. Hortic.* **1993**, *11*, 124–127.
47. Prendergast-Miller, M.; Duvall, M.; Sohi, S. Biochar–root interactions are mediated by biochar nutrient content and impacts on soil nutrient availability. *Eur. J. Soil Sci.* **2014**, *65*, 173–185. [CrossRef]
48. Rellán-Álvarez, R.; Lobet, G.; Dinneny, J.R. Environmental control of root system biology. *Annu. Rev. Plant Biol.* **2016**, *67*, 619–642. [CrossRef] [PubMed]

© 2019 by the authors. Licensee MDPI, Basel, Switzerland. This article is an open access article distributed under the terms and conditions of the Creative Commons Attribution (CC BY) license (http://creativecommons.org/licenses/by/4.0/).

Article

Zinc Adsorption by Activated Carbon Prepared from Lignocellulosic Waste Biomass

Sari Tuomikoski [1,*], Riikka Kupila [2], Henrik Romar [1,2], Davide Bergna [2], Teija Kangas [1], Hanna Runtti [1] and Ulla Lassi [1,2]

1. Research Unit of Sustainable Chemistry, University of Oulu, P.O. Box 4300, FI-90014 Oulu, Finland; henrik.romar@chydenius.fi (H.R.); teija.kangas@oulu.fi (T.K.); hanna.runtti@oulu.fi (H.R.); ulla.lassi@oulu.fi (U.L.)
2. Unit of Applied Chemistry, Kokkola University Consortium Chydenius, University of Jyväskylä, Talonpojankatu 2B, FI-67100 Kokkola, Finland; riikka.kupila@chydenius.fi (R.K.); davide.bergna@chydenius.fi (D.B.)
* Correspondence: sari.tuomikoski@oulu.fi; Tel.: +358-294-481644

Received: 2 October 2019; Accepted: 22 October 2019; Published: 28 October 2019

Featured Application: In this study, activated carbon from lignocellulosic waste biomass was prepared and it was used to the zinc removal. The adsorption capacity obtained for the prepared material was compared favorably to other adsorbents used in zinc removal presented in literature. Therefore, the suitable adsorbent for zinc removal from waste biomass was developed by using non-hazardous materials. The adsorbent can be used with wide temperature range and with quite high concentration range.

Abstract: Sawdust was used as a precursor for the production of biomass-based activated carbon. Carbonization and activation are single-stage processes, and steam was used as a physical activation agent at 800 °C. The adsorption capacity towards zinc was tested, and the produced activated carbon proved effective and selectively adsorbent. The effects of pH, initial concentration, adsorbent dosage, time, temperature, and regeneration cycles were tested. The adsorption capacity obtained in this study was compared favorably to that of the materials reported in the literature. Several isotherms were applied to describe the experimental results, with the Sips isotherm having the best fit. Kinetic studies showed that the adsorption follows the Elovich kinetic model.

Keywords: lignocellulosic biomass; adsorption; carbonization; adsorbent; zinc; regeneration

1. Introduction

Water purification and the removal of toxic substances from wastewaters are increasing global challenges. Both surface and ground waters worldwide are reported to be contaminated with impurities deriving from natural and human-involved sources [1]. Metals such as cadmium, chromium, copper, nickel, and zinc (Zn) are commonly associated with pollution problems [2]. Significant anthropogenic sources of Zn in the environment include: metalliferous mining [2]; agricultural sources, such as fertilizers and manure; sewage sludge [3]; metallurgical industries, such as the production of special alloys and steel [4]; landfill leachates [5]; electroplating in the metal finishing industry [6]; and miscellaneous sources, such as batteries [7]. For humans, Zn is an important micronutrient, but too-high amounts of Zn can cause depression, lethargy, and neurologic issues such as seizures and ataxia [8–11]. Therefore, the effective removal of Zn from water is vital.

Various methods and combinations of methods exist for the removal of Zn, including electrocoagulation [7], ion exchange [5], membrane filtration [6], coagulation–flocculation, flotation, and chemical precipitation [8]. The most frequently used methods are ion exchange and membrane filtration [8]. In addition, adsorption is a suitable method for removing heavy metals such as Zn because it is simple, highly efficient, and cost-effective and has regeneration potential [11,12]. Activated carbon is the most widely applied adsorbent for water purification and is traditionally prepared using, for example, coconut shells or coal as a precursor [13]. The literature contains several studies about the preparation of activated carbon from different agro-based biomass raw materials, such as cherry stones, eucalyptus bark saw dust, bagasse, and peanut husks [14–18].

The preparation of activated carbon basically consists of two stages, carbonization and activation. Carbonization and activation can be performed in one single reaction or in a two-stage reaction where carbonization and activation are separated in time and space [19]. The carbonization yield (degree of conversion) varies widely as a function of the amount of carbon being removed as CO_x or hydrocarbons. The production of activated carbon is typically dominated by a compromise between porosity development and process yield [20]. The typical yield for lignocellulosic materials in the carbonization step is 20 wt%–30 wt% and after the activation step the overall yield is approximately 10 wt% [21]. The pore structure of the carbon after the carbonization step is insufficient, and therefore the activation step is needed. In the activation process, spaces between aromatic sheets are cleared of various carbonaceous compounds and disorganized carbon that might have filled the interstices during carbonization. Carbonized material is converted into a form that contains a large number of randomly distributed pores of various shapes and sizes, giving rise to a product with an extended and extremely high surface area. The carbonized material is further treated using chemical or physical activation to increase the surface area and the porosity [22,23].

Physical activation is a two-stage process in which the carbonization step is followed by physical activation. Physical activation is typically done between 600 °C and 900 °C, the temperature range in which the selective gasification of individual carbon atoms take place. Steam and carbon dioxide are the most commonly used physical activation agents [22]. The loss of sample mass during the activation step occurs due to the combined effect of devolatilization and the loss of fixed carbon that remains after the release of volatiles [20,21]. The porosity is generated in the slow oxidation process where carbon atoms react with oxygen generating carbon dioxide [21].

The aim of this work is to develop eco-innovative biomass-derived water treatment material from a renewable source and to study the conditions in which the material can be used effectively. Due to the bioeconomy and circular economy potential, renewable sources, waste materials, and locally available materials are the future in the production of activated carbon. There is huge potential in terms of the use of raw materials, such as saw dust from sawmills and other locally available waste biomass materials. To that end, sawdust waste from sawmills was used as a precursor in the production of activated carbon. The carbonization and activation conditions of this lignocellulosic biomass were optimized, and the material produced was characterized and tested in terms of Zn adsorption based on wide range of pH levels, initial Zn concentration, adsorbent mass, temperature, and time ranges. Reuse of the material was also studied by testing the material regeneration using 0.1 M hydrogen chloride (HCl). Several isotherm models were used, such as Langmuir, Freundlich, and Sips adsorption isotherms. Pseudo-first-order, pseudo-second-order, and Elovich kinetic models were also used.

2. Materials and Methods

2.1. Preparation of Activated Carbons

A dried and sieved lignocellulosic biomass composed of birch sawdust from a Swedish saw-mill was carbonized and steam activated in a one-step process in a rotating quartz reactor (Nabertherm GmbH RSRB 80-750/11, Lilienthal, Germany). The single-stage thermal process was divided into two sub-stages, a carbonization step in which the temperature was raised to 800 °C and an activation stage. During the activation, the temperature was set to hold at 800 °C for 120 min with a stream of steam to ensure proper surface activation. During the whole process, the reactor was flushed with inert N_2 gas.

The carbons were crushed and sieved, and a 100–425 µm fraction was characterized and used for the Zn adsorption tests.

2.2. Characterization Methods

The pore size, pore volume, and specific surface area of the produced activated carbons were analyzed based on nitrogen adsorption–desorption isotherms at the temperature of liquid nitrogen using a Micromeritics ASAP 2020 system (Norcross, GA, USA). Prior to measurements, samples were pretreated at low pressures and high temperatures to clean the surfaces. Sample tubes were immersed in liquid nitrogen (−197 °C), nitrogen gas was added to the samples in small steps, and the resulting isotherms were obtained. Specific surface areas were calculated from adsorption isotherms according to the Brunauer–Emmett–Teller (BET) method [24] and pore size distributions were calculated using the Barrett–Joyner–Halenda (BJH) algorithm [25]. Sample degassing was performed by elevating the temperature to 50 °C under restricted pressure drop to 15 mm Hg followed by a 10-min hold at 50 °C. After the hold period, the samples were heated 10 °C/min up to 140 °C under free evacuation. Finally, samples were degassed at 140 °C for 3 h. This procedure gives a final, constant pressure of 2 µm Hg.

The content of carbon present in each sample, given as total carbon (TC) percent, was measured using a solid-phase carbon analyzer (Skalar Primacs MCS, Breda, The Netherlands). Dried samples were weighted in quartz crucibles and combusted at 1100 °C in an atmosphere of pure oxygen, and the formed CO_2 was analyzed by the infrared (IR) detector of the instrument. Carbon content values were obtained by reading the signal of the IR analyzer from a calibration curve derived from known masses of a standard substance, oxalic acid. The total mass of carbon was calculated as a percent of the mass initially weighted and was measured after the carbonization and activation steps.

2.3. Adsorption and Desorption Experiments

The effect of pH on the Zn adsorption properties of the activated carbons was tested using 25 mL sample volume, 100 mg/L Zn solution prepared from $ZnCl_2$, adsorbent dosage 5 g/L, and pH range 2–7. Experiments were done in an Erlenmeyer flask, and the mixture was agitated continuously for 24 h to achieve equilibrium between the adsorption and regeneration of Zn. After the pH optimization experiments were done, the effect of the initial Zn concentration was studied at the optimum pH with initial Zn concentration range 10 mg/L–500 mg/L by using 24 h adsorption time. After that, adsorbent dosage optimization tests were done with an adsorbent mass of 0.5–10 g/L at under the previously determined optimum conditions. After these parameters were optimized, the effect of time was studied, with time ranging from 1 min to 24 h. The effect of temperature was studied using temperatures of 10 °C, room temperature, 22 °C, and 40 °C. All the adsorption experiments have been done duplicated. Regeneration experiments were done with a 2.5 g/L dosage of activated carbon adsorbed with Zn in optimized conditions and mixed in 0.1 M HCl. Samples were taken within time range 1 min and 4 h.

Because the pH of the activated carbon prepared from the lignocellulosic biomass is quite alkaline (pH of approximately 10), the pH of the solution was adjusted in all experiments with 0.1 M or 0.01 M HCl to ensure the correct pH value and to prevent the formation of unwanted precipitation. pH adjustments were done after mixing the adsorbent and the Zn solution. Before taking the Zn concentration measurements, the sample was filtered and diluted. The concentration of the Zn in the solution was measured using atomic absorption spectroscopy (AAnalyst 200 atomic absorption spectrometer; Perkin–Elmer, Waltham, MA, USA).

2.4. Isotherm Analysis

Below, the results of three non-linear adsorption models are reported. The non-linear form of Langmuir's Equation [26] is:

$$q_e = \frac{b_L q_m C_e}{1 + b_L C_e}, \tag{1}$$

where b_L (L/mg) represents the energy of sorption, q_m (mg/g) is the Langmuir constant that is related to the sorption capacity, and C_e (mg/L) is the Zn concentration of the solution in equilibrium. The Freundlich model [27] can be written as:

$$q_e = K_F C_e^{1/n_F}, \tag{2}$$

where K_f (L/g) and n_F are Freundlich constants related to the sorption capacity and intensity, respectively. The Sips isotherm [28] combines properties of both the above mentioned models and is given as:

$$q_e = \frac{q_m (b_S C_e)^{n_s}}{1 + (b_S C_e)^{n_s}}. \tag{3}$$

In this equation constant b_S (L/mg) is a constant related to sorption energy, the q_m (mg/g) value relates to the sorption capacity, and C_e (mg/L) is the Zn concentration of the solution in equilibrium. Isotherm parameters were obtained using non-linear regression with OriginPro 2018.

2.5. Kinetic Studies

Three non-linear models, the pseudo first-order [29], pseudo second-order [30], and Elovich models [31], were used to describe the kinetic behavior of the sorption process.

The equation of the pseudo first-order model can be written as:

$$q_t = q_e \left(1 - e^{-k_1 \cdot t}\right) \tag{4}$$

where k_1 (min^{-1}) is the pseudo-first-order rate constant, q_t (mg/g) is the amount of Zn ions adsorbed at time t (min), and q_e (mg/g) is the amount of Zn ions adsorbed at equilibrium [29].

The pseudo-second-order equation is:

$$q_t = \frac{q_e^2 k_2 t}{q_e k_2 t + 1}. \tag{5}$$

where the parameter k_2 is the pseudo-second-order rate equilibrium constant (g/mg min) [30].

The non-linear form of the Elovich equation is:

$$q = \frac{1}{\beta} \ln\left(v_0 \beta + \frac{1}{\beta} \ln t\right), \tag{6}$$

where β (g/mg) is the desorption constant and v_0 (mg/g min) is the initial sorption rate [31]. Kinetic parameters were obtained using non-linear regression with OriginPro 2018.

For both the kinetic and isotherm models, the residual root mean square error (RMSE) was determined to evaluate the fit of the isotherm equations to the experimental data. The smaller the error function value, the better the curve fit. The calculated expressions of error function can be defined as follows:

$$\text{RMSE} = \sqrt{\frac{1}{n-p} \sum_{i=1}^{n} \left(q_{e(exp)} - q_{e(calc)}\right)^2}, \tag{7}$$

where n is the number of experimental data points, p is the number of parameters in the isotherm model, and $q_{e(exp)}$ (mg/g) and $q_{e(calc)}$ (mg/g) are the experimental and calculated values of sorption capacity in equilibrium, respectively.

2.6. Weber–Morris Model

The diffusion mechanism was analyzed using the intraparticle diffusion model introduced by Weber and Morris, as follows:

$$q_t = k_{id} t^{1/2} + C, \tag{8}$$

where k_{id} (mg g^{-1} min$^{-1/2}$) is the intraparticle diffusion in the rate-determining step, and C is the intercept related to the thickness of the boundary layer [32].

2.7. Thermodynamics

The thermodynamic parameters calculated from the experimental result reveals the spontaneity of the process. To solve these parameters, the Van't Hoff equation was applied. The equation is expressed as follows:

$$\ln K_d = -\frac{\Delta H^0}{RT} + \frac{\Delta S^0}{R}, \quad (9)$$

where R is the universal gas constant (8.314 J/mol K), T (K) is the temperature used in the experiments, and K_d (L/g) is the equilibrium constant calculated from the experimental results in equilibrium conditions using the following equation:

$$K_d = \frac{q_e}{C_e} \quad (10)$$

In addition, ΔH^0 where calculated by Van't Hoff equations:

$$\Delta G^0 = -RT \ln K_d \quad (11)$$

3. Results and Discussion

Biomass-based activated carbon was used in the removal of Zn. The effects of pH, competitive ions, initial concentration, adsorbent dosage, temperature, time, and desorption were studied. Isotherm analysis, kinetic modeling, and thermodynamic calculations were done. Experiments were also done for commercial activated carbon as a reference sample, but the results are not presented because the removal percentage was 10–30% depending on the initial Zn concentration and the adsorbent dosage (1 g/L–10 g/L).

3.1. Characterization of Activated Carbon

Table 1 presents the yield of the carbonization and activation steps, TC content, surface area, pore volume, and mean pore sizes of the prepared activated carbon. Based on the literature, the carbon content for biomass-based activated carbons typically varies between 65% and 97%. Therefore, the TC content obtained in this study is very high [33], specific surface area can vary between 336 m^2/g and 1080 m^2/g for biomass-based activated carbon, and the pore volume can vary between 0.02 cm^3/g and 0.80 cm^3/g [33]. Therefore, the results obtained in this study are in the same order of magnitude [33]. The total yield of the carbonization and activation steps is 10%, which is typical for a lignocellulosic biomass [21].

Table 1. Characterization results of produced activated carbon sample.

Sample	Yield [%]	TC [%]	Brunauer–Emmett–Teller (BET)			Barrett–Joyner–Halenda (BJH)		
			Specific Surface Area [m^2/g]	Pore Volume [cm^3/g]	Average Pore Size [nm]	Micropore [%]	Mesopore [%]	Macropore [%]
Birch carbon	10	97	860	0.61	2.8	14	84	2

3.2. Effect of pH and Competitive Ion

pH optimization tests were done with the pH ranging between 2 and 7; the results are presented in Figure 1. The results clearly show that the optimum pH value for adsorption is 4 and that removal efficiency is slightly lower with pH values of 6 and 7. There is no removal capacity with a pH value of 2.

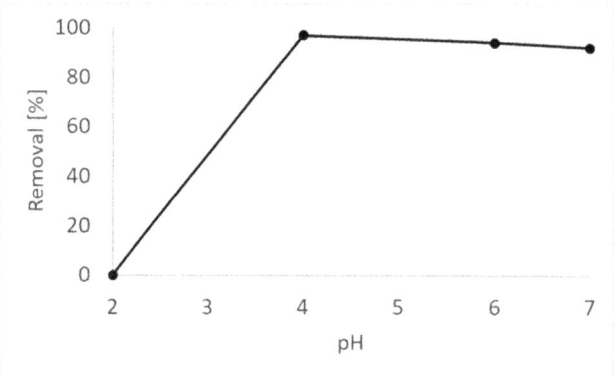

Figure 1. The removal percentage as a function of pH. Experiments were done in duplicate and the range of results were at maximum 3 percentage points.

The effect of competitive ions was studied using a Zn solution prepared from $ZnCl_2$ or $ZnSO_4$ within a pH range of 2–6. The results clearly show that the adsorbed amount of Zn and the removal percentage seem to be dependent only on pH value and not on the competitive ions (Cl^- or SO_4^{2-}). Thus, adsorption towards the biomass-based activated carbon is selective.

3.3. Effect of Initial Concentration

The effect of the initial Zn concentration was studied at the optimum pH value of 4. Other concentrations were also studied: 10 mg/L, 25 mg/L, 50 mg/L, 75 mg/L, 100 mg/L, 125 mg/L, 150 mg/L, 200 mg/L, and 500 mg/L. The results of the initial concentration experiments are presented in Figure 2. Biomass-based activated carbon works quite well in the concentration range of 0 mg/L–150 mg/L. The removal percentage increases as a function of the initial Zn concentration to 75 mg/L and then starts to decrease sharply. Therefore, the optimum Zn concentration, 75 mg/L, was selected for further experiments. The optimal adsorption sites are occupied first at low concentrations. When the concentration increases, there may exist some driving forces to increase the removal percentage. [34].

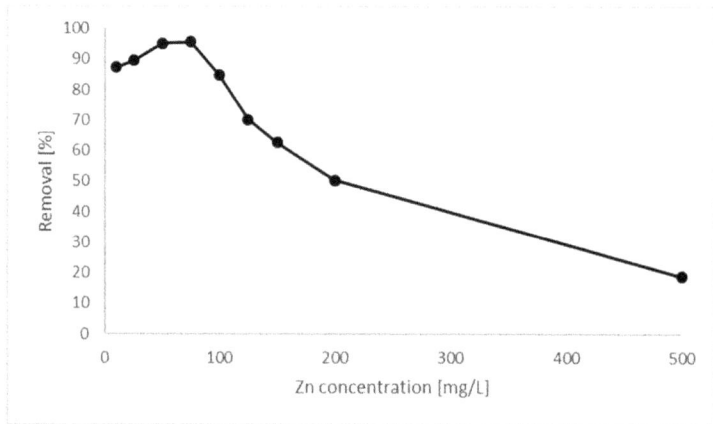

Figure 2. Removal percentages as a function of Zn initial concentration. Experiments were done duplicated and the removal % varied in the range of 3 percentage points.

Adsorption capacity increases as a function of Zn concentration. The adsorption capacity was compared to other adsorbents used in Zn removal, shown in Table 2. The adsorption capacities presented in the literature vary between 2.21 and 52.09 mg/g. The adsorption capacity (21.44 mg/g) obtained in this study was in the same order of magnitude. The amount of Zn adsorbed as a function of Zn concentration is presented in Figure 3. Katsou et al. used the initial metal concentration of 320 mg/L [35] and Mohan and Singh reported the initial metal concentration between 1–1000 mg/L [18]. The initial metal concentration used in this study was 10–500 mg/L. The initial metal concentration may have effect to the adsorption capacity.

Table 2. Adsorption capacities of zinc (Zn) on various adsorbents.

Adsorbent	Adsorption Capacity [mg/g]	Reference
Natural zeolite	2.21	[36]
Vermiculite	3.88	[37]
Defatted rice brans	5–17	[38]
Solid residue of olive mill products	5.40	[39]
Activated carbon from almond shells	6.65	[40]
Natural zeolite	13.02	[35]
Apricot stones carbon	13.21	[41]
Biomass-based carbon	21.44	This study
Lignite	22.83	[42]
Activated carbon derived from bagasse	31.11	[18]
Peat	52.09	[43]

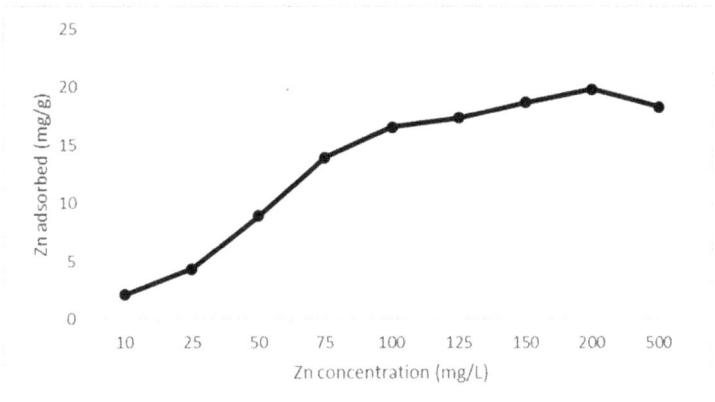

Figure 3. Zinc (Zn) adsorption capacity to carbon surface in different initial concentrations. Experiments were done duplicated and the range of results were at maximum 0.3 mg/g.

3.4. Effect of Adsorbent Dose

Figure 4 shows the effect of adsorbent dosage on Zn removal % and adsorption capacity. The highest removal percentage was obtained with doses of 3 g/L and 5 g/L. The adsorption capacities, q [mg/g], increase when the adsorbent dose increases. This is a logical result because the higher amount of adsorbent can uptake more Zn ions from the solution. The higher the dosage, the greater the number of available adsorption sites there are to protect against impurities [18,44].

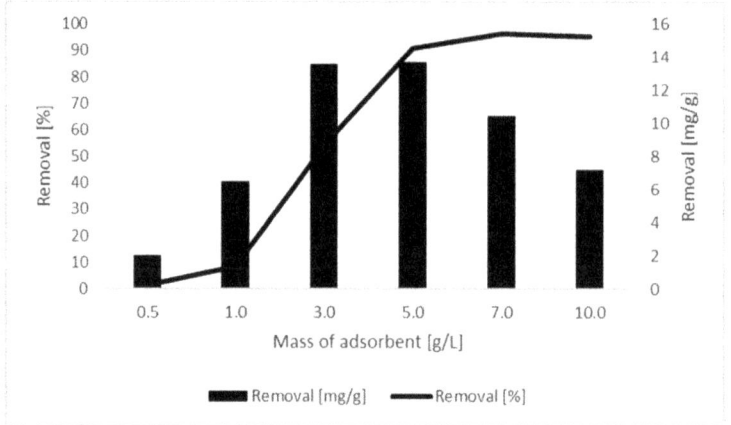

Figure 4. Zinc removal [%] and adsorption capacity [mg/g] as a function of the mass of the adsorbent. Experiments were done duplicated and the range of results were at maximum 3 percentage points in values of removal % and at maximum 0.5 mg/g in adsorption capacity determinations.

3.5. Isotherms

Several non-linear isotherm models (Langmuir, Freundlich, Sips, Bi-Langmuir, Toth, Temkin, and Dubinin–Radushkevich) were applied to the experimental data. However, according to the R^2 values and RMSEs, the Sips model was clearly the best fit (R^2 = 0.95, RMSE = 1.81). For that reason, only the parameters of the most traditional Langmuir and Freundlich isotherms, in addition to the Sips isotherm, are presented in Table 3. Corresponding fits are illustrated in Figure 5. The applicability of the Sips model is a very reasonable result, as it is generally known to represent adsorption well on heterogenous surfaces, which is the case with biomass-based activated carbon materials.

Table 3. Isotherm parameters of Langmuir, Freundlich, and Sips models.

	Parameters and Errors	Value
Experiment	Max. q_e [mg/g]	21.44
Langmuir	q_m [mg/g]	20.78
	b_L [L/mg]	0.29
	R^2	0.87
	RMSE	2.68
Freundlich	n_F	5.86
	K_F [(mg/g)/(mg/L)^n]	8.50
	R^2	0.71
	RMSE	4.06
Sips	q_m [mg/g]	19.10
	b_S [L/mg]	0.41
	n_S	4.02
	R^2	0.95
	RMSE	1.81

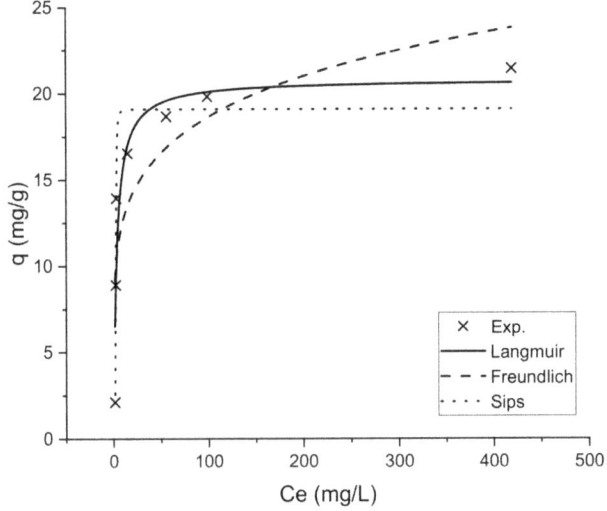

Figure 5. Langmuir, Freundlich, and Sips models applied to the experimental results.

3.6. Kinetic Studies

Zn adsorption towards biomass-based carbon at room temperature is quite fast, as can be seen in Figure 6. Almost 80% removal can be achieved within the first 60 min. The adsorption equilibrium was attained at 240 min, and it remained constant thereafter. Maximum Zn removal and adsorption capacity were 95% and 14.4 mg/g, respectively. The fast removal of impurities is typical for carbonaceous adsorbents [34,44].

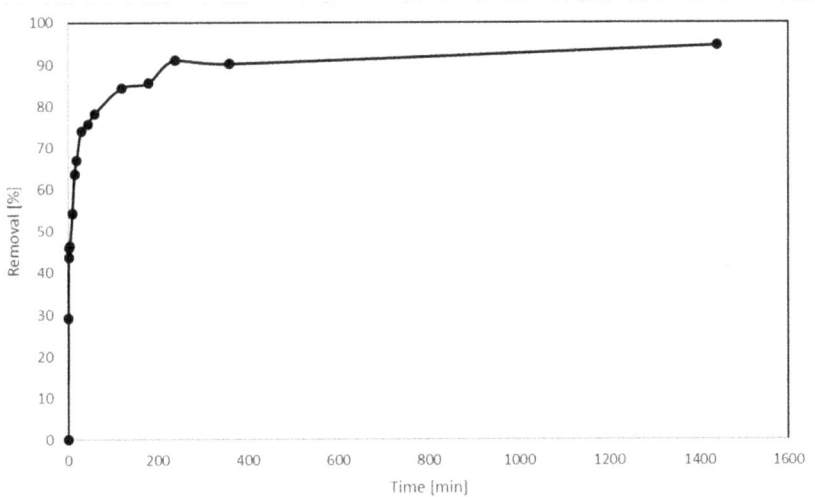

Figure 6. Zinc (Zn) removal [%] as a function of time (minutes). Experiments were done duplicated and the range of results were at maximum 3 percentage points.

Pseudo first-order, pseudo second-order, and Elovich kinetic models were applied to the experimental data. The obtained kinetic parameters are listed in Table 4, and fits are shown in

Figure 7. The best fitting model was the Elovich model, with a correlation coefficient of 0.97 and an RMSE of 0.68. However, the correlation of the pseudo second-order model was also rarely good as it typically is for carbonaceous adsorbents [34,44]. The R^2 value was 0.94, and the calculated equilibrium removal (13.01 mg/g) was quite similar to the experimental one (14.36 mg/g). It can be seen in Figure 7 that the sorption process of the Zn removal did not take place immediately; rather, equilibrium was reached after a few hundred minutes. This is typical of processes following the Elovich model.

Table 4. Kinetic parameters of pseudo-first-order, pseudo-second-order, and Elovich models.

Model	Parameter	Value
Experiments	$q_{e(exp)}$ [mg/g]	14.36
Pseudo-1-order	$q_{e(cal)}$ [mg/g]	12.15
	k_1 [min^{-1}]	0.22
	R^2	0.83
	RMSE	1.66
Pseudo-2-order	$q_{e(cal)}$ [mg/g]	13.01
	k_2 [g/ mg min]	0.023
	R^2	0.94
	RMSE	1.04
Elovich	β [g/mg]	0.69
	v_0 [mg/g min]	60.28
	R^2	0.97
	RMSE	0.68

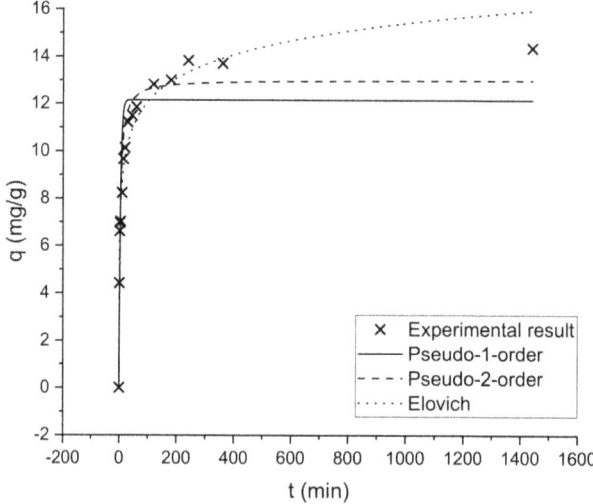

Figure 7. Pseudo-first-order, pseudo-second-order, and Elovich models applied to the experimental results.

3.7. Weber–Morris Model

The Weber–Morris intraparticle diffusion model was applied to the kinetic data of biomass-based carbon against the Zn^{2+}. If a plot of q_t versus $t^{1/2}$ presents a straight line from the origin, the adsorption mechanism only follows the intraparticle diffusion process. As can be seen in Figure 8, the data of Zn^{2+} sorption on biomass-based carbon shows four plots that do not pass through the origin. The first stages can be attributed to the instantaneous or external surface adsorption. The majority of Zn^{2+} ions are diffused through the solution to the external surface of the adsorbent in which the instantaneous sorption takes place. In the second stage, the adsorption capacity increases only slightly. This stage can be attributed to the slow diffusion of Zn^{2+} ions from the surface sites into inner pores. In the third stage, the adsorption rate stays almost constant due to the low Zn^{2+} ion concentration left in the solution [45,46].

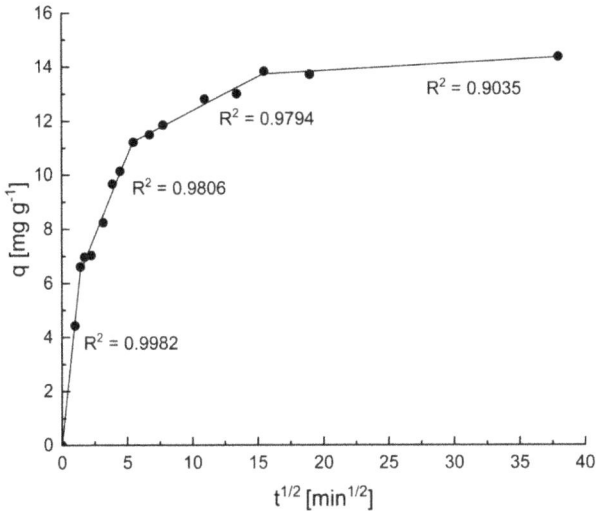

Figure 8. Weber–Morris intraparticle diffusion model.

3.8. Thermodynamic Effect of Temperature

The effect of temperature on adsorption was studied by performing the experiments at three different temperatures (10 °C, 22 °C, and 40 °C) to study how the adsorption system works in colder environments or for higher temperature process waters. The removal percentages at 10 °C, 22 °C, and 44 °C were 88%, 91%, and 92%, respectively. Therefore, temperature has no meaningful effect on Zn removal, and the produced material can be used effectively in a wide temperature range.

Experiments were performed at three different temperatures, 10 °C, 20 °C, and 40 °C, and the results were used to calculate the parameters of the Van't Hoff plot (paragraph 2.6, Equation (8)). ΔH^0 was calculated from the slope of the plot, and ΔS^0 was calculated from the intercept of the plot. In addition, Gibbs free energies ΔG^0 were calculated utilizing Equation (10). All the values are shown in Table 5, and the plot is presented in Figure 9.

The obtained ΔG^0 values are negative, indicating that the adsorption process is spontaneous. The positive value of ΔS^0 indicates that entropy is increasing in the process, and the positive ΔH^0 value, which is smaller than 40 kJ/mol, refers to endothermic physisorption [47,48].

Table 5. Thermodynamic parameters of Zinc (Zn) sorption on the biomass-based activated carbon.

Temperature (°C)	$\Delta G°$ (kJ/mol)	$\Delta S°$ (J/mol K)	$\Delta H°$ (kJ/mol)
10	−17.06		
23	−18.50	110.15	14.00
40	−20.42		

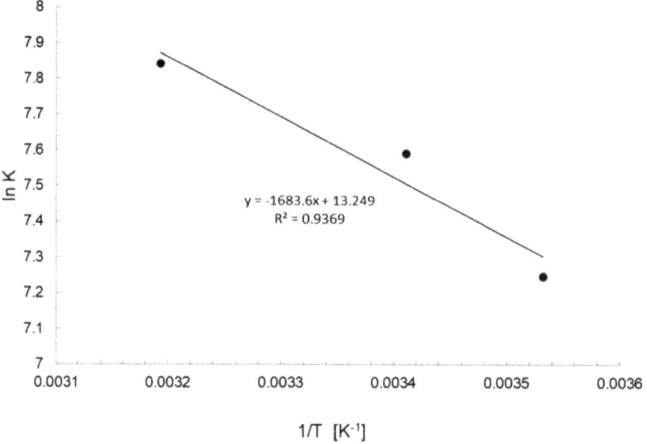

Figure 9. Van't Hoff plot of zinc (Zn) sorption on the biomass based activated carbon.

3.9. Desorption Experiments

After adsorption, the used adsorbents can be disposed of or regenerated; in both cases, there will be secondary pollution. Metal-loaded, used adsorbents are toxic for humans and the environment. In regeneration, metals are recovered in the solution form, and secondary pollution is caused by the regeneration solution [1]. Therefore, the regeneration solution used plays an important role, and the use of non-hazardous regeneration solutions is preferred. The regeneration solutions typically used are HCl [49,50], H_2SO_4 [36], 1 M KCl [35], 0.01 M $NaNO_3$ [51], and NaCl [52]. The regenerability of adsorbents is vital to enable cost-efficiency. In this study, Zn desorption was done using 0.1 M HCl; the results of the desorption experiments are presented in Figure 10. The desorption occurred immediately after the used adsorbent came into contact with the desorption solution. The desorption percentage was 80%, which was not related to the time. Therefore, it is possible to regenerate biomass-based carbon quickly using non-hazardous chemicals.

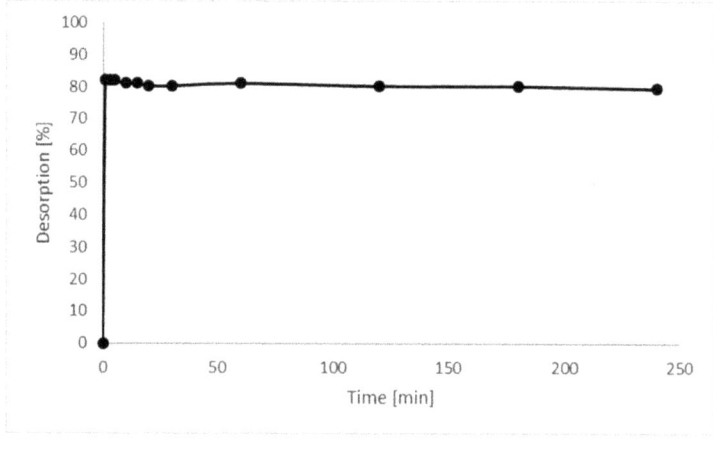

Figure 10. Desorption [%] as a function of time [min].

4. Conclusions

The adsorption of Zn(II) ions on biomass-based activated carbon was studied as a function of pH, solution metal–ion concentration, adsorbent dosage, time, and temperature. The amount (expressed as a %) of adsorbed Zn was the highest at pH 4. The removal percentage increased to 75 mg/L as a function of the initial Zn concentration, but the studied carbon adsorbent worked quite well at the 150 mg/L concentration level. The optimum adsorbent dosage was 3 g/L or 5 g/L. Zn adsorption onto the biomass-based activated carbon was not temperature-dependent. The adsorption isotherm followed the Sips isotherm and the Elovich kinetic model. Thermodynamic calculations showed that the adsorption process is spontaneous (negative ΔG^0), that entropy increases during adsorption (positive ΔS^0), and that the adsorption process consists of endothermic physisorption (ΔH^0 smaller than 40 kJ/mol). The adsorption capacity obtained for biomass-based activated carbon was compared favorably to other adsorbents used in Zn removal presented in literature. As a conclusion, the suitable adsorbent for Zn removal from waste biomass was developed by using non-hazardous materials. The adsorbent can be used with wide temperature range and with quite high concentration range.

Author Contributions: S.T., H.R. (Henrik Romar), D.B., and R.K. were responsible for the conceptualization of the paper; S.T., H.R. (Henrik Romar), D.B., and R.K. were responsible for the methodology; H.R. (Hanna Runtti) and T.K. were responsible for the software used; R.K. was responsible for the investigation; all authors were responsible for the resources; S.T., R.K., T.K., and H.R. (Hanna Runtti) were responsible for the data curation; S.T. was responsible for preparing and writing the original draft; all authors were responsible for writing, reviewing, and editing; R.K., T.K., and H.R. (Hanna Runtti) were responsible for visualization; H.R. (Hanna Runtti) was responsible for supervision; U.L. was responsible for project administration; and U.L. was responsible for funding acquisition.

Funding: This work was done under the auspices of the Waterpro (ERDF project number: A74635, funded by the Central Ostrobothnia Regional Council). The author thanks Maa-ja Vesitekniikan Tuki Ry for the financial support.

Conflicts of Interest: The authors declare no conflict of interest. The funders had no role in the design of the study; in the collection, analyses, or interpretation of data; in the writing of the manuscript, or in the decision to publish the results.

References

1. Tzou, Y.; Wang, S.; Hsu, L.; Chang, R.; Lin, C. Deintercalation of Li/Al LDH and its application to recover adsorbed chromate from used adsorbent. *Appl. Clay Sci.* **2007**, *37*, 107–114. [CrossRef]
2. O'Connell, D.W.; Birkinshaw, C.; O'Dwyer, T.F. Heavy metal adsorbents prepared from the modification of cellulose: A review. *Bioresour. Technol.* **2008**, *99*, 6709–6724. [CrossRef] [PubMed]
3. Nicholson, F.A.; Smith, S.R.; Alloway, B.J.; Carlton-Smith, C.; Chambers, B.J. An inventory of heavy metals inputs to agricultural soils in England and Wales. *Sci. Total Environ.* **2003**, *311*, 205–219. [CrossRef]
4. Rule, K.L.; Comber, S.D.W.; Ross, D.; Thornton, A.; Makropoulos, C.K.; Rautiu, R. Diffuse sources of heavy metals entering an urban wastewater catchment. *Chemosphere* **2006**, *63*, 64–72. [CrossRef]
5. Fernández, Y.; Marañón, E.; Castrillón, L.; Vázquez, I. Removal of Cd and Zn from inorganic industrial waste leachate by ion exchange. *J. Hazard. Mater.* **2005**, *126*, 169–175. [CrossRef]
6. Castelblanque, J.; Salimbeni, F. NF and RO membranes for the recovery and reuse of water and concentrated metallic salts from waste water produced in the electroplating process. *Desalination* **2004**, *167*, 65–73. [CrossRef]
7. Mansoorian, H.J.; Mahvi, A.H.; Jafari, A.J. Removal of lead and zinc from battery industry wastewater using electrocoagulation process: Influence of direct and alternating current by using iron and stainless steel rod electrodes. *Sep. Purif. Technol.* **2014**, *135*, 165–175. [CrossRef]
8. Kurniawan, T.A.; Chan, G.Y.S.; Lo, W.; Babel, S. Physico-chemical treatment techniques for wastewater laden with heavy metals. *Chem. Eng. J.* **2006**, *118*, 83–98. [CrossRef]
9. Laura, M. Plum, Lothar Rink and Hajo Haase the essential toxin: Impact of zinc on human health. *Int. J. Environ. Res. Public Health* **2010**, *7*, 1342–1365. [CrossRef]
10. Fu, F.; Wang, Q. Removal of heavy metal ions from wastewaters: A review. *J. Environ. Manag.* **2011**, *92*, 407–418. [CrossRef]
11. Demirbas, A. Heavy metal adsorption onto agro-based waste materials: A review. *J. Hazard. Mater.* **2008**, *157*, 220–229. [CrossRef] [PubMed]

12. Runtti, H.; Tuomikoski, S.; Kangas, T.; Lassi, U.; Kuokkanen, T.; Rämö, J. Chemically activated carbon residue from biomass gasification as a sorbent for iron(II), copper(II) and nickel(II) ions. *J. Water Process Eng.* **2014**, *4*, 12–24. [CrossRef]
13. Ahmadpour, A.; Do, D.D. The preparation of activated carbon from macadamia nutshell by chemical activation. *Carbon* **1997**, *35*, 1723–1732. [CrossRef]
14. Lussier, M.G.; Shull, J.C.; Miller, D.J. Activated carbon from cherry stones. *Carbon* **1994**, *32*, 1493–1498. [CrossRef]
15. Ricordel, S.; Taha, S.; Cisse, I.; Dorange, G. Heavy metals removal by adsorption onto peanut husks carbon: Characterization, kinetic study and modeling. *Sep. Purif. Technol.* **2001**, *24*, 389–401. [CrossRef]
16. Kumar, J.; Balomajumder, C.; Mondal, P. Application of agro-based biomasses for zinc removal from wastewater—A review. *Clean Soil Air Water* **2011**, *39*, 641–652. [CrossRef]
17. Mishra, V.; Majumder, C.B.; Agarwal, V.K. Sorption of Zn(II) ion onto the surface of activated carbon derived from eucalyptus bark saw dust from industrial wastewater: Isotherm, kinetics, mechanistic modeling, and thermodynamics. *Desalin. Water Treat.* **2012**, *46*, 332–351. [CrossRef]
18. Mohan, D.; Singh, K.P. Single- and multi-component adsorption of cadmium and zinc using activated carbon derived from bagasse—An agricultural waste. *Water Res.* **2002**, *36*, 2304–2318. [CrossRef]
19. Bergna, D.; Varila, T.; Romar, H.; Lassi, U. Comparison of the properties of activated carbons produced in one-stage and two-stage processes. *C* **2018**, *4*, 41. [CrossRef]
20. Azargohar, R.; Dalai, A.K. Production of activated carbon from Luscar char: Experimental and modeling studies. *Microporous Mesoporous Mater.* **2005**, *85*, 219–225. [CrossRef]
21. Marsh, H.; Rodríguez-Reinoso, F. (Eds.) CHAPTER 5—Activation processes (thermal or physical). In *Activated Carbon*; Elsevier Science Ltd.: Oxford, UK, 2006; pp. 243–321. [CrossRef]
22. Bansal, R.C.; Goyal, M. *Activated Carbon Adsorption*; CRC Press: Boca Raton, FL, USA, 2005. [CrossRef]
23. Williams, P.T.; Reed, A.R. Development of activated carbon pore structure via physical and chemical activation of biomass fibre waste. *Biomass Bioenergy* **2006**, *30*, 144–152. [CrossRef]
24. Brunauer, S.; Emmett, P.H.; Teller, E. Adsorption of gases in multimolecular layers. *J. Am. Chem. Soc.* **1938**, *60*, 309–319. [CrossRef]
25. Barrett, E.P.; Joyner, L.G.; Halenda, P.P. The determination of pore volume and area distributions in porous substances. I. computations from nitrogen isotherms. *J. Am. Chem. Soc.* **1951**, *73*, 373–380. [CrossRef]
26. Langmuir, I. The adsorption of gases on plane surfaces of glass, mica and platinum. *J. Am. Chem. Soc.* **1918**, *40*, 1361–1403. [CrossRef]
27. Freundlich, H. Over the adsorption in solution. *J. Phys. Chem.* **1906**, *57*, 385–471.
28. Sips, R. On the structure of a catalyst surface. *J. Chem. Phys.* **1948**, *16*, 490–495. [CrossRef]
29. Lagergren, S. About the theory of so-called adsorption of soluble substances. *K. Sven. Vetensk. Handl.* **1898**, *24*, 1–39.
30. Ho, Y.S.; McKay, G. Pseudo-second order model for sorption processes. *Process Biochem.* **1999**, *34*, 451–465. [CrossRef]
31. Zeldowitsch, J. Über den mechanismus der katalytischen oxydation von CO an MnO2 [About the mechanism of catalytic oxidation of CO over MnO2]. *Acta Physicochim URSS* **1934**, *1*, 364–449.
32. Weber, W.J.; Morris, J.C. Kinetics of adsorption on carbon from solutions. *J. Sanit. Eng. Div. Am. Soc. Civil Eng.* **1963**, *89*, 31–60.
33. Riikka, L.; Davide, B.; Henrik, R.; Tero, T.; Tao, H.; Ulla, L. Physico-chemical properties and use of waste biomass-derived activated carbons. *Chem. Eng. Trans.* **2017**, *57*, 43–48.
34. Kilpimaa, S.; Runtti, H.; Kangas, T.; Lassi, U.; Kuokkanen, T. Physical activation of carbon residue from biomass gasification: Novel sorbent for the removal of phosphates and nitrates from aqueous solution. *J. Ind. Eng. Chem.* **2015**, *21*, 1354–1364. [CrossRef]
35. Katsou, E.; Malamis, S.; Tzanoudaki, M.; Haralambous, K.J.; Loizidou, M. Regeneration of natural zeolite polluted by lead and zinc in wastewater treatment systems. *J. Hazard. Mater.* **2011**, *189*, 773–786. [CrossRef] [PubMed]
36. Motsi, T.; Rowson, N.A.; Simmons, M.J.H. Adsorption of heavy metals from acid mine drainage by natural zeolite. *Int. J. Miner. Process.* **2009**, *92*, 42–48. [CrossRef]
37. Covelo, E.F.; Vega, F.A.; Andrade, M.L. Competitive sorption and desorption of heavy metals by individual soil components. *J. Hazard. Mater.* **2007**, *140*, 308–315. [CrossRef]

38. Marshall, W.E.; Johns, M.M. Agricultural by-products as metal adsorbents: Sorption properties and resistance to mechanical abrasion. *J. Chem. Technol. Biotechnol.* **1996**, *66*, 192–198. [CrossRef]
39. Gharaibeh, S.H.; Abu-el-sha'r, W.Y.; Al-Kofahi, M.M. Removal of selected heavy metals from aqueous solutions using processed solid residue of olive mill products. *Water Res.* **1998**, *32*, 498–502. [CrossRef]
40. Ferro-García, M.A.; Rivera-Utrilla, J.; Rodríguez-Gordillo, J.; Bautista-Toledo, I. Adsorption of zinc, cadmium, and copper on activated carbons obtained from agricultural by-products. *Carbon* **1988**, *26*, 363–373. [CrossRef]
41. Budinova, T.K.; Petrov, N.V.; Minkova, V.N.; Gergova, K.M. Removal of metal ions from aqueous solution by activated carbons obtained from different raw materials. *J. Chem. Technol. Biotechnol.* **1994**, *60*, 177–182. [CrossRef]
42. Allen, S.J.; Brown, P.A. Isotherm analyses for single component and multi-component metal sorption onto lignite. *J. Chem. Technol. Biotechnol.* **1995**, *62*, 17–24. [CrossRef]
43. McKay, G.; Porter, J.F. Equilibrium parameters for the sorption of copper, cadmium and zinc ions onto peat. *J. Chem. Technol. Biotechnol.* **1997**, *69*, 309–320. [CrossRef]
44. Hanna, R.; Sari, T.; Teija, K.; Toivo, K.; Jaakko, R.; Ulla, L. Sulphate removal from water by carbon residue from biomass gasification: Effect of chemical modification methods on sulphate removal efficiency. *BioResources* **2016**, *11*, 3136–3152.
45. Caliskan, N.; Kul, A.R.; Alkan, S.; Sogut, E.G.; Alacabey, İ. Adsorption of Zinc(II) on diatomite and manganese-oxide-modified diatomite: A kinetic and equilibrium study. *J. Hazard. Mater.* **2011**, *193*, 27–36. [CrossRef] [PubMed]
46. Arias, F.; Sen, T.K. Removal of zinc metal ion (Zn2+) from its aqueous solution by kaolin clay mineral: A kinetic and equilibrium study. *Colloids Surf. A Physicochem. Eng. Asp.* **2009**, *348*, 100–108. [CrossRef]
47. Katal, R.; Baei, M.S.; Rahmati, H.T.; Esfandian, H. Kinetic, isotherm and thermodynamic study of nitrate adsorption from aqueous solution using modified rice husk. *J. Ind. Eng. Chem.* **2012**, *18*, 295–302. [CrossRef]
48. Bhatnagar, A.; Kumar, E.; Sillanpää, M. Nitrate removal from water by nano-alumina: Characterization and sorption studies. *Chem. Eng. J.* **2010**, *163*, 317–323. [CrossRef]
49. Iqbal, M.; Saeed, A.; Akhtar, N. Petiolar felt-sheath of palm: A new biosorbent for the removal of heavy metals from contaminated water. *Bioresour. Technol.* **2002**, *81*, 151–153. [CrossRef]
50. Saeed, A.; Akhter, M.W.; Iqbal, M. Removal and recovery of heavy metals from aqueous solution using papaya wood as a new biosorbent. *Sep. Purif. Technol.* **2005**, *45*, 25–31. [CrossRef]
51. Hu, J.; Shipley, H.J. Evaluation of desorption of Pb (II), Cu (II) and Zn (II) from titanium dioxide nanoparticles. *Sci. Total Environ.* **2012**, *431*, 209–220. [CrossRef]
52. Xu, W.; Li, L.Y.; Grace, J.R. Regeneration of natural Bear River clinoptilolite sorbents used to remove Zn from acid mine drainage in a slurry bubble column. *Appl. Clay Sci.* **2012**, *55*, 83–87. [CrossRef]

© 2019 by the authors. Licensee MDPI, Basel, Switzerland. This article is an open access article distributed under the terms and conditions of the Creative Commons Attribution (CC BY) license (http://creativecommons.org/licenses/by/4.0/).

Article

Evaluation of Biochar and Compost Mixes as Substitutes to a Commercial Propagation Mix

Lan Huang [1], Ping Yu [2] and Mengmeng Gu [3,*]

[1] Institute of Urban Agriculture, Chinese Academy of Agricultural Sciences, Chengdu 610000, China; huanglan_92@163.com
[2] Department of Horticultural Sciences, Texas A&M University, College Station, TX 77843, USA; yuping520@tamu.edu
[3] Department of Horticultural Sciences, Texas A&M AgriLife Extension Service, College Station, TX 77843, USA
* Correspondence: mgu@tamu.edu; Tel.: +1-979-845-8567

Received: 26 September 2019; Accepted: 14 October 2019; Published: 17 October 2019

Abstract: The effects of biochar (BC) on seed propagation depend on the type of BC, BC incorporation rate, base substrate, and plant seed species. Limited research tested BC-compost mixes for seed propagation. High percentages (70% or 80%, by volume) of BC with vermicompost (VC) or chicken manure compost (CM) were evaluated to substitute a commercial propagation mix (control) in three experiments. Seeds, including basil, coleus, edamame, marigold, okra, petunia, radish, salvia, tomato, vinca, and zinnia in Experiments 1 and 2 had similar or higher emergence percentages (EPs) and emergence indexes (EIs) in both BC:VC mixes, while celosia, cowpea, corn, and pumpkin had lower EPs or EIs in either 8BC:2VC or 7BC:3VC mixes compared to the control. Seedling fresh weights in both BC:VC mixes were similar to the control except for vinca, pumpkin, marigold, and salvia. The BC:VC mixes had no negative effects on plant dry weights at 7 weeks after transplanting. In Experiment 3, BC:CM mixes suppressed the seed germination or seedling growth of coleus, corn, cowpea, marigold, petunia, pumpkin, radish, salvia, vinca, watermelon, and zinnia due to high pH and CM's high electrical conductivity. Therefore, 7BC:3VC and 8BC:2VC can be used as seed propagation mix, while 7BC:3CM and 8BC:2CM are not recommended.

Keywords: biochar; germination; propagation mix; vermicompost; chicken manure compost

1. Introduction

Seed germination refers to the process, which starts with the seed imbibition of water and terminates with the emergence of radicle [1]. In order for the seeds to germinate, it is important for the seeds to be in touch with adequate propagation substrates to provide optimal conditions. Generally, commercial propagation mix includes different percentages of peat moss, vermiculite, and perlite to provide appropriate physical properties to hold water and provide air for seeds to grow. However, the harvest of peat moss has caused ecological concerns about damage to the peatlands and greenhouse gas emissions [2,3], and the cost of the commonly used germination mix components was continuously increasing [4,5]. All these factors have led to the necessity to find alternative propagation mix for seed germination.

Biochar (BC) is a carbonaceous material obtained from the pyrolysis of biomass, including plant-based materials, such as wood, grasses, or crop residues [6,7], and animal-based materials, such as crab shell [8] or manure [9–11]. Biochar has the potential to reduce greenhouse gas emissions [12], sequester environmental contaminants [13] and be used as a valuable substrate component for plant production [14–18]. Biochar is regarded as a sustainable product, which turns agriculture waste products into valuable materials. As the byproduct of the pyrolysis process, the use of BC in plant production adds extra value to the bioenergy process [19]. Research has shown the potential of BC to be used in seed propagation substrate for agricultural production [20]. It was shown that coconut shell BC (5%, 10%, 20%, or 40%, by volume) could be mixed with an adapted cornell seed germination mix and did not affect final germination percentages [21]. Additionally, the rate of BC incorporation, base substrate, and the plant species all have effects on the final seed germination and seedling growth. Prasad et al. [22] showed tomato (*Solanum lycopersicum*) seed germination and plant growth were higher in peat mixes with 10% woodchips BC than those in mixes with 50% BC. Similarly, Margenot et al. [23] evaluated the softwood BC substitution for peat in a 70:30 (v/v) peat:perlite mixture on marigold (*Tagetes erecta*) seed germination and seedling growth. They concluded that mixes with high BC substitution rate (50%, 60%, or 70%, by vol) had lower germinations than the mixes with low BC rates (0, 10%, 20%, 30%, or 40%, by vol) with all substrate pHs adjusted to 5.8. Compared to the 70:30 (v/v) peat:perlite mixture, mixes with 10% or 70% BC led to lower plant height while mixes with 20%, 30%, 40%, 50%, or 60% BC had no effect. Fan et al. [24] also pointed out that the lower germination rate and plant height of water spinach (*Ipomoea aquatica* Forsk) were found in substrates with more wheat straw BC, and that the seed germination rate and plant height in substrates with 16% (by volume) BC with 0.8 g L^{-1} super absorbent polymer were higher than the substrates without super absorbent polymer, indicating that both the BC incorporation rate and base substrate affect seed germination and seedling establishment. Gascó et al. [25] showed that five different seeds, including lentil (*Lens culinaris* Medikus cv. Pardina), cress (*Lepidium sativum*), cucumber (*Cucumis sativus* cv. Wisconsin SMR-58), tomato (cv. Alicante), and lettuce (*Lactuca sativa* cv. Great Lakes) seeds had different responses to three different BC mixes prepared from wood, paper sludge plus wheat husks, and sewage sludge. Therefore, the impact of substrates amended with BC on seed propagation and plant growth had been variable. Mixes with high percentages of BC tend to have less positive effects on seed germination and seeding growth compared to those with low percentages.

However, the hardwood BC was suitable to be used as container substrates to grow plants. Research has shown positive results for using a high rate (80%, by volume) of the hardwood BC with composts when growing tomato and basil (*Ocimun basilicum*) in containers [26,27]. It may be suitable as a propagation mix at a high rate, if incorporated with substrate components with fine textures, such as composts.

Composts are valuable alternative components for soilless substrates. Composting and vermicomposting are generally considered as environmentally-friendly and sustainable management processes to recycle plant or animal organic wastes [28]. Vermicompost (VC) is the end product of using earthworms to degrade organic wastes [29,30], which has fine particulate structures and a lot of nutrients [31,32]. Biochar mixes with VC may increase nutrient retention and have a positive impact on plant growth [7]. Chicken manure compost (CM) is also fine-textured and is cheaper than VC. Properly treated CM could have nutrients readily available to plants [33,34].

Most research testing the effect of BC in germination substrates usually choses to use BC to substitute certain percentages of the commercial propagation mix. There is limited research testing propagation substrate with BC and composts instead of commercial propagation mix. More research is needed to find mixes with BC, especially at high percentages, on seed germination with comparable results to commercial germination mix.

Due to previous success of using the mixed hardwood BC in container production trials, the objective of this study was to evaluate the mixes of high percentages (70% or 80%, by volume) of the mixed hardwood BC with the rest being composts (VC or CM) on seed germination and seedling growth, compared to a commercial seed propagation mix (the control).

2. Materials and Methods

2.1. Germination Substrate Selection and Characterization

The BC used in this study was a byproduct of the fast pyrolysis of mixed hardwood (Proton Power, Inc., Lenior City, TN, USA). The nutrient analysis (N, P, K, Ca, Mg, S, B, Ca, Cu, Fe, Mn, Na, and Zn) of the BC was tested by the Texas A&M AgriLife Extension Service Soil, Water, and Forage Testing Laboratory in College Station, TX, USA and is shown in Table 1. The total N was determined spectrophotometrically by a combustion process [35] and the minerals (including P, K, Ca, Mg, S, B, Ca, Cu, Fe, Mn, Na, and Zn) were determined by inductively coupled plasma analysis of a nitric acid digest [36]. Two composts, including VC (Pachamama Earthworm Castings; Lady Bug Brand, Conroe, TX, USA) and CM (Back to Nature, Inc., Slaton, TX, USA), were chosen to mix with the BC. A commercial propagation mix (Propagation Mix; Sun Gro Inc., Agawam, MA, USA) was used as the control to germinate seeds. The pHs of the BC, VC, and CM were measured using a handheld pH-EC meter (HI 98129, Hanna Instruments, Woonsocket, RI, USA) and electrical conductivity (EC) was measured using the Bluelab Combo Meter (Bluelab Corporation Limited, Tauranga, New Zealand) according to the pour-through extraction method using the same amount of leachate for each test [37]. The pHs of the BC, VC, and CM were 11.18, 4.8, and 7.5, respectively. The ECs of the BC, VC, and CM were 2.0, 6.7, and 32.9 dS m^{-1}, respectively. The physical properties of the BC, VC, and CM were reported previously by Huang et al. [26]. Physical properties, including the total porosity (the percent of pores in a substrate), container capacity (the maximum percent volume a substrate can hold water), air space (the percent of pores filled with air after drainage), and bulk density (the ratio of dry weight to the volume of the substrate) were determined using the North Carolina State University Horticultural Substrates Laboratory porometers [38]. Six replications were tested to determine the physical properties. The total porosities of the BC, VC, and CM were 84.7%, 75.0%, and 64.4%, respectively. The container capacities of the BC, VC, and CM were 60.3%, 72.2%, and 60.0%, respectively. The air spaces of the BC, VC, and CM were 24.4%, 2.8%, and 4.4%, respectively. The bulk densities of the BC, VC, and CM were 0.15, 0.38, and 0.62 g cm^{-3}, respectively.

Table 1. Nutrient analysis of the biochar used in the experiments.

N	P	K	Ca	Mg	S	Fe	B	Cu	Mn	Na	Zn
(%)						(mg kg^{-1})					
0.23	456	6362	27,507	1299	231	2039	15	9	905	107	13

2.2. Experimental Design and Setup

Three experiments were conducted to test the effect of mixes with the BC-compost (VC or CM) on seed germination and seedling growth. Each treatment in the three experiments was replicated 4 times and arranged in a completely randomized design in the greenhouse located on Texas A&M University campus, College Station, TX, USA. Six seeds were sown for each replication with one seed per cell (hexagon with side length of 2.6 cm; height: 4.2 cm; volume: 20 cm^3). The average greenhouse temperature, relative humidity, and dew point for Experiment 1 were approximately 22.4 °C, 46.7%, and 8.7 °C, respectively. The average greenhouse temperature, relative humidity, and dew point for Experiments 2 and 3 were approximately 27.1 °C, 92.4%, and 25.6 °C, respectively.

2.2.1. Experiment 1

Both Experiment 1 and 2 evaluated the potential of mixes of the mixed hardwood BC (70% or 80% by volume) with the rest being VC to substitute the commercial seed propagation mix (the control) on different seeds but were conducted at different times. In Experiment 1, seven types of flower and vegetable seeds, from Morgan County Seeds (Barnett, MO, USA; tomato *Solanum lycopersicum*, okra *Abelmoschus esculentus*), Ball Horticultural Company (West Chicago, IL, USA; celosia *Celosia plumosa*, vinca *Catharanthus roseus* 'Vitalia Pink,' marigold *Tagetes patula*), Plantation Products LLC. (Norton, MA, USA; zinnia *Zinnia elegans*), and Johnny's Selected Seeds (Winslow, ME, USA; basil *Ocimum basilicum*), were sown on 14 October 2016 and observed for 2 weeks. Four types of seeds from Botanical Interests (Broomfield, CO, USA), including radish (*Raphanus sativus*) and three relatively bigger seeds (cowpea

Vigna unguiculata, corn *Zea mays* var. rugosa, pumpkin *Cucurbita pepo*) were sown on 21 October 2016 and observed for 10 days.

2.2.2. Experiment 2

Experiment 2 was conducted with the same treatments in Experiment 1: 7BC:3VC and 8BC:2VC (by volume) to compare with the commercial propagation mix (the control). During germination test, six types of flower seeds, from Floranova Ltd. (Pierceton, IN, USA; salvia *Salvia splendens* 'Sizzler Red,' petunia Petunia 'Espresso Sweet Pink'), Ball Horticultural Company (West Chicago, IL, USA; marigold *Tagetes patula*, vinca *Catharanthus roseus* 'Vitalia Pink'), and PanAmerican Seed Co. (West Chicago, IL, USA; zinnia *Zinnia angustifolia*, coleus *Solenostemon scutellarioides* 'Wizard Golden'), were sown on 17 May 2017 and observed for 20 days. Five types of seeds, from Producers Cooperative Association (Bryan, TX, USA; corn *Zea mays* 'Bodacious') and Botanical Interests (Broomfield, CO, USA; cowpea *Vigna unguiculata*, edamame *Glycine max* 'Butterbean,' pumpkin *Cucurbita pepo*, radish *Raphanus sativus*), were sown on 17 May 2017 and observed for 16 days.

2.2.3. Experiment 3

The BC was mixed with CM at the ratio of 7BC:3CM and 8BC:2CM (by volume) to compare with the commercial propagation mix (the control). Experiment 3 was conducted on 22 April 2017. During the germination test, twelve types of seeds, from Producers Cooperative Association (Bryan, TX, USA; watermelon *Citrullus lanatus* 'Black Diamond,' corn *Zea mays* 'Bodacious'), Botanical Interests (Broomfield, CO, USA; pumpkin *Cucurbita pepo*, edamame *Glycine max* 'Butterbean,' cowpea *Vigna unguiculata*, radish *Raphanus sativus*), Floranova Ltd. (Pierceton, IN, USA; salvia *Salvia splendens* 'Sizzler Red,' petunia *Petunia* x *hybrida* F1 'Espresso Sweet Pink'), Ball Horticultural Company (West Chicago, IL, USA; marigold *Tagetes patula*, vinca *Catharanthus roseus* 'Vitalia Pink'), and PanAmerican Seed Co. (West Chicago, IL, USA; zinnia *Zinnia angustifolia*, coleus *Solenostemon scutellarioides* 'Wizard Golden'), were sown and observed for 20 days.

2.3. Measurements

For Experiment 1, 2, and 3, the number of seed germinations was monitored every day. Seeds were counted as germinated when two cotyledons were visible. The emergence percentage (EP) and emergence index (EI) were calculated to reflect the overall germination rate and germination speed, respectively.

Emergence percentage was calculated using the following formula:

$$EP = \left(\frac{\text{No. of emerged seedlings}}{\text{Total No. of seeds}}\right) \times 100\%$$

Emergence index was calculated as:

$$EI = \sum_{i=1}^{n} \left(\frac{EP_i}{T_i}\right)$$

where EP_i is EP on day i, and T_i is the number of days after sowing.

For Experiments 1 and 3, at the end of the germination test, the seedlings in each replication were harvested from the medium surface to measure the mean fresh weight (FW) of each seedling in each replication. For Experiment 2, after the germination test, one randomly-selected seedling from each replication was transplanted to a commercial growing mix (Professional growing mix; Sun Gro Inc., Agawam, MA, USA) in a 6-inch pot (depth: 10.8 cm; top diameter: 15.5 cm; bottom diameter: 11.3 cm; volume: 1330 cm^3) and randomly arranged in the greenhouse. The other remaining seedlings in each replication were harvested from the medium surface and the mean FW of each seedling in each replication measured. The transplanted plants were harvested after 7 weeks to test the effect of the mixes with the BC and VC on seedling growth. The dry weight (DW) of plants was measured after oven-dried at 80 °C until constant weight.

2.4. Statistical Analysis

Data were analyzed with one-way analysis of variance (ANOVA) using JMP Statistical Software (version Pro 12.2.0; SAS Institute, Cary, NC, USA) to test the effect of different germination mix on seed germination and seedling growth. The type of germination mix was the main factor. Treatments with the BC and composts (either VC or CM) were compared to the control (the commercial propagation mix) using Dunnett's test and means were separated when treatments were significant at $p < 0.05$.

3. Results and Discussion

3.1. The Effect of the Biochar and Compost Mix on Seed Germination Stage

In Experiment 1, it was observed that all seeds in the BC and VC mixes had similar EPs to the control's except for cowpea and corn in the mix with 70% BC and 30% VC (by volume; 7BC:3VC) and pumpkin in both BC:VC mixes, which were lower than the control (Figure 1a). Celosia in 8BC:2VC and pumpkin in both BC:VC mixes had lower EIs than the control's, with the other seeds having better or equal EIs (Figure 1b). In Experiment 2, it was observed that all seeds had similar EPs and EIs in both BC:VC mixes when compared to the control's (Figure 2). This indicated that all the seeds in both BC:VC mixes had similar germination rates and speeds compared to the ones in the commercial propagation mix. Most of the results were similar in Experiments 1 and 2 except for the EP of cowpea, corn, and pumpkin and the EI of pumpkin. The different cultivars of the cowpea seeds could have led to different EPs. For corn and pumpkin, the different result could be due to the different batches of the seeds, temperatures of the days when conducting Experiments 1 and 2, and different amounts of water applied for germination.

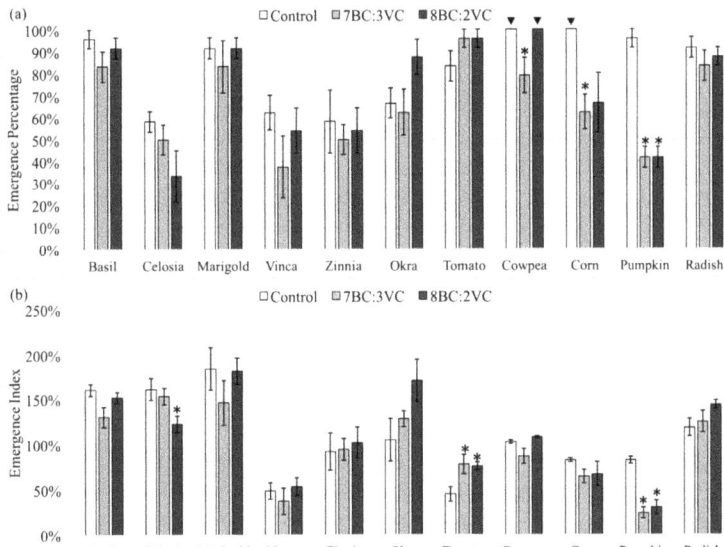

Figure 1. Seeds' emergence percentages (**a**) and emergence indexes (**b**) (mean ± standard error) in commercial propagation mix (control), 70% biochar:30% vermicompost (7BC:3VC, by volume), or 8BC:2VC in Experiment 1. The asterisks (*) indicate differences from the control using Dunnett's test at $p < 0.05$ ($n = 4$). Means indicated by "▼" are 100% ± 0.

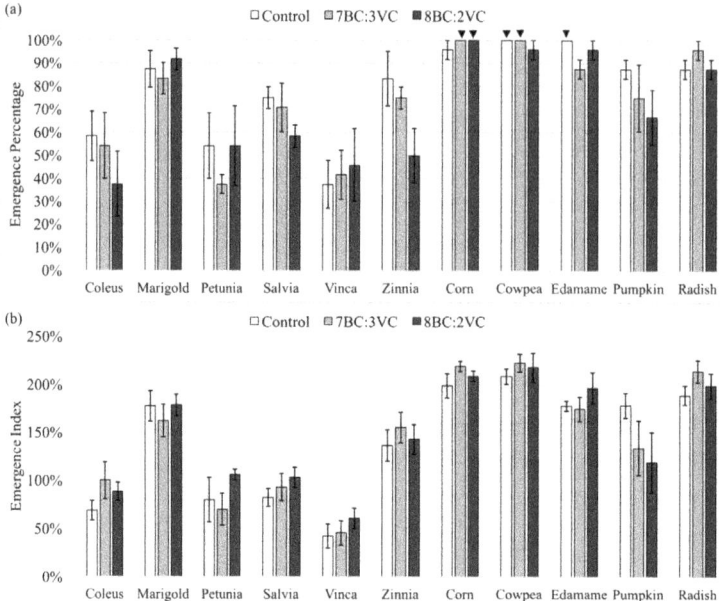

Figure 2. Seeds' emergence percentages (**a**) and emergence indexes (**b**) (mean ± standard error) in commercial propagation mix (control), 70% biochar:30% vermicompost (7BC:3VC, by volume), or 8BC:2VC in Experiment 2. Means indicated by "▼" are 100% ± 0.

The results from Experiments 1 and 2 demonstrate that mixes of 70% or 80% (by volume) BC with the rest being VC have the potential to be substituted into the commercial seed propagation mixes for basil, coleus, edamame, marigold, okra, petunia, radish, salvia, tomato, vinca, and zinnia seeds with similar or higher EPs and EIs compared to the control's. Similar to our results, Nieto et al. [39] tested and indicated that the germination rate of the lettuce seeds in mixes with 75% BC (from pruning waste at 300 or 500 °C) and peat were similar to the 100% peat control. Margenot et al. [23] also showed that the substrate with 70% (by volume) softwood BC and 30% perlite without pH adjustment had no negative effect on the germination rate of marigold seeds compared to the 70:30 peat:perlite mixture. Webber III et al. [20] concluded that the percentage of seedling establishment for green beans (*Phaseolus vulgaris* var. 'Bowie') in mixes of 75% sugarcane bagasse BC or pine BC, with the rest being commercial growing media, was similar to that of the control. The reason for the lower EPs and EIs for cowpea, corn, and pumpkin seeds in Experiment 1 could be that these relatively bigger seeds do not have close contact with relatively big BC particles in the mixes. The uptake of the water, available from substrate particles in the case of container substrate, is required for the seeds to germinate [40].

In Experiment 3, all the seeds in mixes with the BC and CM had lower EPs than the ones in the control except for the edamame and watermelon seeds in 8BC:2CM, which were similar to the control (Figure 3). Similarly, the EIs of all the seeds in BC:CM mixes were lower than the ones in the control except for the edamame seeds in 8BC:2CM and corn in both BC:CM mixes. Therefore, all the seeds in either 7BC:3CM or 8BC:2CM mixes had lower EPs or EIs than those of the control, except for EI of corn. The reason for the suppressed seed germination could be the high pH of the germination mix (pH of the BC: 11.18; pH of the CM: 7.5) and/or high EC (32.9 dS m^{-1}) of the CM, which led to salinity stress for the seeds. A similar result has been found in other research, which showed slow and poor germination of the seeds induced by a pig slurry compost due to the detrimental effect of high salinity [41]. Due to the addition of a high percentage of the mixed hardwood BC, the nutrient retention of the CM could be improved due to BC's absorption ability and porous structure [42]. High pH or high salt stresses would suppress seed germination since the radicles are sensitive, with more deleterious effects under combined salt-alkaline stress [43]. Edamame and watermelon seeds may have higher tolerance, compared to the other seeds.

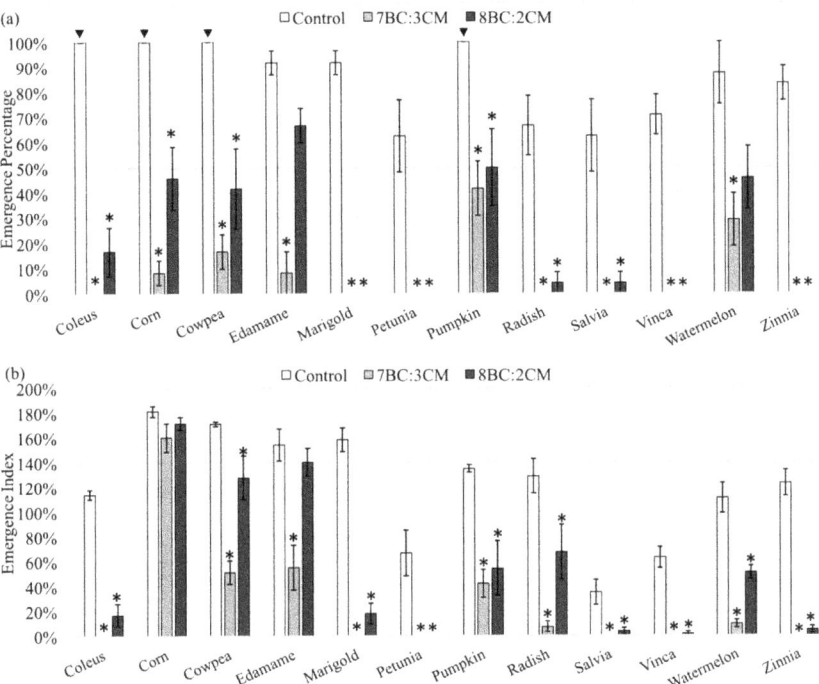

Figure 3. Seeds' emergence percentages (**a**) and emergence indexes (**b**) (mean ± standard error) in commercial propagation mix (control), 70% biochar:30% chicken manure compost (7BC:3CM, by volume), or 8BC:2CM in Experiment 3. The asterisks (*) indicate differences from the control using Dunnett's test at $p < 0.05$ ($n = 4$). Means indicated by "▼" are 100% ± 0.

3.2. The Effects of the Biochar and Compost Mix on Seedling Growth

In Experiment 1, the FWs of all the seedlings in the BC:VC mixes were similar to the ones in the control except for the vinca and pumpkin seedlings in 7BC:3VC (Figure 4). Similarly, all harvested seedling FWs in BC:VC mixes in Experiment 2 were similar to control, except those of marigold and salvia in 8BC:2VC, which were lower than that of the control (Figure 5a). All plant DWs at 7 weeks after transplanting were similar to or higher than those of the controls (Figure 5b). Research has shown that the FWs and DWs of the squash (*Cucurbita pepo* var. 'Enterprise') seedlings in mixes of 75% (by volume) standard bagasse BC with the rest being commercial growing mix were similar to the ones in the control while those of the cantaloupe (*Cucumis melo* var. 'Magnum 45') seedlings were lower than the ones in the control [44]. Similarly, the FWs and DWs of green bean var. 'Bowie' seedlings in mixes of 75% (by volume) sugarcane bagasse BC or pine BC with the rest being commercial growing media were similar to the control [20]. Sáez et al. [41] also showed that mixes of 60% or 80% (by volume) holm oak BC (with the rest being pig slurry compost) had similar or higher shoot DWs compared to the mixes with 60% or 80% coir (the resting being pig slurry compost), respectively.

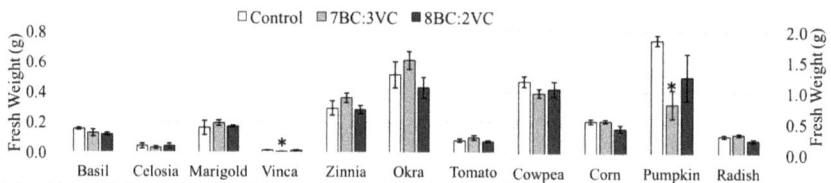

Figure 4. Fresh weight (mean ± standard error) of the seedlings in commercial propagation mix (control), 70% biochar:30% vermicompost (7BC:3VC, by volume), or 8BC:2VC in Experiment 1. The asterisks indicate differences from the control using Dunnett's test at $p < 0.05$ ($n = 4$).

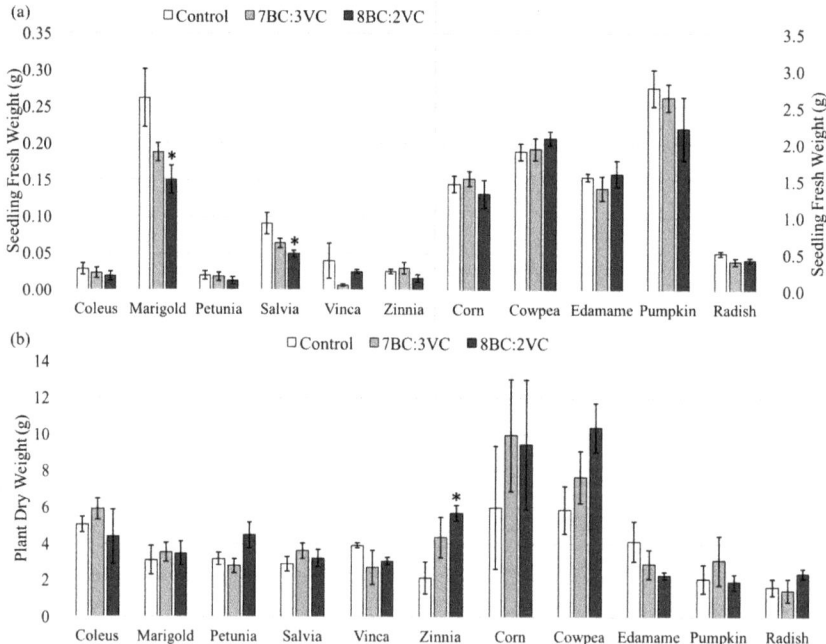

Figure 5. Fresh weights of the seedlings (**a**) and the dry weights of the plants harvested at 7 weeks after transplantation (**b**) (mean ± standard error), which were germinated in commercial propagation mix (control), 70% biochar:30% vermicompost (7BC:3VC, by volume), or 8BC:2VC in Experiment 2. The asterisks indicate differences from the control using Dunnett's test at $p < 0.05$ ($n = 4$).

In Experiment 3, similar to the results on EPs and EIs, the FWs of all the seedlings in the BC:CM mixes were lower than those in the control except for the edamame seedling in 8BC:2CM and cowpea seedlings in both BC:CM mixes (Figure 6). The salinity and alkaline stresses caused by BC:CM mixes not only caused low and slow seed germination but also suppressed plant growth. The same CM and mixed hardwood BC has been used in the other research and the results also showed that the growth index, shoot DW and FW, and root and total DW of basil plants in mixes of the BC (80% or 90%, by volume) and 5% CM with the rest being commercial growing mix were reduced due to high salt and high pH [26].

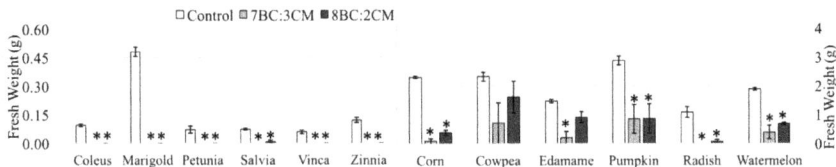

Figure 6. Fresh weights (mean ± standard error) of the seedlings in commercial propagation mix (control), 70% biochar:30% chicken manure compost (7BC:3CM, by volume), or 8BC:2CM in Experiment 3. The asterisks indicate differences from the control using Dunnett's test at $p < 0.05$ ($n = 4$).

4. Conclusions

The results have shown that the high-percentage (70% or 80%, by volume) BC mixes, with the rest being VC, have the potential to be used as germination mixes to substitute the commercial peat-based seed propagation mix, while mixes with 70% or 80% (by volume) BC, with the rest being CM, are not recommended for seed germination or plant production. Basil, marigold, vinca, zinnia, okra, tomato, radish, coleus, petunia, salvia, and edamame seeds had similar or higher EPs and EIs in both BC:VC mixes compared to the control. The FWs of all the seedlings in the BC:VC mixes were similar to the ones in the control except for the vinca and pumpkin in 7BC:3VC in Experiment 1 and marigold and salvia in 8BC:2VC in Experiment 2. The DWs of all plants grown in both BC:VC mixes at 7 weeks after transplantation were similar to or higher than the control in Experiment 2. Due to the high pHs of the BC:CM mixes and high salinity of CM, all the seeds in either 7BC:3CM or 8BC:2CM mixes had lower EPs or EIs than those of the controls, except for the EI of corn. The FWs of all the seedlings in the BC:CM mixes were lower than those in the control except for the edamame in 8BC:2CM and cowpea in both BC:CM mixes. Therefore, 7BC:3VC and 8BC:2VC could be used as seed propagation mixes, while 7BC:3CM or 8BC:2CM are not recommended.

These results would be only suitable for these specific BCs, VCs, and CMs, since different materials have different physical and chemical properties. More plant materials should be tested for the broader use of the BC:VC mixes for seed germination, or a small trial needs to be conducted before using such mixes in production.

Author Contributions: This review is a product of the combined effort of all the authors. M.G. conceived the experiments; L.H. conducted the experiment, analyzed the data, wrote the original draft, and improved it based on other authors' advice, mainly M.G.'s.; M.G. reviewed, edited, and revised the manuscript; P.Y. conducted some measurements of the parameters and reviewed the manuscript.

Funding: This research received no external funding.

Acknowledgments: The authors thank the Agriculture Women Excited to Share Opinions, Mentoring, and Experiences (AWESOME) faculty group of the College of Agriculture and Life Sciences at Texas A&M University for assistance with editing the manuscript.

Conflicts of Interest: The authors declare no conflict of interest.

References

1. Bewley, J.D.; Bradford, K.; Hilhorst, H. *Seeds: Physiology of Development, Germination and Dormancy*, 3rd ed.; Springer: New York, NY, USA, 2012; p. 133.
2. Carlile, B.; Coules, A. Towards sustainability in growing media. *Acta Hortic.* **2013**, *1013*, 341–349. [CrossRef]
3. Leifeld, J.; Menichetti, L. The underappreciated potential of peatlands in global climate change mitigation strategies. *Nat. Commun.* **2018**, *9*, 1071. [CrossRef] [PubMed]
4. Landis, T.D.; Morgan, N. Growing Media Alternatives for Forest and Native Plant Nurseries. In *National Proceedings: Forest and Conservation Nursery Associations, Missoula, MT, USA, 23–25 June 2008*; Dumroese, R.K., Riley, L.E., Eds.; USDA Forest Service, Rocky Mountain Research Station: Fort Collins, CO, USA, 2009; pp. 26–31.

5. Jackson, B.E.; Wright, R.D.; Barnes, M.C. Pine tree substrate, nitrogen rate, particle size, and peat amendment affect poinsettia growth and substrate physical properties. *HortScience* **2008**, *43*, 2155–2161. [CrossRef]
6. Lehmann, J. A handful of carbon. *Nature* **2007**, *447*, 143–144. [CrossRef]
7. Liu, R.; Gu, M.; Huang, L.; Yu, F.; Jung, S.K.; Choi, H.S. Effect of pine wood biochar mixed with two types of compost on growth of bell pepper (*Capsicum annuum* L.). *Hortic. Environ. Biotechnol.* **2019**, *60*, 313–319. [CrossRef]
8. Cho, M.S.; Meng, L.; Song, J.H.; Han, S.H.; Bae, K.; Park, B.B. The effects of biochars on the growth of *Zelkova serrata* seedlings in a containerized seedling production system. *Sci. Technol.* **2017**, *13*, 25–30. [CrossRef]
9. Uzoma, K.; Inoue, M.; Andry, H.; Fujimaki, H.; Zahoor, A.; Nishihara, E. Effect of cow manure biochar on maize productivity under sandy soil condition. *Soil Use Manag.* **2011**, *27*, 205–212. [CrossRef]
10. Jin, Y.; Liang, X.; He, M.; Liu, Y.; Tian, G.; Shi, J. Manure biochar influence upon soil properties, phosphorus distribution and phosphatase activities: A microcosm incubation study. *Chemosphere* **2016**, *142*, 128–135. [CrossRef]
11. Lei, O.; Zhang, R. Effects of biochars derived from different feedstocks and pyrolysis temperatures on soil physical and hydraulic properties. *J. Soils Sediments* **2013**, *13*, 1561–1572. [CrossRef]
12. Woolf, D.; Amonette, J.E.; Street-Perrott, F.A.; Lehmann, J.; Joseph, S. Sustainable biochar to mitigate global climate change. *Nat. Commun.* **2010**, *1*, 56. [CrossRef]
13. Ippolito, J.A.; Laird, D.A.; Busscher, W.J. Environmental benefits of biochar. *J. Environ. Q.* **2012**, *41*, 967–972. [CrossRef] [PubMed]
14. Guo, Y.; Niu, G.; Starman, T.; Volder, A.; Gu, M. Poinsettia growth and development response to container root substrate with biochar. *Horticulturae* **2018**, *4*, 1. [CrossRef]
15. Guo, Y.; Niu, G.; Starman, T.; Gu, M. Growth and development of Easter lily in response to container substrate with biochar. *J. Hortic. Sci. Biotechnol.* **2018**, *94*, 80–86. [CrossRef]
16. Peng, D.; Gu, M.; Zhao, Y.; Yu, F.; Choi, H.S. Effects of biochar mixes with peat-moss based substrates on growth and development of horticultural crops. *Hortic. Science Technol.* **2018**, *36*, 501–512.
17. Choi, H.S.; Zhao, Y.; Dou, H.; Cai, X.; Gu, M.; Yu, F. Effects of biochar mixtures with pine-bark based substrates on growth and development of horticultural crops. *Horti. Environ. Biotechnol.* **2018**, *59*, 345–354. [CrossRef]
18. Huang, L.; Gu, M. Effects of biochar on container substrate properties and growth of plants—A review. *Horticulturae* **2019**, *5*, 14. [CrossRef]
19. Laird, D.A. The charcoal vision: A win–win–win scenario for simultaneously producing bioenergy, permanently sequestering carbon, while improving soil and water quality. *Agron. J.* **2008**, *100*, 178–181. [CrossRef]
20. Webber, C.L., III; White, P.M., Jr.; Gu, M.; Spaunhorst, D.J.; Lima, I.M.; Petrie, E.C. Sugarcane and pine biochar as amendments for greenhouse growing media for the production of bean (*Phaseolus vulgaris* L.) seedlings. *J. Agric. Sci.* **2018**, *10*, 58–68. [CrossRef]
21. Hoover, B.K. Herbaceous perennial seed germination and seedling growth in biochar-amended propagation substrates. *HortScience* **2018**, *53*, 236–241. [CrossRef]
22. Prasad, M.; Tzortzakis, N.; McDaniel, N. Chemical characterization of biochar and assessment of the nutrient dynamics by means of preliminary plant growth tests. *J. Environ. Manag.* **2018**, *216*, 89–95. [CrossRef]
23. Margenot, A.J.; Griffin, D.E.; Alves, B.S.; Rippner, D.A.; Li, C.; Parikh, S.J. Substitution of peat moss with softwood biochar for soil-free marigold growth. *Ind. Crop. Prod.* **2018**, *112*, 160–169. [CrossRef]
24. Fan, R.; Luo, J.; Yan, S.; Zhou, Y.; Zhang, Z. Effects of biochar and super absorbent polymer on substrate properties and water spinach growth. *Pedosphere* **2015**, *25*, 737–748. [CrossRef]
25. Gascó, G.; Cely, P.; Paz-Ferreiro, J.; Plaza, C.; Méndez, A. Relation between biochar properties and effects on seed germination and plant development. *Biol. Agric. Hortic.* **2016**, *32*, 237–247. [CrossRef]
26. Huang, L.; Niu, G.; Feagley, S.E.; Gu, M. Evaluation of a hardwood biochar and two composts mixes as replacements for a peat-based commercial substrate. *Ind. Crop. Prod.* **2019**, *129*, 549–560. [CrossRef]
27. Huang, L. Effects of Biochar and Composts on Substrates Properties and Container-Grown Basil (*Ocimum basilicum*) and Tomato (*Solanum lycopersicum*). Master's Thesis, Texas A&M University, College Station, TX, USA, 2018.

28. Lim, S.L.; Lee, L.H.; Wu, T.Y. Sustainability of using composting and vermicomposting technologies for organic solid waste biotransformation: Recent overview, greenhouse gases emissions and economic analysis. *J. Clean. Prod.* **2016**, *111*, 262–278. [CrossRef]
29. Chan, P.L.; Griffiths, D. The vermicomposting of pre-treated pig manure. *Biol. Wastes* **1988**, *24*, 57–69. [CrossRef]
30. Manna, M.; Jha, S.; Ghosh, P.; Ganguly, T.; Singh, K.; Takkar, P. Capacity of various food materials to support growth and reproduction of epigeic earthworms on vermicompost. *J. Sustain. For.* **2005**, *20*, 1–15. [CrossRef]
31. Atiyeh, R.; Subler, S.; Edwards, C.; Bachman, G.; Metzger, J.; Shuster, W. Effects of vermicomposts and composts on plant growth in horticultural container media and soil. *Pedobiologia* **2000**, *44*, 579–590. [CrossRef]
32. Sinha, R.K.; Agarwal, S.; Chauhan, K.; Valani, D. The wonders of earthworms & its vermicompost in farm production: Charles darwin's friends of farmers, with potential to replace destructive chemical fertilizers. *Agric. Sci.* **2010**, *1*, 76.
33. Urra, J.; Alkorta, I.; Lanzén, A.; Mijangos, I.; Garbisu, C. The application of fresh and composted horse and chicken manure affects soil quality, microbial composition and antibiotic resistance. *Appl. Soil Ecol.* **2019**, *135*, 73–84. [CrossRef]
34. Chen, J.; Shi, L.; Liu, H.; Yang, J.; Guo, R.; Chen, D.; Jiang, X.; Liu, Y. Screening of fermentation starters for organic fertilizer composted from chicken manure. *Asian Agric. Res.* **2017**, *9*, 92–98.
35. Parkinson, J.; Allen, S. A wet oxidation procedure suitable for the determination of nitrogen and mineral nutrients in biological material. *Commun. Soil Sci. Plant. Anal.* **1975**, *6*, 1–11. [CrossRef]
36. Havlin, J.L.; Soltanpour, P. A nitric acid plant tissue digest method for use with inductively coupled plasma spectrometry. *Commun. Soil Sci. Plant. Anal.* **1980**, *11*, 969–980. [CrossRef]
37. LeBude, A.; Bilderback, T. *The Pour-Through Extraction Procedure: A Nutrient Management Tool for Nursery Crops*; AG-717-W: 2009; North Carolina Cooperative Extension: Raleigh, NC, USA, 2009.
38. Fonteno, W.; Hardin, C.; Brewster, J. *Procedures for Determining Physical Properties of Horticultural Substrates Using the NCSU Porometer*; Horticultural Substrates Laboratory, North Carolina State University: Raleigh, NC, USA, 1995.
39. Nieto, A.; Gascó, G.; Paz-Ferreiro, J.; Fernández, J.; Plaza, C.; Méndez, A. The effect of pruning waste and biochar addition on brown peat based growing media properties. *Sci. Hortic.* **2016**, *199*, 142–148. [CrossRef]
40. Manz, B.; Müller, K.; Kucera, B.; Volke, F.; Leubner-Metzger, G. Water uptake and distribution in germinating tobacco seeds investigated in vivo by nuclear magnetic resonance imaging. *Plant. Physiol.* **2005**, *138*, 1538–1551. [CrossRef] [PubMed]
41. Sáez, J.; Belda, R.; Bernal, M.; Fornes, F. Biochar improves agro-environmental aspects of pig slurry compost as a substrate for crops with energy and remediation uses. *Ind. Crop. Prod.* **2016**, *94*, 97–106. [CrossRef]
42. Laird, D.; Fleming, P.; Wang, B.; Horton, R.; Karlen, D. Biochar impact on nutrient leaching from a midwestern agricultural soil. *Geoderma* **2010**, *158*, 436–442. [CrossRef]
43. Li, R.; Shi, F.; Fukuda, K. Interactive effects of salt and alkali stresses on seed germination, germination recovery, and seedling growth of a halophyte *Spartina alterniflora* (Poaceae). *S. Afr. J. Bot.* **2010**, *76*, 380–387. [CrossRef]
44. Webber, C.L., III; White, P.M., Jr.; Spaunhorst, D.J.; Lima, I.M.; Petrie, E.C. Sugarcane biochar as an amendment for greenhouse growing media for the production of cucurbit seedlings. *J. Agric. Sci.* **2018**, *10*, 104. [CrossRef]

© 2019 by the authors. Licensee MDPI, Basel, Switzerland. This article is an open access article distributed under the terms and conditions of the Creative Commons Attribution (CC BY) license (http://creativecommons.org/licenses/by/4.0/).

Article

Biochar as a Stimulator for Germination Capacity in Seeds of Virginia Mallow (*Sida hermaphrodita* (L.) *Rusby*)

Bogdan Saletnik *, Marcin Bajcar, Grzegorz Zaguła, Aneta Saletnik, Maria Tarapatskyy and Czesław Puchalski

Department of Bioenergy Technology, Faculty of Biology and Agriculture, Rzeszow University, Ćwiklińskiej 2D, 35-601 Rzeszow, Poland
* Correspondence: bogdan.saletnik@urz.pl

Received: 11 July 2019; Accepted: 4 August 2019; Published: 7 August 2019

Abstract: This article presents the findings of a laboratory study investigating the stimulation and conditioning of seeds with biochar and the effects observed in the germination and emergence of Virginia mallow (*Sida hermaphrodita* (L.) *Rusby*) seedlings. The study shows that biochar, applied as a conditioner added to water in the process of seed hydration, improves their germination capacity. When the processed plant material was added to water at a rate of 5 g (approx. 1250 seeds) per 100 mL, the rate of germination increased to 45.3%, and was 23.3% higher when compared to the control group, and 7.3% higher than in the seeds hydrated without biochar. The beneficial effects of biochar application were also reflected in the increased mass of Virginia mallow seedlings. The mass of seedlings increased by 73.5% compared to the control sample and by 25.9% compared to the seeds hydrated without biochar. Given the low cost of charcoal applied during the hydro-conditioning process, the material can be recommended as a conditioner in large-scale production of Virginia mallow.

Keywords: *Sida hermaphrodita* L. *Rusby*; biochar; seed conditioning

1. Introduction

Crop yield and seed quality are among the basic parameters reflecting the effectiveness of plant production. They are determined by a number of factors, such as genotype, environmental conditions during crop growth and maturation, as well as the storage, selection, germination and enhancement of seed material. Given the processes of climate change, various stress-inducing factors influence seeds during their growth and maturation in significant and various ways. By selecting plants as well as breeding new cultivars to achieve enhanced qualities, it is possible to acquire a high yield and fine quality of seeds resistant to disease and stress. Irrespective of the effects of such breeding projects, the yield and quality of seeds may be improved by applying biological methods during plant growth or maturation and in the processes of storing and preparing the seeds for sowing [1].

Virginia mallow (*Sida hermaphrodita* L. *Rusby*) is a plants successfully grown for its biomass, which is used as a raw material in the production of renewable energy. It is a fast-growing perennial plant cultivated for its ornamental, melliferous and medicinal properties and as a fibre and fodder crop; it also produces a large quantity of biomass, comparable to or greater than common osier. However, the poor germination rate, frequently amounting to only 30–50%, poses a challenge in the cultivation of the crop. Virginia mallow production is also limited due to the difficulties in establishing plantations. The problems related to Virginia mallow seed germination and the disproportions between germination rate and seedling emergence in the field make it impossible to plan crop density [2]. In order to eliminate this obstacle, the seeds can be subjected to a pre-sowing treatment. The method is designed to stimulate the germination of seeds and the emergence of seedlings in the shortest possible time under a wide range of environmental conditions. Most importantly, the processes of germination and emergence to a large degree determine seedling vigour, their susceptibility to disease, their capacity for further growth and development and, ultimately, crop productivity [3].

Poor germination rate due to dormancy, linked with imperviousness of seed coats to water and gases, may be improved with such treatments as scarification, i.e., the mechanical alteration of seed coats, or short-term refrigeration of seeds at low temperature, periodical chilling (vernalisation) as well as growth stimulation treatments, variability to drought and saline stress through [4,5]. Seed germination and plant growth may also be improved with a number of operations, e.g., the decontamination and screening of seeds for their physical properties and treatments based on such factors as temperature or magnetic fields, as well as their irradiation with light of varying wave length and intensity [1,6]. Enhanced germination rate may also be achieved if, prior to sowing, the seeds are soaked in aqueous monocultures of algae or *Cyanobacteria* [7]; on the other hand the exposure of adequately wetted seeds to high or low temperatures for a specific length of time results in the seedlings having higher resistance to cold weather and the plants having greater metabolic activity and growth [8].

The germination potential of seeds can also be effectively improved by the controlled wetting of seeds up to a specified level, followed by incubating for some time at an appropriate temperature. The related methods include hydro-conditioning, osmo-conditioning and matri-conditioning. They differ in terms of the wetting techniques applied:

- hydro-conditioning involves direct seed humidification with water, without a medium;
- matri-conditioning involves water carried by a solid inorganic substance with high negative water potential; and
- osmo-conditioning involves water carried by osmotically active substances with low osmotic potential.

A specific amount of water imbibed by the seeds initiates most of the metabolic processes preceding germination, but it is insufficient for initiating the growth and emergence of radicles. The seeds are primed this way, at a temperature of 15 or 20 °C as a rule, for a time ranging from a few hours to several days. Prior to planting, adequately conditioned seeds may be dried and stored, preferably at low temperatures and mild relative air humidity (40%). The priming treatment leads to a faster germination and emergence of seedlings, and decreased vulnerability to environmental conditions during initial plant growth. The effectiveness of the conditioning may be increased by adding growth regulators to the water [1,9,10].

The quality of the seeds is also greatly dependent on the mother plants. Under field conditions the quality of the seeds may be increased by the biological fertilisation of mother plants with substances improving the environmental conditions, keeping mother plants a suitable distance away from each other, introducing microorganisms, algae or non-toxic *Cyanobacteria* that produce allelopathic interactions within the soil, and finally by treating mother plants with a bioformula mainly stimulating metabolic processes [11–17].

Biochar is a fine-grained product of carbonisation, obtained through the pyrolysis of plant biomass and organic waste, and containing aromatic, aliphatic and oxidised compounds of carbon. The chemical composition of biochar is determined by the composition of the feedstocks and the way the process of pyrolysis is conducted. Biochars contain stable organic carbon, washed out carbon and ash. The mineral fraction of biochar comprises macro- and micro-elements which may act as a source of minerals for micro-organisms living in the soil. Biochar introduced into the soil is highly resistant to degradation and microbial decay, which means it preserves the chemical stability of the environment to which it is introduced [18–22]. Biochar may protect plants against infection [23]. Added to soil, it improves soil fertility and productivity and may lead to an increase in the land's water capacity, as well as decreasing soil acidity. Biochar also has the capacity to retain and exchange nutrients, which results in a better availability of nutrients in plants [19,24–26].

In view of the growing body of evidence regarding the beneficial effects of biochar as a soil amendment, already successfully used in the production of various crops, the authors of this paper decided to expand the scope of this research and to apply biochar in seed conditioning. The purpose of the current study was to assess the usefulness of biochar in enhancing the viability of Virginia mallow seeds. It was designed as a preliminary laboratory study in order to verify whether further research would be justified, and if it would be practicable to apply the relevant method under field conditions.

2. Materials and Methods

2.1. Preparation and Characterisation of Biochars

Biochar used for the experiment was produced from barley straw. The material was examined for pH values, and contents of macro- and micro-elements. Biochar was refined in a mechanical lab mill, to obtain grains of a size a fraction below 5 mm. It was then sieved through a mesh with 1 mm openings, and rinsed with demineralised water several times to remove any contaminants. The biochar was then subjected to drying at 75 °C for 12 h.

The biochar used in the study was characterised by a pH value of 6.59 with total contents of carbon and nitrogen amounting to 74.35% and 0.93%, respectively. The tests also showed the water and ash content of the biochar being at a level of 9.11% and 11.57%, respectively. The biochar was characterised by a high content of volatile substances, i.e., 66.42% (Table 1).

Table 1. The pH value of the biochar, with the content of the absorbable forms of macro-elements and the percent content of carbon and nitrogen in the biochar. x—mean, SD—standard deviation.

pH (KCl)	Carbon	Nitrogen	P_2O_5	K_2O	Mg
	%		$mg\ kg^{-1}$		
			$x \pm SD$		
6.59 ± 0.21	74.35 ± 0.24	0.93 ± 0.07	1382 ± 41	5752 ± 63	645 ± 22

Table 2 presents the contents of selected macro- and micro-elements in the biochar generated from plant biomass. In view of the fact that many fertilisers used in agriculture today may contain toxic metals, the biochar used in the experiment was subjected to appropriate tests. The findings showed no aluminium, arsenic, cadmium or lead content in the material.

Table 2. The contents of selected macro- and micro-elements in the biochar generated from biomass. x—mean, SD—standard deviation.

Al	As	Ca	Cd	Cr	Cu	Mn
			$mg\ kg^{-1}$			
			$x \pm SD$			
<0.01	<0.01	1852 ± 21	<0.01	<0.01	10 ± 0.8	240 ± 2.5
Mo	**Na**	**Ni**	**Pb**	**S**	**Sr**	**Zn**
			$mg\ kg^{-1}$			
			$x \pm SD$			
<0.01	<0.01	<0.01	<0.01	880 ± 12	<0.01	130 ± 11.5

2.2. Plant Sample Collection

The study used commercially available Virginia mallow (*Sida hermaphrodita* R.) seeds, from a plantation located in the Warmia-Masuria Province, Poland.

2.3. Experimental Design

In the first stage of the experiment, 5 g of biochar was added to 100 mL of demineralised water and carefully mixed. Subsequently 0035 g of Virginia mallow seeds, with a moisture level of 5%, were introduced into the solution (1000 seeds of Virginia mallow weight approx. 4 g). The seeds were hydrated for four hours at 20 °C, after which they were transferred to hermetic containers and incubated at a temperature of 20 °C for two days. The seeds were also aerated every day. At the next stage, a new 5 g sample of Virginia mallow seeds was subjected to hydration in 100 mL of water, with no biochar added to the solution, following the same procedure as described above. All the seeds conditioned in the above way were then dried on filter paper in laboratory conditions, to achieve the initial moisture level, and their viability was assessed. For this purpose, 3×50 seeds were planted (representing all the variants, i.e., seeds hydrated with biochar, hydrated without biochar, and not hydrated) at a temperature of 20 °C, onto wet paper placed in Petri dishes. Each day the germinating specimens were counted, i.e., the seeds in which the radicle had emerged and reached a minimum length of 1 mm. After 14 days of the experiment, the plant material was collected for further testing. The seedlings (overground parts) collected for this purpose were weighed and rinsed several times with demineralised water, to remove any possible contamination.

2.4. Examination of Samples

The samples of biochar and the Virginia mallow seedlings were subjected to laboratory analyses using applicable analytical standards (Table 3).

Table 3. The parameters analysed with the research methods used.

Item	Parameter	Research Method
1.	pH in KCl	PN-ISO 10390:1997 [27]
2.	Content of absorbable forms of phosphorus (P_2O_5)	PN-R-04023:1996 [28]
3.	Content of absorbable forms of potassium (K_2O)	PN-R-04022:1996/Az1:2002 [29]
4.	Content of absorbable form of magnesium (Mg)	PN-R-04020:1996/Az1:2004 [30]
5.	Content of carbon and nitrogen	PN-EN 15104:2011 [31]
6.	Total content of selected macro- and microelements	Method using atomic emission spectrometry with excitation in argon plasma (ICP-OES)

The pH of the biochar was determined by measuring the concentration of hydrogen ions, i.e., the activity of hydrogen ions (H^+), with the use of the potentiometric method. The analysis was performed in a KCl solution with a concentration of 1 mol dm^{-3}, assuming a mass of study sample to solution volume ratio of 1:2.5. All measurements were taken using the Nahita pH meter, model 907 (AUXILAB, Beriáin, Spain).

The total carbon and nitrogen content was tested using a TrueSpec CHN analyser from LECO (LECO Corporation, Saint Joseph, MI, USA).

The mineralisation of the material was repeated three times. The elemental content of the samples was determined using a method based on inductively-coupled plasma atomic emission spectroscopy (ICP-OES), using an iCAP Dual 6500 analyser from Thermo Scientific (Schaumburg, IL, USA). The mineralisation of the samples was performed in Teflon containers using a mixture of acids under specific conditions (Table 4). In each case, the sample obtained in the mineralisation process was supplemented with mineralised water to make up a volume of 50 mL. In the calibration step, standard solutions for all elements were prepared using a spectroscopic grade reagent (Thermo) with a three-step curve. The curve fit factor for all elements was over 0.99. The selection of a measuring line of appropriate length was validated by the method of standard additions. The recovery on selected lines was above 98.5% for each of the elements. Every time, the CRM 1515 (Certified Reference Material) was used and the selection of appropriate lines was carried out using the standard addition method.

Every time we also used internal standards for matrix curve correction; these were yttrium (Y) and ytterbium (Yb), two elements not detected in the samples. The detection limit of the analytical method used for the elements studied was no worse than 0.01 mg kg^{-1}.

Table 4. The parameters of the mineralisation process.

Material	Acid		Temperature and Time	Power	Application Note
Biochar	7 mL HNO$_3$ 65% 1 mL H$_2$O$_2$ 30%	-	temperature increase to 200 °C, time: 15 min;	1500 W	HPR-PE-19 [32]
Plants	6 mL HNO$_3$ 65% 2 mL H$_2$O$_2$ 30%	-	maintaining at temperature of 200 °C, time: 15 min		HPR-AG-02 [33]

2.5. Statistical Analyses

All parts of the experiment were independently repeated three times. The results obtained were subjected to statistical analyses using Statistica ver. 10.0. The results were statistically analysed with multiway ANOVA and differences between the means were assessed using the Duncan test.

3. Results and Discussion

Figure 1 shows the number of germinating seeds of Virginia mallow, at the relevant time, in the control sample and relative to the method of seed conditioning (hydro-conditioning alone, and hydro-conditioning in water with biochar added).

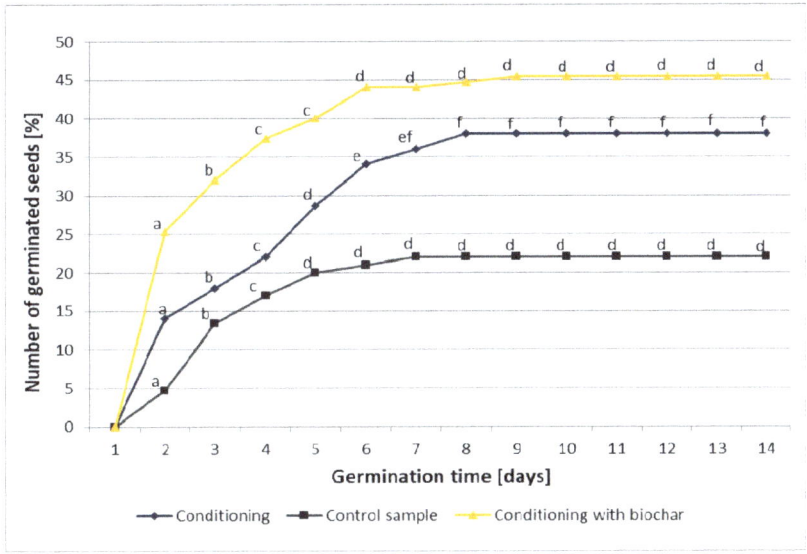

Figure 1. Germination rate in Virginia mallow seeds in the control sample, the sample subjected to hydro-conditioning alone, and the sample subjected to hydro-conditioning with biochar added. According to the Duncan's test, the means marked with the same letter are not significantly different at $\alpha \leq 0.05$ (differences between time germination inside each treatment).

The germination process in the three samples started at the same time—two days after the seeds were planted. The seeds in the control sample were found to have a germination rate of 22%. Poor germination rates observed in Virginia mallow are associated with the fact that the plant produces hard seeds, with a calcareous coat impervious to water. The impervious layer regulates seed dormancy, as well as the speed at which water penetrates the seeds to reach a sufficient hydration level [34].

The poor germination rate of the control sample could also have been affected by the climatic conditions occurring in the year the plants were harvested, the threshing method used and the duration of storage [35]. The seeds subjected to hydro-conditioning showed a better germination rate than the control sample, reaching 38%. Hydro-conditioning is rather commonly applied in order to improve the ability of hard seeds to germinate. The positive impact of this process on the germination and vigour of crops and garden plants has been reported by numerous researchers. Grzesik et al. [36] demonstrated the positive effects of hydration as a factor in stimulating growth and metabolic processes in Virginia mallow. The highest metabolic activity was observed in the seeds hydrated to achieve a 35–40% water content, and incubated at 20 °C for six days. Janas and Grzesik [37] showed that a pre-sowing hydro-conditioning applied to dill seeds can effectively enhance their viability and stimulate the faster emergence and growth of seedlings. The same authors determined the optimum conditions for hydro-conditioning applied to seedlings of root parsley. Parsley seeds hydrated to a hydration level of 45% and incubated at 20 °C for three to six days were found to germinate earlier and in larger numbers than non-hydrated seeds. The seedlings emerging from the former developed faster and contained more chlorophyll [38]. Similar findings were reported by Mendonça et al. [39] who found that seedlings from carrot seeds subjected to hydro-conditioning were faster to emerge and grow compared to the control sample. Xavier Fernanda da Motta et al. [40] established that hydro-conditioning is the most effective treatment to improve the germination and vigour of onion seeds. The positive effects of hydration, observed in germination, emergence and plant growth, are associated with metabolic activity, which is more rapidly initiated and increased in seeds before they are planted in soil, and in seedlings at the very beginning of their growth [36].

The addition of processed plant material (biochar from biomass) produced an increase in the number of germinating seeds. Biochar added at the rate of 5 g per 100 mL of water increased the speed of germination by 20.7%, and the germination capacity of the seeds increased to 45.3%, which was 23.3% higher than the control group; conversely, these values increased by 11.3% and 7.3%, respectively, when compared to the seeds subjected to hydration. The method applied may present several advantages over the methods of pre-sowing seed treatment and enhancement currently used. In the literature we can find numerous seed scarification methods, more or less effective, and varied in terms of their feasibility for use in mass production. A number of authors have investigated methods of breaking seed dormancywith the use of concentrated sulphuric acid [41–47]. Doliński et al. [47] observed an increase in the germination capacity of Virginia mallow seeds subjected to scarification with sulphuric acid. The germination capacity following the scarification varied depending on the year the seeds were harvested and on the duration of their storage [35,47]. Although it produces satisfying results, this method is hazardous to seeds (prolonged exposure to the acid leads to embryo death) and harmful to people carrying out the scarification process. Furthermore, scarification with sulphuric acid may lead to a decrease in seed vigour [41,43,47]. Mechanical processing is also frequently used to interrupt seed dormancy. However, such methods may prove troublesome for production on a large scale; examples include nicking lupin seed coats with a razor blade [48], manually rubbing burclover *Medicago polymorpha* seeds with sand paper, and shaking seeds of clover *Trifolium subterraneum* with sand in a laboratory shaker [49]. Germination rates may also be improved by a treatment involving the stimulation of seeds with a magnetic field, however, these techniques are rather costly and not as effective as other known methods of seed scarification [50,51]. Effective methods for breaking seed dormancy include thermal scarification. Dormancy was successfully broken in seeds of burclover, *Medicago polimorpha,* and clover, *Trifolium subterraneum,* following a short-term heat treatment with a blowlamp. As a result, seed germination rate increased from 10 to 79% for burclover, and from 46 to 95% for clover [49]. Thermal scarification may also be performed by immersing seeds in hot water or treating them with liquid nitrogen. Doliński et al. [35,47] observed a significant increase in the germination capacity of Virginia mallow seeds following a short treatment with hot water. The temperature required for interrupting the dormancy of a Virginia mallow seed depends on the length of time the seed has spent in storage. Sharma et al. [52] reported even a ten-fold increase in the viability of seeds of selected fast-growing tropical tree species, i.e., *Albizzia lebbek, Albizzia procera, Peltophorum pterocarpum, Acacia auriculiformis* and *Leucaena leucocephala*, following a short-term (1–10 min) treatment with hot water (100 °C). The dormancy of fodder galega seeds was interrupted following a treatment with liquid nitrogen (−196.5 °C) [53]. Notably, the duration and temperature of the treatments applied to break seed dormancy depends on the plant species.

The experiment described in this paper is the first to apply biochar as an additional conditioner in the process of seed hydration. Rogovska et al. [54] assessed the effects of biochars on seedling growth and absorption of allelochemicals present in corn (*Zea mays* L.) residues. Corn seeds were germinated in aqueous extracts of six varied biochars. Extracts from the six biochars had no effect on percent germination; however, extracts from three biochars produced at high conversion temperatures significantly inhibited shoot growth by an average of 16% relative to control (deionized water). In the literature we can find studies assessing the effects of biochar as a soil fertiliser on seed germination. Solaiman et al. [55] carried out a laboratory experiment which showed a positive association between biochar, obtained from various biological feedstocks, and increased germination rates in wheat seeds. Generally, the beneficial effects were produced by five types of biochar applied in doses <50 t h^{-1}. Conversely, higher doses failed to stimulate the germination process, or even inhibited it. Shamim et al., [56] assessed the effects of biochars on seed germination, early growth of *Oryza sativa* L. In this study, the germination percentage of paddy increased in case of two of three biochar was above the control level but the difference was not significant. The third type of biochar shows decrease in germination percentage than control. Other reports on the related study found that there is no negative impact of biochar on the germination of paddy seeds [57]. The study based on forest seeds found that biochar application increases germination of seeds [58]. Saletnik et al. [19] showed that by supplementing soil with the most effective dose of biochar defined for that species, it is possible to achieve a 20% increase in the germination rate of Virginia mallow seeds. It was also observed that biochar from plant biomass, introduced to the soil, led to a significant increase in the mass obtained by plants in the early stages of their development [19].

Figure 2 presents the total mass of the seedlings harvested after 14 days of the experiment, relative to the sample type. The total mass of the plants collected from the control sample was 0.275 g. The mass of the plants collected from seeds subjected to conditioning with biochar was significantly greater than the mass of the seedlings from seeds subjected to hydroconditioning alone, or those from the control sample. The total mass of the seedlings increased by 73.1% compared to the control sample, and by 25.6% compared to the hydroconditioned sample.

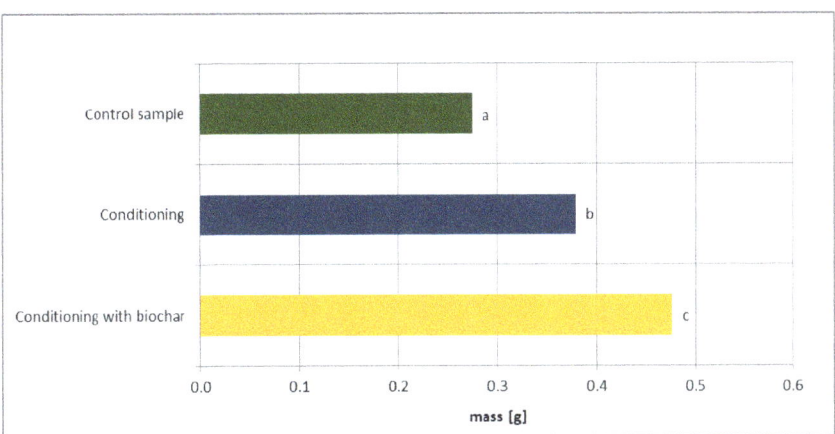

Figure 2. The total masses of Virginia mallow seedlings in the experimental variants. According to Duncan's test, means marked with the same letter are not significantly different at $\alpha \leq 0.05$.

The literature on this topic suggests that biochar beneficially affects crop yields, for instance Robertson et al. [58] planted lodgepole pine (*Pinus contorta* var. *latifolia*) or sitka alder (*Alnus viridis* ssp. *sinuata*) seeds in pots containing field-collected forest soils amended with 0.5, or 10% (dry mass basis) biochar with and without urea fertilizer (150 mg N kg^{-1}). Pine had greater biomass in biochar fertilizer treatments compared to control and fertilizer-only treatments. Alder seedlings had greater shoot biomass when grown in biochar-amended soils compared with unamended control. According to Liu et al. [59] biochar amendments used in fields at a rate below 30 t h^{-1} resulted in increased productivity of these specific crops: legumes by 30%, vegetables—29%, grasses—13%, wheat—11%, maize—8%, rice—7% [59]. Liu et al. [60] demonstrated the positive effects of biochar made from wheat straw in the productivity of rapeseed and sweet potato. A biochar amendment applied at a rate of 40 t ha^{-1} resulted in a yield increase of 36.02% and 53.77%, respectively [60]. Hossain et al. [61] used wastewater sludge biochar as a fertiliser. With the amendment used at a rate of 10 t h^{-1} the yield of cherry tomato increased by 64% compared to control conditions. Chan et al. [62] reported the positive effects of poultry litter biochar, applied at a rate of 10 t h^{-1}, in the increased mass of radish. Uzoma et al. [63] investigated the impact of cow manure biochar on maize yield. Biochar amendment applied at rates of 15 t and 20 t h^{-1} increased maize grain yield by 150% and 98%, respectively, compared to the control. The findings of the study suggested that cow manure biochar contained important nutrients which significantly contributed to increased maize productivity. Numerous authors point out that there is a threshold value for the application of biochar soil amendment when it comes to crop productivity. It was shown that the use of biochar at higher rates does not produce statistically significant effects on crop yields [64]. Saletnik et al. [19] observed a significant increase in the mass of Virginia mallow at an early stage of growth if biochar was added to soil. Furthermore, the study showed that a combination of biochar and ash from biomass, added at an optimum rate, improves the germination capacity of seeds, while increasing the potassium content of the overground parts of plants and reducing their concentrations of phosphorus and calcium [19]. Biochar has been also studied as method to reduce the greenhouse gas emissions from agriculture soils [65–67]. Cayuela et al. [66] found that biochar reduced soil N_2O emissions by 54% in laboratory and field studies. The biochar feedstock, pyrolysis conditions and C/N ratio were shown to be key factors influencing emissions of N_2O while a direct correlation was found between the biochar application rate and N_2O emission reductions.

Table 5 presents the results reflecting the total contents of selected macro- and micro-elements in overground parts of Virginia mallow plants, in both the control and experimental samples, 14 days after the experimental culture was established.

In each case the samples analysed were characterised by having the highest potassium, calcium and phosphorus contents. The lowest levels of macro- and micro-elements were identified in plants grown from seeds conditioned with biochar. This may have been caused by the faster development and intensified metabolic processes, associated with a greater nutrient requirement, at the early stage of growth [36,68]. The samples differed most significantly in their potassium content. Potassium is absorbed extremely rapidly by plants, second only to nitrogen, though by young ones possessing the fast-growing meristematic tissue responsible for the development of roots and stems in particular. This element is an essential plant nutrient; in the case of a potassium deficiency it is sent to apical meristems and young leaves first of all. At the early stages of plant development, phosphorus is necessary for the normal growth of the root system, and calcium is an important factor in both the regulation of cellular metabolism and structural functions, as well as being a universal carrier of information [68–71].

Table 5. Contents of selected macro- and micro-elements in overground parts of Virginia mallow plants in the control sample, in the hydro-conditioned sample, and the sample hydroconditioned with biochar added. X–mean, SD–standard deviation. According to Duncan's test, means marked with the same letter are not significantly different at $\alpha \leq 0.05$.

	Al	As	Ca	Cd	Cr	Cu	K
				mg 100 g^{-1}			
				x ± SD			
Control sample	2.26 ^{b} ± 0.06	<0.01	32.7 ^{b} ± 0.64	<0.01	<0.01	0.17 ^{b} ± 0.01	303.57 ^{c} ± 3.75
Conditioning	0.14 ^{a} ± 0.06	<0.01	30.84 ^{a} ± 0.70	<0.01	<0.01	0.18 ^{b} ± 0.01	264.34 ^{b} ± 4.83
Conditioning with biochar	<0.01	<0.01	29.90 ^{a} ± 0.70	<0.01	<0.01	0.07 ^{a} ± 0.01	208.20 ^{a} ± 3.56
	Mg	Mn	Na	Ni	P	Pb	S
				mg 100 g^{-1}			
				x ± SD			
Control sample	38.52 ^{c} ± 0.56	0.21 ^{b} ± 0.02	35.80 ^{c} ± 0.29	<0.01	66.00 ^{b} ± 0.34	<0.01	33.41 ^{c} ± 0.21
Conditioning	25.36 ^{b} ± 0.55	0.12 ^{a} ± 0.03	32.07 ^{b} ± 0.83	<0.01	72.24 ^{c} ± 0.48	<0.01	30.21 ^{b} ± 0.18
Conditioning with biochar	18.27 ^{a} ± 0.34	0.11 ^{a} ± 0.01	7.71 ^{a} ± 0.29	<0.01	55.73 ^{a} ± 0.57	<0.01	23.77 ^{a} ± 0.24

4. Conclusions

The current study shows that by adding biochar to water in the process of hydro-conditioning it is possible to achieve an over 20% increase in the germination capacity of Virginia mallow seeds. It was noted that the hydro-conditioning of seeds with biochar added resulted in a significantly greater mass of the seedlings collected at the early stage of plant development.

These findings provide a justification for further research focusing on the use of biochar as a stimulator for seed germination. Subsequent laboratory tests should be expanded, e.g., to include analyses of the enzymes involved in the germination process. The method used should be developed further with respect to optimum temperatures, incubation time and biochar dosage. Research related to these factors should be continued under field conditions in order to confirm the beneficial effects of biochar in the development, growth and productivity of crops. The application of biochar as a biological factor stimulating seed germination, as presented in the experiment, seems to present advantages related to the practicality and simplicity of the treatment as well as to the availability of the conditioning material. This method may also prove to be a cost-efficient alternative to the existing methods of a pre-sowing scarification of seeds used in plant production.

Author Contributions: Conceptualization, B.S.; Data curation, B.S., A.S. and G.Z.; Formal analysis, B.S., G.Z., M.B. and M.T.; Methodology, B.S., G.Z., M.B. and M.T.; Project administration, C.P.; Supervision, C.P.; Writing—original draft, B.S. and A.S.; Writing—review & editing, C.P. and A.S.

Funding: This research received no external funding.

Conflicts of Interest: The authors declare no conflict of interest.

References

1. Grzesik, M.; Janas, R.; Górnik, K.; Romanowska-Duda, Z. Biological and physical methods of seed production and processing. *J. Res. Appl. Agric. Eng.* **2012**, *57*, 147–152.
2. Tworkowski, J.; Szczukowski, S.; Stolarski, M.J.; Kwiatkowski, J.; Graban, Ł. Productivity and properties of virginia fanpetals biomass as fuel depending on the propagule and plant density. *Fragm. Agron.* **2014**, *31*, 115–125.
3. Podleśny, J.; Sowiński, M. The effect of seeds stimulation by magnetic field on growth, development and dynamics of biomass accumulation in pea (*Pisum sativum* L.). *Agric. Eng.* **2005**, *9*, 103–110.
4. Duczmal, K.; Tucholska, H. *Nasiennictwo [Seed Production]*; PWRiL: Poznań, Poland, 2000; Volume 1, pp. 205–234.
5. Cavallaro, V.; Barbera, A.C.; Maucieri, C.; Gimma, G.; Scalisi, C.; Patanè, C. Evaluation of variability to drought and saline stress through the germination of different ecotypes of carob (*Ceratonia siliqua* L.) using a hydrotime model. *Ecol. Eng.* **2016**, *95*, 557–566. [CrossRef]
6. Ciupak, A.; Szczurowska, I.; Gładyszewska, B.; Pietruszewski, S. Impact of laser light and magnetic field stimulation on the process of buckwheat seed germination. *Tech. Sci.* **2007**, *10*, 1–10. [CrossRef]
7. Karthikeyanb, N.; Prasannaa, R.; Nainb, L.; Kaushik, B.D. Evaluating the potential of plant growth promoting *Cyanobacteria* as inoculants for wheat. *Eur. J. Soil Biol.* **2007**, *43*, 23–30. [CrossRef]
8. Górnik, K. The effect of temperature treatments during 'Wielkopolski' sunflower seed imbibition and storage on plant tolerance to chilling. *Folia Hortic.* **2011**, *23*, 83–88. [CrossRef]
9. Grzesik, M.; Janas, R. Effects of hydropriming on metabolic activity, seed germination and seedling emergence of carrot. *J. Res. Appl. Agric. Eng.* **2011**, *56*, 127–132.
10. Badek, B.; van Duijn, B.; Grzesik, M. Effects of water supply methods and incubation on germination of China aster (*Callistephus chinensis*) seeds. *Seed Sci. Technol.* **2007**, *35*, 569–576. [CrossRef]
11. Nahm, M.; Morhart, C. Virginia mallow (*Sida hermaphrodita* (L.) *Rusby*) as perennial multipurpose crop: Biomass yields, energetic valorization, utilization potentials, and management perspectives. *GCB Bioenergy* **2018**, *10*, 393–404. [CrossRef]
12. Błażewicz-Woźniak, M. Effect of no-tillage and mulching with cover crops on yield of parsley. *Folia Hortic.* **2005**, *17*, 3–10.

13. Górnik, K.; Grzesik, M. Effect of Asahi SL on China aster 'Aleksandra' seed yield, germination and some metabolic events. *Acta Physiol. Plant.* **2002**, *24*, 378–383. [CrossRef]
14. Janas, R. Effect of tytanit on yield and quality of onion seeds. *Postępy Nauk Rol.* **2009**, *541*, 133–139.
15. Janas, R.; Grzesik, M. Pro-ecological methods of improving horticultural plant seeds quality. *Adv. Agric. Sci. Probl. Issues* **2006**, *510*, 213–221.
16. Grzesik, M.; Romanowska-Duda, Z.B. Technologia hydrokondycjonowania nasion ślazowca pensylwańskiego (*Sida hermaphrodita*) w aspekcie zmian klimatycznych [Technology for hydro-conditioning of Virginia mallow (*Sida hermaphrodita*) seeds in view of climate changes]. *Prod. Biomasy Wybrane Probl.* **2009**, *VII*, 63–69.
17. Grzesik, M.; Romanowska-Duda, Z.B. New Technologies of the energy plant production in the predicted climate changed conditions. *Bjuleten Djerżawnowo Nikitsk. Bot. Sada. Ukr. Akad. Agrar. Nauk* **2009**, *99*, 65–68.
18. Saletnik, B.; Zaguła, G.; Bajcar, M.; Tarapatskyy, M.; Bobula, B.; Puchalski, C. Biochar as a Multifunctional Component of the Environment—A Review. *Appl. Sci.* **2019**, *9*, 1139. [CrossRef]
19. Saletnik, B.; Bajcar, M.; Zaguła, G.; Czernicka, M.; Puchalski, C. Influence of biochar and biomass ash applied as soil amendment on germination rate of Virginia mallow seeds (*Sida hermaphrodita* R.). *Econtechmod Int. Q. J.* **2016**, *5*, 71–76.
20. Lehman, J. Bio-energy in the black. *Front. Ecol. Environ.* **2007**, *5*, 381–387. [CrossRef]
21. Malińska, K. Biochar - a response to current environmental issues. *Eng. Prot. Environ.* **2012**, *15*, 387–403.
22. Lehmann, J.; Rilling, M.C.; Thies, J.; Masiello, C.A.; Hockaday, W.C.; Crowley, D. Biochar effects on soil biota—A review. *Soil Biotechnol. Biochem.* **2011**, *43*, 1812–1836. [CrossRef]
23. Nigussie, A.; Kissi, E.; Misganaw, M.; Ambaw, G. Effect of biochar application on soil properties and nutrient uptake of lettuces (*Lactuca sativa*) grown in chromium polluted soils. *Am. Eur. J. Agric. Environ. Sci.* **2012**, *12*, 369–376.
24. Karhu, K.; Mattila, T.; Bergstrom, I.; Regina, K. Biochar addition to agricultural soil increased CH4 uptake and water holding capacity—Results from a short-term pilot field study. *Agric. Ecosyst. Environ.* **2011**, *140*, 309–313. [CrossRef]
25. Hossain, M.K.; Strezov, V.; Chan, K.Y.; Ziolkowski, A.; Nelson, P.F. Influence of pyrolysis temperature on production and nutrient properties of wastewater sludge biochar. *J. Environ. Manag.* **2011**, *92*, 223–228. [CrossRef]
26. Lehman, J.; Joseph, S. *Biochar for Environmental Management: Science and Technology*; Earthscan: London, UK, 2009; Volume 2, pp. 13–30.
27. Polish Committee for Standardization. *Soil Quality—Determination of Ph*; Polish Committee for Standardization: Warsaw, Poland, 1997.
28. Polish Committee for Standardization. *Chemical and Agricultural Analysis of the Soil—Determination of the Content of Absorbable Phosphorus in Mineral Soils*; Polish Committee for Standardization: Warsaw, Poland, 1996.
29. Polish Committee for Standardization. *Chemical and Agricultural Analysis of the Soil—Determination of the Content of Potassium in Mineral Soils*; Polish Committee for Standardization: Warsaw, Poland, 2002.
30. Polish Committee for Standardization. *Chemical and Agricultural Analysis of the Soil—Determination of the Content of Magnesium in Mineral Soils*; Polish Committee for Standardization: Warsaw, Poland, 2004.
31. British Standards Institution. *Solid Biofuels—Determination of Total Carbon, Hydrogen and Nitrogen Content—Instrumental Methods*; British Standards Institution: London, UK, 2011.
32. Milestone. *SK-10 High Pressure Rotor; HPR-PE-19 Carbon Black*; Milestone: Shelton, CT, USA, 2019.
33. Milestone. *SK-10 High Pressure Rotor; HPR-AG-02 Dried Plant Tissue*; Milestone: Shelton, CT, USA, 2019.
34. Woodstock, L.W. Seed imbibition: A critical period for successful germination. *J. Seed Technol.* **1988**, *12*, 1–15.
35. Doliński, R. Influence of treatment with hot water, chemical scarification and storage time on germination of Virginia fanpetals (*Sida hermaphrodita* (L.) *Rusby*) seeds. *Biul. Inst. Hod. Aklim. Roślin* **2009**, *257*, 293–303.
36. Grzesik, M.; Janas, R.; Romanowska-Duda, Z. Stimulation of growth and metabolic processes in Virginia mallow (*Sida hermaphrodita* L. *Rusby*) by seed hydroconditioning. *Probl. Agric. Eng.* **2011**, *4*, 81–89.
37. Grzesik, M.; Janas, R. Effect of conditioning on dill (*Anethum graveolens* L.) seed germination and plant emergence. *J. Res. Appl. Agric. Eng.* **2013**, *58*, 188–192.
38. Grzesik, M.; Janas, R. Physiological method for improving seed germination and seedling emergence of root parsley in organic systems. *J. Res. Appl. Agric. Eng.* **2014**, *59*, 80–86.

39. Mendonça, S.R.; Silva Pereira, J.C.; Teles da Cruz, A. Emergence of carrot seeds cv. Brasília submitted to hydro-conditioning. *Ipê Agron. J.* **2018**, *2*, 18–25.
40. Xavier, F.M.; Brunes, A.P.; Cavalcante, J.A.; Meneghello, G.E.; Radke, A.K.; Noguez Martins, A.B.; Winke Dias, L.; Revers Meneguzzo, M.R. Germination of *Allium cepa* L. seeds subjected to physiological conditioning and drying. *Rev. Ciênc. Agrár.* **2017**, *40*, 1–10.
41. Imani, A.F.; Salehi Sardoei, A.; Shahdadneghad, M. Effect of H2SO4 on Seed Germination and Viability of *Canna indica* L. Ornamental Plant. *Int. J. Adv. Biol. Biomed. Res.* **2014**, *2*, 223–229.
42. Rostami, A.A.; Shasavar, A. Effects of Seed Scarification on Seed Germination and Early Growth of Olive Seedlings. *J. Biol. Sci.* **2009**, *9*, 825–828. [CrossRef]
43. Kheloufi, A.; Mansouri, L.M.; Boukhatem, F.Z. Application and use of sulphuric acid pretreatment to improve seed germination of three acacia species. *Reforesta* **2017**, *3*, 1–10. [CrossRef]
44. Zare, S.; Tavili, A.; Darini, M.J. Effects of different treatments on seed germination and breaking seed dormancy of Prosopis koelziana and Prosopis Juliflora. *J. For. Res.* **2011**, *22*, 35–38. [CrossRef]
45. Tanaka-Oda, A.; Kenzo, T.; Fukuda, K. Optimal germination condition by sulfuric acid pretreatment to improve seed germination of Sabina vulgaris. *J. For. Res.* **2009**, *14*, 251–256. [CrossRef]
46. Saied, A.S.; Gebauer, J.; Buerkert, A. Effects of different scarification methods on germination of Ziziphus spina-christi seeds. *Seed Sci. Technol.* **2008**, *36*, 201–205. [CrossRef]
47. Doliński, R.; Kociuba, W.; Kramek, A. Influence of short treatment with hot water, chemical scarification and gibberellic acid on germination of Virginia mallow (*Sida hermaphrodita* (L.) Rusby) seeds. *Adv. Agric. Sci. Probl. Issues* **2007**, *517*, 139–147.
48. Mackay, W.A.; Davis, T.D.; Sankhla, D. Influence of scarification and temperature treatments on seed germination of *Lupinus havardiiv*. *Seed Sci. Technol.* **1995**, *23*, 815–821.
49. Martin, I.; De la Caudra, C. Evaluation of different scarification methods to remove hard-seediness in *Trifolium subterraneum* and *Medicago polimorpha* accessions of the Spanish base genebank. *Seed Sci. Technol.* **2004**, *32*, 671–681. [CrossRef]
50. Pietruszewski, S.; Kania, K. Effect of magnetic field on germination and yield of wheat. *Int. Agrophys.* **2010**, *24*, 297–302.
51. Komarzyński, K.; Pietruszewski, S. Influence of alternating magnetic field on the germination of seeds with low germination capacity. *Acta Agrophys.* **2008**, *11*, 429–435.
52. Sharma, S.; Naithani, R.; Vargtrese, B.; Keshavkant, S.; Naithani, S.C. Effect of hot-water treatment on seed germination of some fast growing tropical tree species. *J. Trop. For.* **2008**, *24*, 49–53.
53. Tworkowski, J.; Szczukowski, S.; Jakubiuk, P. Skaryfikacja a wartość siewna nasion rutwicy wschodniej (*Galega orientalis* Lam.) [Scarification versus viability of galega (*Galega orientalis* Lam.) seeds]. *Adv. Agric. Sci. Probl. Issues* **1999**, *468*, 233–240.
54. Rogovska, N.; Laird, D.; Cruse, R.M.; Trabue, S.; Heaton, E. Germination Tests for Biochar Quality. *J. Environ. Qual.* **2012**, *41*, 1014–1022. [CrossRef]
55. Solaiman, Z.M.; Murphy, D.V.; Abbot, L.K. Biochars influence seed germination and early growth of seedlings. *Plant Soil* **2012**, *353*, 273–287. [CrossRef]
56. Shamim, M.; Saha, N.; Hye, F.B. Effect of biochar on seed germination, early growth of *Oryza sativa* L. and soil nutrients. *Trop. Plant Res.* **2018**, *5*, 336–342. [CrossRef]
57. Kamara, A.; Kamara, A.; Mansaray, M.; Sawyerr, P. Effects of biochar derived from maize stover and rice straw on the germination of their seeds. *Am. J. Agric. For.* **2014**, *2*, 246–249. [CrossRef]
58. Robertson, S.J.; Rutherford, P.M.; Lo' pez-Gutie´rrez, J.C.; Massicotte, H.B. Biochar enhances seedling growth and alters root symbioses and properties of sub-boreal forest soils. *Can. J. Soil Sci.* **2012**, *92*, 329–340. [CrossRef]
59. Liu, X.; Zhang, A.; Ji, C.; Joseph, S.; Bian, R.; Li, L.; Pan, G.; Paz-Ferreiro, J. Biochar's effect on crop productivity and the dependence on experimental conditions—A meta-analysis of literature data. *Plant Soil.* **2013**, *373*, 583–594. [CrossRef]
60. Liu, Z.; Chen, X.; Jing, Y.; Li, Q.; Zhang, J.; Huang, Q. Effects of biochar amendment on rapeseed and sweet potato yields and water stable aggregate in upland red soil. *Catena* **2013**, *123*, 45–51. [CrossRef]
61. Hossain, M.K.; Strezov, V.; Chan, K.Y.; Nelson, P.F. Agronomic properties of wastewater sludge biochar and bioavailability of metals in production of cherry tomato. *Chemosphere* **2010**, *78*, 1167–1171. [CrossRef]

62. Chan, K.Y.; Van Zwieten, L.; Meszaros, I.; Downie, A.; Joseph, S. Using poultry litter biochars as soil amendments. *Aust. J. Soil Res.* **2008**, *46*, 437–444. [CrossRef]
63. Uzoma, K.C.; Inoue, M.; Andry, H.; Fujimaki, H.; Zahoor, A.; Nishihara, E. Effect of cow manure biochar on maize productivity under sandy soil condition. *Soil Use Manag.* **2011**, *27*, 205–212. [CrossRef]
64. Lehmann, J.; Gaunt, J.; Rondon, M. Biochar sequestration in terrestrial ecosystems–A review. *Mitig. Adapt. Strateg. Glob. Chang.* **2006**, *11*, 395–419. [CrossRef]
65. Zhang, Y.; Wang, H.; Maucieri, C.; Liu, S.; Zou, J. Annual nitric and nitrous oxide emissions response to biochar amendment from an intensive greenhouse vegetable system in southeast China. *Sci. Hortic.* **2019**, *246*, 879–886. [CrossRef]
66. Cayuela, M.L.; Van Zwieten, L.; Singh, B.P.; Jeffery, S.; Roig, A.; Sánchez-Monedero, M.A. Biochar's role in mitigating soil nitrous oxide emissions: A review and meta-analysis. *Agric. Ecosyst. Environ.* **2014**, *191*, 5–16. [CrossRef]
67. Maucieri, C.; Zhang, Y.; McDaniel, M.D.; Borin, M.; Adams, M.A. Short-term effects of biochar and salinity on soil greenhouse gas emissions from a semi-arid Australian soil after re-wetting. *Geoderma* **2017**, *307*, 267–276. [CrossRef]
68. Szweykowska, A. *Plant Physiology*; Wydawnictwo Naukowe: Poznań, Poland, 1999; pp. 67–78.
69. Krzywy, E. *Fertilization of Soils and Plants*; Akademia Rolnicza: Szczecin, Poland, 2000; p. 177.
70. Wińska-Krysiak, M. Calcium transporting proteins in plants. *Acta Agrophysica* **2006**, *7*, 751–762.
71. Bezak-Mazur, E.; Stoińska, R. The importance of phosphorus in the environment—Review article. *Arch. Waste Manag. Environ. Prot.* **2013**, *15*, 33–42.

© 2019 by the authors. Licensee MDPI, Basel, Switzerland. This article is an open access article distributed under the terms and conditions of the Creative Commons Attribution (CC BY) license (http://creativecommons.org/licenses/by/4.0/).

Article

Effect of Biochar Application Depth on Crop Productivity Under Tropical Rainfed Conditions

Juana P. Moiwo [1,*], Alusine Wahab [1], Emmanuel Kangoma [1], Mohamed M. Blango [1], Mohamed P. Ngegba [2] and Roland Suluku [3]

[1] Department of Agriculture and Biosystems Engineering, School of Technology, Njala University, PMB, Freetown 47235, Sierra Leone
[2] Department of Extension and Rural Sociology, School of Agriculture, Njala University, PMB, Freetown 47235, Sierra Leone
[3] Department of Animal Science, School of Agriculture, Njala University, PMB, Freetown 47235, Sierra Leone
* Correspondence: jupamo2001@yahoo.com; Tel.: +232-76542008

Received: 11 January 2019; Accepted: 26 February 2019; Published: 27 June 2019

Featured Application: Biochar is an exciting discovery in the search for a solution to the deteriorating environmental and climatic conditions in the age. When activated and applied, it has the potential to adsorb hold on both carbon and nutrients and remain fertile for a considerably long period of time. Unfortunately, farming in especially western Africa is still dominated by the slash-burn shifting cultivation. Here, a considerable amount of charcoal from a mix of biomass stock is left behind after burning and plowed into the soil during cultivation. This study targets this system of cultivation. It is the long journey to the transformation of the burning to charcoal to pyrolysis to biochar to enrichen the soil.

Abstract: Although inherently fertile, tropical soils rapidly degrade soon after cultivation. The period of time for which crops, mulch, compost, and manure provide nutrients and maintain mineral fertilizers in the soil is relatively short. Biochar, on the other hand, has the potential to maintain soil fertility and sequester carbon for hundreds or even thousands of years. This study determined the effect of biochar application depth on the productivity of NERICA-4 upland rice cultivar under tropical rainfed conditions. A fixed biochar–soil ratio of 1:20 (5% biochar) was applied in three depths—10 cm (TA), 20 cm (TB), and 30 cm (TC) with a non-biochar treatment (CK) as the control. The study showed that while crop productivity increased, root penetration depth decreased with increasing biochar application depth. Soil moisture was highest under TA (probably due to water logging in sunken-bed plots that formed after treatment) and lowest under TC (due to runoff over the raised-bed plots that formed too). Grain yield for the biochar treatments was 391.01–570.45 kg/ha (average of 480.21 kg/ha), with the potential to reach 576.47–780.57 kg/ha (average of 695.73 kg/ha) if contingent field conditions including pest damage and runoff can be prevented. By quantifying the effect of externalities on the field experiment, the study showed that biochar can enhance crop productivity. This was good for sustainable food production and for taking hungry Africa off the donor-driven food ration the nation barely survives on.

Keywords: biochar; tropical rainfed condition; crop productivity; root-zone soil; application depth

1. Introduction

Biochar is charcoal produced by burning biomass under controlled temperature and oxygen conditions, mainly for the purpose of soil amendment [1–4]. As a porous material with active functional groups, high pH and cation exchange capacity [5], biochar can serve as a natural pool of organic carbon and plant nutrient in the soil [6–8]. There is a widespread interest in biochar research because of its rediscovered potential to mitigate climate change, amend soils and increase crop productivity [9,10]. However, this potential is strongly influenced by biochar production (temperature, oxygen supply, etc.), application (rate, mode, time, soil type, etc.) and feedstock (plant, animal, domestic waste, etc.) conditions [11–15].

The highly fertile dark earths (the so-called *terra preta*) left behind centuries ago by the prehistoric Amerindians are characteristically rich in carbon [16–18]. The origin of the rich dark earths is traced to as far back as 1542 when Orellana first reported to the Spanish Court on the well-established networks of agricultural settlements along a tributary of the Amazon River [19,20]. Unlike the predominantly *terra mulata* (brown earths), which is the natural soil in the Amazon, *terra preta* dark earths is anthropogenic or anthropic creation of humans [20–22]. Typically, *terra preta* has three or more times total soil organic carbon and far more phosphorus, calcium and humus than the surrounding *terra mulata* brown soils. The Amazonian *terra preta* has remained highly fertile centuries after its formation.

In addition to the potential to sequester carbon, the porous structure, large surface area, and high particle charge of biochar [23] interact with the soil [24–26] that results in beneficial effects on the ecosystem [27]. Among the countless benefits of the use of biochar are enhanced plant growth [28], improved soil water-holding capacity [29], reduced crop diseases [30], limited bioavailability of heavy metals [31], reduced soil nitrogen oxide emission [32], and reduced leaching of nutrients and fertilizers [29]. Schmidt et al. [33] and Schmidt [34] have documented a more exhaustive list of the beneficial uses of biochar. Biochar is hailed as an all-win solution to global energy, food, and environmental challenges [11].

In the tropics where soil fertility is depleted soon after cultivation in the predominantly slash-burn shifting cultivation [8,20,35], there is the need to switch to slash-char sedentary farming for sustainable crop production [32,36]. Slash-burn shifting cultivation destroys up to 97% of soil organic carbon as against less than 50% under slash-char farming, of which most is retained as highly stable organic carbon [16,24]. With an average residence time of over 1000 years [37], biochar can sustain crop productivity for as long as it remains active in the soil [38].

In a meta-analysis including 116 studies from 21 countries around the world, Liu et al. [39] noted that biochar increases crop productivity by 11% on average. In another meta-analysis including 371 independent studies, Biederman and Harpole [40] noted considerable variations in the degree of enhancement of crop yield components and soil properties. Biochar application rates below 30 tons/ha typically result in 30% more productivity of legume and 29% more of vegetable crops than of cereal crops such as corn (8%) and rice (7%) [38,40]. Studies, including literature data analyses, also show that biochar application rates above 55 tons/ha can lead to reductions in crop productivity [16,40].

The *terra preta* soil, which has so much reignited biochar research, occurs in varying depths and rates across the Amazon [20]. This forms the basis for investigation of the variations in soil/crop response to biochar application rates in terms of biochar–soil mixture, activation mode and/or application depth. In this study, a field experiment was conducted to determine the effect of biochar application depth on rice crop productivity under tropical rainfed conditions. The results of the study will add to the efforts in understanding the potential of biochar to sustainable crop production, improve soil properties and mitigate global climate change.

2. Materials and Methods

2.1. Study Area

This study was conducted at the experimental site at the Department of Agricultural Engineering, Njala University. The site is located at 012.07762° W, 08.10958° N and is 80 m +MSL (above mean sea level); which is in Kori Chiefdom in the southern province of Sierra Leone (Figure 1). The study area has a typical tropical climate, with gravelly brown soil and shrub vegetation [32,41].

Here, there are two distinct seasons—the wet season (WS) and dry season (DS). The WS (May to Oct.) is characterized by cloudiness, heavy rains, and by low temperatures, solar radiation, photoperiod, and sunshine hours, but high humidity. In the DS (Nov. to Apr.), the climatic condition is the near-reverse of that of the WS. The unimodal average annual rainfall is 2526 mm and the temperature is 31 °C. While over 80% of the rain falls in the WS, solar radiation is highest in March (604 MJ/m^2/day).

The soils at the experimental site are typical Oxisols and Ultisols and therefore generally slightly acidic [42], but nonetheless good for rice cultivation. Biochar is reported to have pH range of 10–12 [43] and therefore suitable for application in such soils.

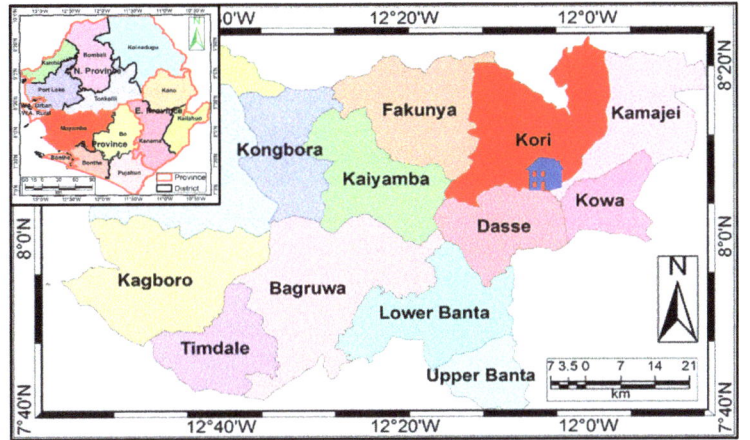

Figure 1. A map depicting Moyamba District (red) in the southern province of Sierra Leone (inset) and then an expanded map of the province (main plate) depicting Kori Chiefdom (red) and the location of the Njala University study area (blue house) in the chiefdom.

2.2. Biochar Preparation

The wood biomass used in the biochar production was a random collection in the bush nearby (1–2 km) the experimental site. The random wood mix was as much as possible similar to the normal conditions under which charcoal is unintentionally added to the soil under the conventional slash-burn shifting cultivation practiced in the region [35]. The dry wood biomass was fed into a locally-built Elsa Stove (retort kiln) of 55 cm in diameter and 90 cm in height and allowed to burn at 350–550 °C under low oxygen condition [32]. The pyrolyzed wood biomass (called biochar) was then collected, ground into fine dust and stored for later application in the field.

2.3. Site Preparation

The experimental site was cleared and leveled in July 2016 using simple farm tools including machete, hoe and shovel. A total of 24 plots (each 1.0 m × 1.0 m with 0.5 m footpath) were constructed and separated out by bamboo canes to prevent flow of water, soil and nutrient across the plots (Figure 2). Also, a drainage was constructed around the entire field site to prevent runoff on to the plots. The dimension of the experimental field site was 12.5 m × 5.0 m (length by width), with a total area of 62.5 m^2.

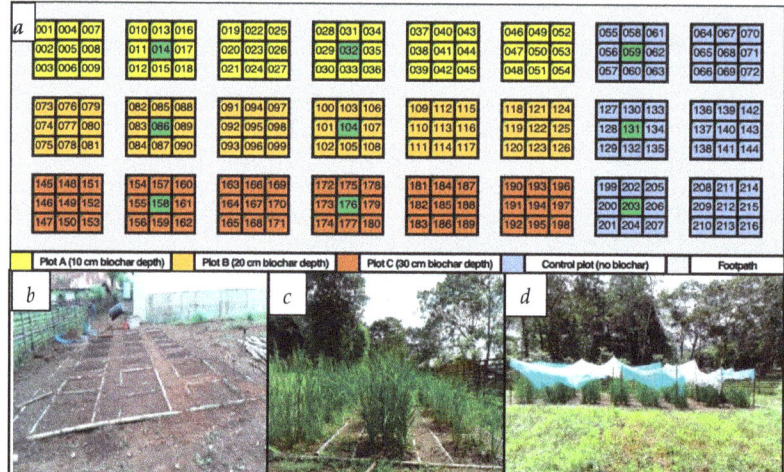

Figure 2. A sketch of the experimental field layout depicting the order of the rice plant hills in each plot (plate *a*). Note that the big box is the plot and each of the small boxes with the numbers represent a rice hills in the plot. The different colors in (plate *a*) indicate different biochar application depths and the green shades (plate *a*) are the sampling points for soil moisture measurement. Plate (*b*) is on-the-ground picture of the sketch in plate (*a*). The next plates (*c*) and (*d*) show respectively the heading and milking growth stages of the NERICA-4 rice. Also note that the milking stage in plate (*d*) is covered with mosquito net to prevent bird damage.

2.4. Biochar-Soil Mixture and Depth Application

In a previous experiment involving biochar treatment of the same soil, the 1:20 (5%) biochar–soil ratio had the best effect on rice yield [32]. Thus in this study, the 5% biochar–soil ratio (by volume) was used for three different application depths: 0-10 cm (TA), 0-20 cm (TB), and 0-30 cm (TC). These depths were used because rice root hardly penetrates beyond 30 cm in the soil. Also the plow depth and depth of application of fertilizers or organic manure are mostly within this depth range.

For TA, the soil in the 1.0 m × 1.0 m plot was dug out to the depth of 0.1 m; i.e., a soil volume of 0.1 m^3. The dug-out 0.1 m^3 of soil was mixed with 0.005 m^3 of biochar dust to have the 5% (1:20) biochar–soil mixture. The process was repeated for TB, where 0.2 m^3 of soil dug out from the pit was mixed with 0.010 m^3 biochar dust to have 5% (1:20) biochar–soil ratio. It was again repeated for TC, where 0.3 m^3 of soil dug out from the pit was mixed with 0.015 m^3 of biochar dust to get 1:20 (5%) biochar–soil ratio. No biochar was used in the control treatment (CK). Each treatment was replicated 6 times to get a total of 24 treatment plots and each plot planted in 9 hills of rice (a total of 216 hills). This meant that the study was a 1 × 3 factorial treatment (with one biochar–soil mixture applied in three depth variations) for a total of 3 treatments and then the control (Figure 2a).

2.5. Agronomic Practices

New Rice for Africa (NERICA) is a cultivar group of interspecific hybrid rice developed from the African *Oryza glaberrima* (high biomass) and Asian *Oryza sativa* (high yield) varieties by the Africa Rice Center (WARDA) to improve the yield of African rice cultivars. NERICA-4 rice cultivar is a high-tillering and high-yielding upland rice hybrid. Its strong vigor at seedling and vegetative stages with moderate tillering, intermediate height, high water logging resistance, and short duration (120 days) helps to successfully suppress weed. The seed rice was manually planted at equal row and column spacing of 25 cm and planting depth of 2 cm (Figure 2b). The experiment was done under rainfed conditions and therefore irrigation was not at any point considered in the study. The seed rice was planted on the 21st of July 2016 at 16:00 pm (Sierra Leone local time) and at a seeding rate of 4 seeds per hill and planting density of 9 hills per plot for each of the 24 plots. Then 4 weeks after planting, an organic growth booster (trade named D. I. Grow Fertilizer with N1.85%:P1.85%:K3.31% + Te) and the insecticide (Lara Force) were applied on the rice seedlings to boost growth and prevent pest infestation. The growth-booster fertilizer and pesticide treatments were repeated in the 6th and 8th weeks after planting.

Hand weeding was done at the end of the 5th week and then tinning done to one seedling per hill in the 6th week. During thinning, only the most vigorous seedling was retained in each hill. Then NPK (15:15:15) fertilizer was applied in the row method, followed by mulching with grass straw. Because the experiment was conducted under rainfed condition, mulching moderated soil water loss through evaporation, suppressed weed growth, and prevented soil hardening and cracking during intermittent spells of no rainfall. Also fertilizers were applied to further boost the generally porous and poor soil for good crop growth, especially for the control treatment. The pesticide was also used to prevent pest damage because pests extremely strive under hot, humid tropical conditions.

Booting started at the end of the 9th week, followed by heading at the end of the 10th week (Figure 2c). During milking, the entire rice field was covered with mosquito tent (knitted together) by hooking it over bamboo cane poles to prevent bird damage and grain loss (Figure 2d). The rice was not ready for harvest until the end of the 14th week, probably due the cloudy and raining weather conditions.

At harvest, the panicle was first cut at the sheath point of the last leaf using penknife. Then the remaining rice stalk was uprooted from the soil for further treatment (including measurement of the various parameters).

2.6. Data Collection

Data were collected for various growth and yield parameters of the Nerica-4 rice plant, including both belowground (root length and weight) and aboveground (shoot height and weight, tiller number, panicle length, and grain weight) parameters. All the variables were measured at hill-scale. Soil moisture was gravimetrically measured at harvest, with samples taken from 2 plots per treatment, except for the control treatment where 3 plots were used (Figure 2a).

For crop height, measurements started at the end of the 4th week (using a graduated PVC box pipe) and repeated every 2 weeks until harvest at the end of the 14th week (which was on 4th November 2016). Tiller number count started at the end of the 6th week and also repeated every 2 weeks until harvest. With tillering, only the height of the tallest tiller in a hill was recorded. For the measurement involving weight, both fresh and dry weights were taken after harvest, although only dry weight was presented in this work. Again the data were taken at hill scale and since there was a large number of treatment hills (216 hills), samples were taken from the weighed fresh variables for dry weighing. The samples were tagged serially from 001 to 216, corresponding to the hill sequence in Figure 2a. The samples (of root, stem, panicle and grain) were again weighed fresh, oven-dried at 105 °C for 92 h (to a constant weight) and then reweighed to get the dry weight. Then the per-hill dry weight was calculated (by wet weight) from the sample as:

$$V_{dwh} = \left(\frac{S_{dw}}{S_{fw}}\right) \times V_{fwh} \qquad (1)$$

where V_{dwh} is variable dry weight per hill [g]; S_{dw} is sample dry weight [g]; S_{fw} is sample fresh weight [g]; and V_{fwh} is variable fresh weight per hill [g].

2.7. Data Analysis

The results were reported as mean and standard deviations (SD) of the variables. For statistical analysis, the data were first screened using the Dixon Outlier Test [44] and all outliers removed before processing for the mean and SD [45]. The SPSS (Statistical Package for Social Sciences) software, now renamed PASW (Predictive Analytics SoftWare), was used to analyze the data. The PASW software has powerful and efficient focal statistics and a wide global application [46].

Two-way analysis of variance (ANOVA) along with Student–Newman–Keuls (SNK) tests were conducted to isolate treatment (biochar depth) effects on crop productivity at the 99% ($p = 0.01$) confidence level [47,48]. ANOVA F-statistic was used to measure variations within samples and ANOVA p-value to test the significance of the independent on the dependent variables. SNK, a stepwise multi-comparison procedure, was used to identify sample means with significant differences [49]. Furthermore, the interdependence of the crop parameters (grain yield against biomass yield, plant height against tiller number, tillers with panicle against tillers without panicle, etc.) was tested using regression analysis and the equations and determination coefficients (R^2) shown on the plots. Also a table was dedicated to the correlativity of the variables.

3. Results and Analysis

3.1. Crop Tiller and Length Dynamics

Tiller count started at the end of the 6th week and was repeated every 2 weeks until the 14th week, which was the week of harvest. As in Figure 3a, the number of tillers increased with time from week 6 to week 10, after which it dropped. The number of tillers per hill was highest for TB (average of 10.56 tillers/hill), followed by TA, CK, and then TC treatment, in that order (Figure 3b). The average rate of tillering for all the treatments was 0.17 tillers/day, which was also highest for TB (0.19 tillers/day) and lowest for TC (0.15 tillers/day). Based on ANOVA and 2-tailed T-test analysis at $p = 0.05$ confidence level, there was significant correlation ($r > 0.97$) between every paired combinations of the treatments but with significant difference, except for TA and CK. Generally, NERICA rice can grow 7-26 tillers per plant. Specifically, NERICA-4 (which is an upland rice cultivar) grown under rainfed conditions can produce an average of 12 tillers per plant [32,50]. The number of tillers in this study was 3-27, with an average of 10 tillers per plant.

The length of the NERICA-4 rice plant was measured along with that of the yield components, including root, shoot and panicle. Figure 3c plots the measured plant growth from the 4th to the 14th week after planting. Similar to tillering, growth was rapid until the 10th week and almost stagnated until the 14th week. Unlike tillering, however, growth was strongest in CK (132.96 cm), followed by TB (131.56 cm), TC (129.30 cm) and then TA (124.19 cm) in that order (Figure 3d). The average height of the NERICA-4 rice cultivar is 120 cm [50], meaning the performance of the rice in terms of height was above average. The average growth rate was 1.85 cm/day and all the treatments strongly correlated ($r > 0.99$), but significantly different at the 0.05 confidence level (2-tailed T-test); except for TA/TC and TB/TC. From the plots on tiller number and plant height, it was not exactly clear that biochar application depth significantly improved the productivity of the rice crop under tropical rainfed conditions.

To understand the growth dynamics further, the lengths of some of the yield components were measured at harvest and plotted in Figure 4. Note that the trend line in Figure 4a is the sum of all the measured yield components, and that (which is the same as the value for plant length) in the bottom plate is the average across the treatments. From the plot, it was clear that with root length added, TA had the highest productivity in terms of plant length (172.90 cm). This was followed by TB, CK, and then TC had the shortest length (134.48). Generally, root length decreased with increasing biochar application depth, and root length under CK was higher than that under TC (the 30 cm biochar depth treatment). Also along the yield components, TA performed the best and TC the worst (Figure 4b). The length increased across the yield components from the root to the panicle, to the shoot and then to the entire plant (which was the sum of the other 3 yield components). TC had the shortest length, and it was the treatment most affected by insect/pests in the study. There was also mounding in TC, resulting from mixing the biochar with excavated soil to the depth of 30 cm. This affected water infiltration and with possible negative effect on the performance of TC.

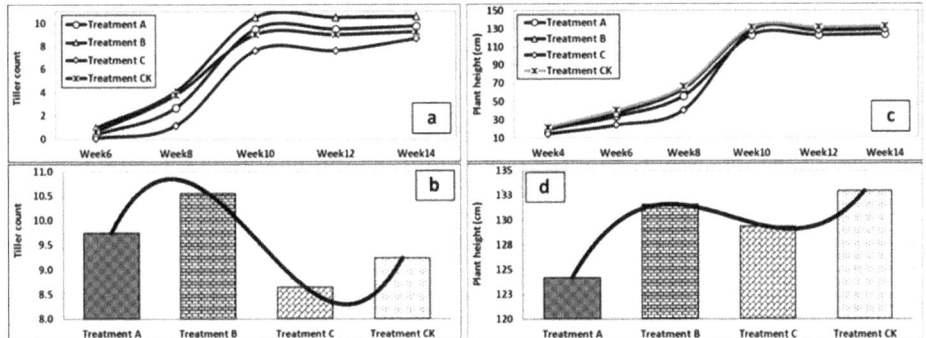

Figure 3. Number of tillers from the 6th to the 14th week (plots **a** & **b**) and plant growth from the 4th to the 14th week (plots **c** & **d**) of planting of New Rice for Africa (NERICA)-4 rice plant in 3 biochar application depths plus the control under tropical rainfed conditions. Treatments A, B, and C respectively denote the 10, 20, and 30 cm biochar application depths and then non-biochar treatment (CK) is the control.

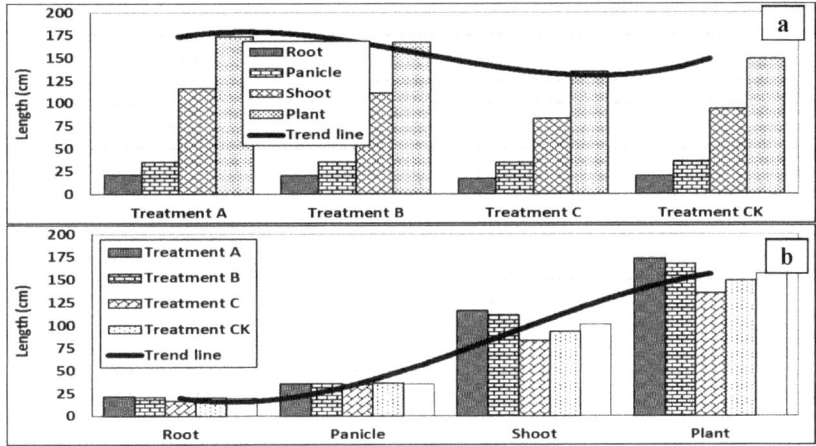

Figure 4. Length of NERICA-4 rice plant along the different treatments (**a**) and yield components (**b**). Treatments A, B, and C respectively denote the 10, 20, and 30 cm biochar application depths and then CK is the control.

3.2. Crop Weight Dynamics

The weights of the various yield components of the NERICA-4 rice plant were taken at harvest, dried, reweighed, and the dry weight plotted in Figure 5. Note that the trend line in Figure 5a is the sum for the yield components (which is the same as the value for the plant), and that in the bottom plate (of Figure 5) is the average across the treatments. Based on the plot in Figure 5, yield weight was smallest at the root scale (1302.56 kg/ha) and highest at the plant scale (16,673.07 kg/ha). At the treatment scale, yield weight was highest under TB (17,799.57 kg/ha) and lowest under TC (14,521.25 kg/ha). Grain weight, which is among the most important yield components of rice, was highest under TB (5704.45 kg/ha), followed by CK (5222.32 kg/ha) and TA (4791.87 kg/ha), and it was lowest under TC (3910.07 kg/ha). For NERICA-4 rice cultivar, average grain weight is 5000 kg/ha [32,50]. This suggested that the plant performed above average for TB and CK, but below average for TA and TC. While the below-average performance of TA could be attributed to the shallow biochar application depth (10 cm, far below rice root penetration depth), that for TC was likely due to pest damage and mounding effect on the yield. In terms of dry weight, root accounted for 7.81%, grain 29.43% and then shoot 62.76% of the NERICA-4 rice plant.

Despite the good correlation ($r > 0.74$), 2-tailed T-test analysis showed significant differences (at $p = 0.05$) among all paired combinations of the parameters along crop yield components. However, there was strong correlation ($r > 0.99$) and no significant differences (at $p = 0.05$) among paired combinations of the treatments for the yield weights. Based on ANOVA analysis, the treatments did not have any significant effect on yield weight, but the yield components were indeed significantly different in terms of weight. It is good to note, however, that yield component is partly a genetic factor that controls root, shoot and grain ratio, irrespective of the treatment.

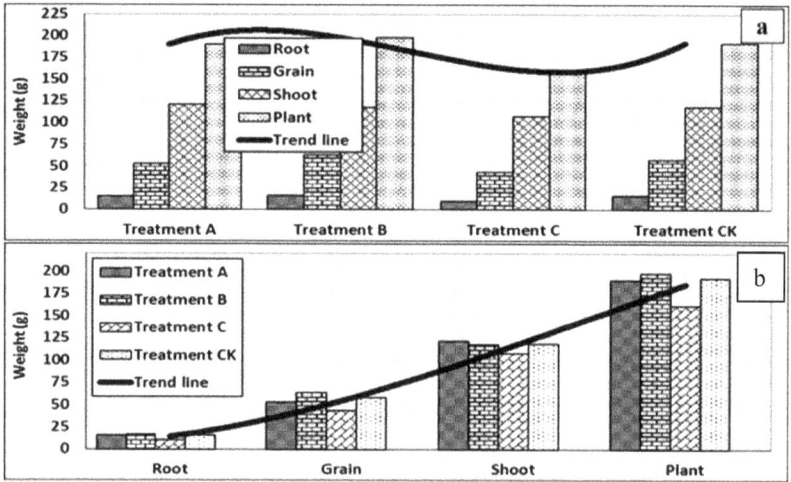

Figure 5. (a) Weight of NERICA-4 rice plant along different treatments; (b) yield components. Treatments A, B, and C respectively denote the 10, 20, and 30 cm biochar application depths and then CK is the control.

3.3. Crop and Soil Moisture Dynamics

From the measured fresh and dry weights of the yield components of the NERICA-4 rice cultivar, crop moisture content was determined using Equation (1) and plotted in Figure 6a,b. Also in Figure 6c is the soil moisture content calculated from soil samples taken from the treatment plots (see the materials and method section). Note that the trend lines in Figure 6 depict the average values across either the treatments or the yield components.

For crop moisture content across the treatments (Figure 6a), moisture content was highest under TB (44.01%), followed by TC (37.14%), CK (36.87%), and then TA (33.75%). Also for crop moisture content across yield components (Figure 6b), it was highest for shoot (49.18%), followed by root (40.18%), the whole plant (37.94%), and then the grain (24.47%). Then for soil moisture content across the treatments (Figure 6c), it was highest under TA (15.79%), followed by TB (15.58%), CK (15.17%), and then TC (14.98%).

For crop moisture content across treatment combinations, there was strong correlation ($r > 0.99$) between every paired combination of the treatments. However, none of the correlations was significant at $p = 0.05$. For crop moisture content across yield component combinations, there was no correlation ($r = 0.45$-0.47) between 2 of the paired yield component combinations (grain, root and grain, shoot) but there was correlation ($r = 0.67$-0.87) between the other paired yield-component combinations. ANOVA analysis showed that except for the root, shoot and root, plant yield-component combinations, there were significant ($p = 005$) differences in crop moisture content at the yield scale. Then at treatment scale for soil moisture content, all the paired combinations were correlated (r = 0.73-0.99) but none was significant at p = 0.05; except for the TA, TC (p = 0.02), and TB, CK (p = 0.03) treatment combinations.

While the differences in both the crop and soil moisture contents was attributable to treatment and genetic factors, also field conditions (that were not part of the treatment) possibly contributed to the differences. For instance, the separation of the treatment plots by bamboo canes caused water logging in TA and CK treatments during rainfall events. Also mixing biochar with dug-out soil caused mounding after backfilling under TC (the treatment with the highest depth, 30 cm), limiting infiltration during rainfall events. These field conditions in combination with treatment effects and genetic factors contributed to the differences in both the soil and crop moisture contents in the study. This necessitated screening for the three factors (genetic, treatment, and field condition) with the highest impact on the observed discrepancies.

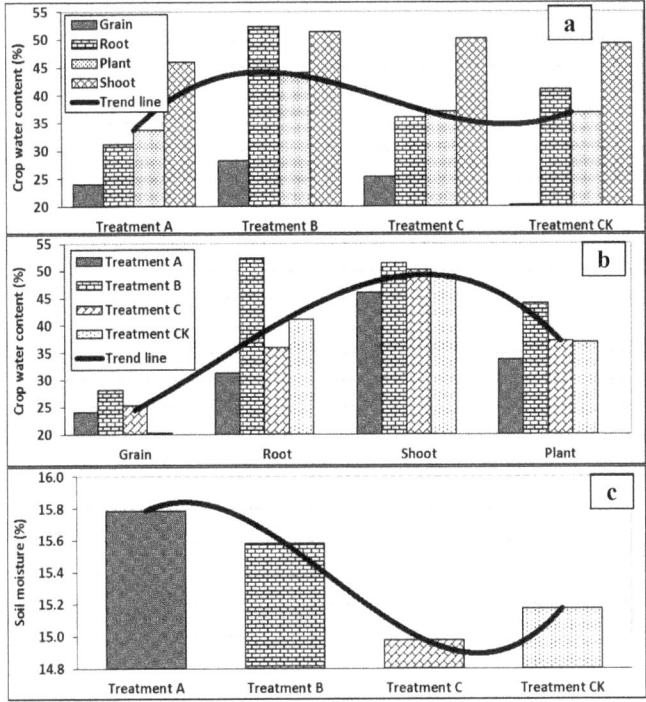

Figure 6. Moisture content of NERICA-4 rice plant along the different treatments (**a**), yield components (**b**) and that of the soil along the treatments (**c**). Treatments A, B, and C respectively denote the 10, 20, and 30 cm biochar application depths and CK is the control.

3.4. Parameter Correlativity

The coefficients of determination (R^2) for paired combinations of the yield components are given in Table 1. With the exception of just 4 paired combinations (TN, SL; TN, RL; TN, PL, and TN, CM; where TN is tiller number, SL shoot length, RL root length, PL panicle length, and CM crop moisture content), all the measured yield components were strongly correlated. In fact, perfect correlations (R^2 = 1.00) existed between 9 of the paired yield components. Despite the strong correlations, however, 2-tailed T-test showed that some of the variables (italicized values) had significant differences (and that was mostly with panicle weight, PW). This suggested that the yield components were largely interdependent. Only 2 of the paired variables (TN, RL and TN, CM; underlined in Table 1) had negative correlation. For instance, the number of tillers (and especially tillers with grains) influenced the quantity of grains produced ($R^2 = 0.89$).

Table 1. A table of correlativity of yield component variables of NERICA-4 upland rice. In the table, TN = tiller number, GW = grain weight, SW = shoot weight, SL = shoot length, RW = root weight, RL = root length, PW = panicle weight, and PL = panicle length and CM crop moisture content.

R^2	TN	GW	SW	SL	RW	RL	PW	PL	CM
TN	1.00								
GW	0.89	1.00							
SW	0.84	0.99	1.00						
SL	0.19	0.91	0.93	1.00					
RW	0.94	0.94	0.99	0.96	1.00				
RL	0.00	0.89	0.91	1.00	0.96	1.00			
PW	0.90	1.00	1.00	0.95	1.00	0.97	1.00		
PL	0.19	0.91	0.93	1.00	0.97	1.00	0.98	1.00	
CM	0.14	0.88	0.92	1.00	0.96	1.00	0.98	1.00	1.00

4. Discussions

4.1. Biochar Research

Maintaining appropriate levels of nutrient cycle is key for successful soil management efforts in humid tropical regions. Cover crop, mulch, compost or manure additions supply nutrients to crops, shorten nutrient cycle by increasing microbial biomass activity and hold onto applied mineral fertilizers [51,52]. However, none of these seem to have a long-term effect like biochar. Biochar research regained momentum with the discovery of the Amazonian *terra preta* soils in the endless search for sustainable agriculture and worsening climate change. The *terra preta* soil is today widely studied for reproduction and application in soil amendment because of its huge potential for nutrient/carbon capture and storage [16]. The biochar-rich *terra preta* soils in the Amazon occur in various depths and spatial extents and are also highly potent in nutrient and carbon, and can therefore support strong vegetation growth. While biochar studies have mostly focused on the production modes and application rates, little has been done on the application depth. This study was set up to determine the effect of biochar application depth on crop productivity under tropical rainfed conditions. A fixed soil-biochar volume ratio of 1:20 (5% biochar) was applied in 3 depths—TA (10 cm), TB (20 cm), and TC (30 cm)—and investigated against a control treatment (CK) with no biochar.

4.2. Treatment Effect on Yield

Based on the results, the number of tillers per hill was highest under TB, followed by TA, CK, and then TC. The increase in the number of tillers in successive time periods (2 weeks in this case) were strongly correlated ($r > 0.85$), but only significantly different ($p = 0.05$) for the period from the 6th to the 8th week and then from the 8th to the 10th week (Figure 3a,b). Tillering was not completely consistent with biochar application depth, probably due to field contingencies (pest damage and water logging) under the tropical rainfed condition.

For the shoot system, increase in plant height in successive time periods (also 2 weeks in this study) was strongly correlated ($r > 0.96$), except for the period between the 8th and 10th weeks ($r = 0.41$). Also 2-tailed T-test showed that the differences in the increase in height between the successive time periods were significant at the 0.05 confidence interval. On average, plant height was highest under CK (control treatment), followed by TB, TC, and TA (Figure 3c,d). Like tillering, plant height was not completely consistent with biochar application depth and probably affected by the same contingent field conditions.

For plant length (which is the total length for root, shoot and panicle), it decreased with increasing biochar application depth (Figure 4). However, plant length under CK (no biochar) was higher than that under TC (30 cm biochar depth). On average, the rice plant increased from root to panicle (cut from the last leaf base) and then to shoot (without panicle). Genetically, rice shoot is longest, followed by panicle and then root (Figure 4). The approximate ratio of panicle-to-root was 1:2, shoot-to-panicle was 1:3, and panicle-to-shoot was 1:5.

The observed field contingencies included water logging, which was highest in CK (where there was no tillage), followed by TA (10 cm tillage depth plus biochar), but no water logging in TB (20 cm tillage depth plus biochar), but even mounding in TC (30 cm tillage depth plus biochar). The differential water logging condition was because with no tillage (CK), the bamboo canes around the plots caused sunken beds to develop. Under TA, the addition of only 1 part of biochar to 20 parts of dug-out soil to the depth of 10 cm increased little the backfill volume and therefore sunken beds also formed after using separation bamboo canes. Under TB, the increase in volume due to the mixing of the 20-cm depth of soil with biochar caused the bed to just level out with the separation canes and there was therefore no water logging. Finally, under TC, the increase in volume due to the mixing of 30 cm depth of excavated soil with biochar cause mounding above the separation canes, causing runoff. As a key factor for rice growth, the differential water conditions of the treatment plots affected the crop yield. Also, because no buffer plots were put around the treatment plots, pest out outbreak affected more the peripheral TA (which bordered on vegetable plots) and TC (which bordered on bush) treatment plots. Pest outbreak was highest in TC, followed by TA, but almost not in TB and CK treatments. The pest damage also contributed to the inconsistencies in the measured plant height and length along the experimental treatments.

The total plant weight (i.e., root, shoot with panicle and grain) was highest for TB, followed by CK, then TA and finally TC (Figure 5). In terms of the measured yield components, weight increased from root to grain and then to shoot. Root accounted for 7.81% of the total plant weight, grain 29.43% and then shoot (with panicle) 62.76%. Especially for grain (which is the most important yield component for rice), it was highest for TB, followed by CK, TA, and then TC. While shoot weight was highest for TA, grain weight and root weight were highest for CK. Weight was not consistent along the treatments (either way from TA to TB to TC) probably because of the negative effect of the contingent field conditions explained above.

For crop moisture content, it was highest under TB, followed by TC, CK, and then TA. For the measured yield components, it was highest under TB and lowest under TA for root and shoot. Then for grain, moisture content was highest under TB and lowest under CK. The percent contributions of grain, root and shoot (plus panicle) to the total moisture content of the rice plant were 21.50%, 35.30%, and 43.21%, respectively. For soil moisture content, it was simply highest for TA, followed by TB, then CK and lowest for TC. Irrespective of the inconsistencies in the measured variables along the treatments (probably due to the contingent field conditions), there were strong correlations among most of the paired combinations of the variables (Table 1).

4.3. Field Contingency Effect on Yield

To further understand the effect of the contingent field conditions on the performance of the measured yield variables, a table of the best- and worst-performing treatments was constructed (Table 2). Note that plant weight (PW) and height (PH) constituted of the sum of the values of all the measured yield components (root, shoot, panicle and grain), and therefore not separately given. Thus, from Table 2, it was apparent that under high soil moisture (SM) content, water was not a factor for the performance of the measured yield variables (MYV), except plant length (PL). Conversely under low soil moisture content, water became critical for the performance of all the measured yield variables, except crop water content (CW). This suggested that rice plant always adjusted its own tillering (TN), growth (measured as PL) and dry matter accumulation (measured as PW) to maintain a certain level of water to keep alive. Soil moisture content was lowest under TC because of mounding (raised bed), implying that most of the soil moisture was absorbed to sustain growth under TC. Furthermore, large plant length was influenced only by soil moisture, whereas small plant length was a function of soil moisture, tiller number, and plant weight. While low tiller number was affected by all the measured yield variables, except crop water content, high tiller number was a function of both crop water content and plant weight.

Table 2. A table depicting the match of the biochar depth treatments (BDT) with the highest and lowest values of the measured yield variables (MYV) of NERICA-4 rice plant under tropical rainfed conditions.

MYV	BDT	Level	MYV Match				
			TN	PW	PL	CW	SM
TN	TB	Highest	√	√	×	√	×
	TC	Lowest	√	√	√	×	√
PW	TB	Highest	√	√	×	√	×
	TC	Lowest	×	√	√	×	√
PL	TA	Highest	×	×	√	×	√
	TC	Lowest	√	√	√	×	√
CW	TB	Highest	v	√	×	√	×
	TA	Lowest	×	×	×	√	×
SM	TA	Highest	×	×	√	×	√
	TC	Lowest	√	√	√	×	√

Note that TA = treatment A (with 10 cm biochar application depth), TB = treatment B (with 20 cm biochar application depth), TC = treatment C (with 30 cm biochar application depth) and CK is control treatment (with no biochar). The measured yield variables include TN = tiller number, PH= plant height, PW = plant weight, PL = plant length, CM = crop moisture content, SM = soil moisture content, √ = a match, and × = no match. Double underline means treatment with water logging (due to the development of sunken) and limited pest infestation whereas italicized bold means treatment with runoff (due to mounding) and high pest infestation.

When related with the contingent field conditions, it was clear that TC (the 30-cm depth treatment with mounding runoff and high pest damage) had the lowest value for the measured yield variables, except crop water content (Table 2). Also, while low crop water content was not really a growth and yield factor, high crop water content influenced both tiller number and plant weight, but not plant length. Here again, soil moisture content had no effect on crop water content. Table 2 also showed that while TA (10-cm depth treatment with high water logging and limited pest damage) was lowest only for crop water content, TB (20-cm depth treatment with no water logging or pest damage) was highest for all the measured yield components, except plant length. This reaffirmed the point that the contingent field conditions affected the results of the experiment. Deductively therefore, it was possible that biochar application depth had a boosting effect on the performance of NERICA-4 rice cultivar. This was because for most of the time, TB (20-cm biochar depth treatment with no pest damage) performed better than CK (control treatment with water logging and no biochar or pest damage).

4.4. Contingent Field Condition and Biochar Treatment Effects

To quantify the effects of the contingent field conditions and the biochar application depth on the measured yield components, simple relational equations were developed. For this purpose, it was assumed that the effect of water logging, wherever it occurred, was the same. Similarly, the effect of pest damage, wherever it occurred, was the same too. Then while water logging and biochar application were yield-gain factors, pest damage was a yield-loss factor. With this, we then take the effect of water logging on the crop performance to be w, that of pest damage be p and that of biochar depth be b. Based on field observations, there was water logging in CK ($CK \to w$), water logging, biochar use and pest damage in TA ($TA \to w + b - p$), biochar use in TB ($TB \to b$) and then biochar use and pest damage in TC ($TC \to b - p$). It then becomes apparent that $TA - TC = w$ and that $CK - w = \overline{CK}$ (where $\overline{CK}$ is crop yield without the effects of water logging, biochar use and pest damage). Also $TC - TB = -p$ (where p is a yield loss factor and therefore negative). With w and p known, then b can be derived from $TA \to w + b - p$; thus $b = TA - \overline{CK} - w + p$. Note that TA, TB, TC and CK are used as defined earlier.

Using the above relations for plant weight (PW), there was 20.30% yield loss under TA due to pest damage, 36.84% yield loss under TB due to runoff (which indicates no water logging), 86.74% yield loss under TC due both to pest damage and runoff and then 14.86% yield gain under CK due to water logging. Similarly for plant length (PL), there was 4.98% yield loss under TA due to pest damage, 38.98% yield loss under TB due to runoff, 83.29% yield loss under TC due both to pest damage and runoff, and then 25.73% yield gain under CK due to water logging. Then finally for tiller number (TN), there was 14.45% yield loss under TA due to pest damage, 36.14% yield loss under TB due to runoff, 73.02% yield loss under TC due both to pest damage and runoff and then 11.82% yield gain under CK due to water logging. Yield loss or gain for crop water content (CW) was not quantified because Table 2 suggested that it had no effect on the output of the rice crop. This implied that the minimum crop water requirement for growth was available throughout the experimental study.

After adjustment for the calculated values (i.e., after removal of the effects of pest damage and water logging on the biochar depth treatments), the effects of only the different biochar application depth treatments on the NERICA-4 rice yield are plotted in Figure 7 for the yield components including plant weight, plant length and tiller number. The plot showed that the rice crop yield increased with increasing biochar application depth and it was generally higher in biochar than in non-biochar treatments. From Figure 4, the measured root length was lowest in TC (19.77 cm) and longest in TA (21.54 cm). It then suggested that the rice root penetrated far beyond the 10-cm depth of biochar in TA, but that only halfway through the 30-cm biochar depth in TC. Thus, biochar generally limited root penetration depth, probably because of sufficient availability of nutrients in the upper layers of the soil where the biochar was concentrated.

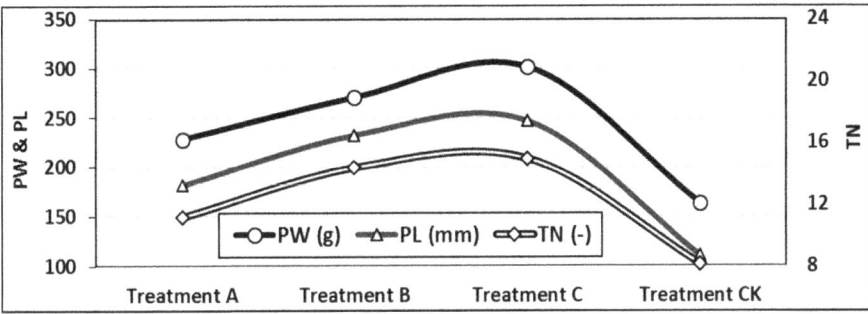

Figure 7. Plot depicting the trends in the effect of biochar depth on selected rice yield components after removal of the effects of contingent field conditions of water logging and pest damage. Note that PW = plant weight, PL = plant length, and TN = tiller number.

5. Conclusions

This study was set up to investigate the effect of biochar application depth on NERICA-4 upland rice cultivar under tropical rainfed conditions. Based on the study, biochar application not only boosted rice growth, but also increased rice yield with increasing application depth. There were different modes of biochar application, either as dust or lump or surface broadcasting or deep tillage. The results of the pilot study reconstructed from the formula developed suggested that the effects of biochar on crop growth and productivity also vary with the mode of application. Based on this study, the appropriate depth of application (which can vary with crop type) was critical for desired effects of biochar on crop productivity. With sufficient biochar in the soil, crop rooting depth can be reduced as biochar can create sufficient reserve of nutrients within the zone of application.

Importantly, however, contingent field conditions (including water logging, runoff, and pest damage) can variously affect rice plant growth. Although contingent field conditions were quantified in this study, sound experimental designs were needed to preclude damage due to any such conditions. For instance, since disease and pest outbreaks often start at peripheral plots, it is necessary to keep reasonable buffers around experimental plots to have sufficient time for effective action in the event of any such outbreaks. Despite all of this, the study showed that deep soil tillage with biochar was a more productive mode of cultivation than shallow or even surface applications under tropical rainfed conditions.

Author Contributions: J.P.M. conceived and designed the experiment and wrote the report; A.W. conducted the experiment and analyzed the data; M.M.B. analyzed and reviewed the data; M.P.N. financed the experiment; and R.S. financed the publication.

Funding: This research received no external funding.

Acknowledgments: We are grateful for the advice from the field technicians in the Department of Agricultural and Biosystems Engineering, Njala University, Sierra Leone.

Conflicts of Interest: The authors declare no conflict of interest.

References

1. Bridgwater, A.V. Renewable fuels and chemicals by thermal processing of biomass. *Chem. Eng. J.* **2003**, *91*, 87–102. [CrossRef]
2. Bridgwater, A.V. Review of fast pyrolysis of biomass and product upgrading. *Biomass Bioenergy* **2012**, *38*, 64–94. [CrossRef]
3. Lehmann, J. Terra preta nova—Where to from here? In *Amazonian Dark Earths: Wim Sombroek's Vision*; Woods, W.I., Wenceslau, G., Teixeira, W.G., Lehmann, J., Steiner, C., Winkler-Prins, A., Rebellato, L., Eds.; Springer Science: Berlin, Germany, 2009; p. 520.
4. Bian, R.; Joseph, S.; Cui, L.; Pan, G.; Li, L.; Liu, X.; Zhang, A.; Rutlidge, H.; Wong, S.; Chia, C.; et al. A three-year experiment confirms continuous immobilization of cadmium and lead in contaminated paddy field with biochar amendment. *J. Hazard. Mater.* **2014**, *272*, 121–128. [CrossRef]
5. Graber, E.R.; Tsechansky, L.; Mayzlish-Gati, E.; Shema, R.; Koltai, H. A humic substances product extracted from biochar reduces Arabidopsis root hair density and length under P-sufficient and Pstarvation conditions. *Plant Soil.* **2015**, *395*, 21–30. [CrossRef]
6. Skjemstad, J.O.; Clarke, P.; Taylor, J.A.; Oades, J.M.; Mcclure, S.G. The chemistry and nature of protected carbon in soil. *Aust. J. Soil Res.* **1996**, *34*, 251–271. [CrossRef]
7. Skjemstad, J.O.; Reicosky, D.C.; Wilts, A.R.; McGowan, J.A. Charcoal carbon in U.S. agricultural soils. *Soil Sci. Soc. Am. J.* **2002**, *66*, 1249–1255. [CrossRef]
8. Barrow, C.J. Biochar: Potential for countering land degradation and for improving agriculture. *Appl. Geogr.* **2012**, *34*, 21–28. [CrossRef]
9. Chan, K.Y.; Van Zwieten, L.; Meszaros, I.; Downie, A.; Joseph, S. Agronomic values of greenwaste biochar as a soil amendment. *Aust. J. Soil Res.* **2007**, *45*, 629–634. [CrossRef]
10. Spokas, K.A.; Cantrell, K.B.; Novak, J.M.; Archer, D.W.; Ippolito, J.A.; Collins, H.P.; Boateng, A.A.; Lima, I.M.; Mekuria, W.; Noble, A. The role of biochar in ameliorating disturbed Soils and sequestering soil carbon in tropical agricultural production systems. *Appl. Environ. Soil Sci.* **2013**, 1–10. [CrossRef]
11. Laird, D.A. The Charcoal Vision: A win-win-win scenario for simultaneously producing bioenergy, permanently sequestering carbon, while improving soil and water quality. *Agron. J.* **2008**, *100*, 178–181. [CrossRef]
12. Novak, J.M.; Lima, I.; Gaskin, J.W.; Steiner, C.; Das, K.C.; Ahmedna, M.; Rehrah, D.; Watts, D.W.; Busscher, W.J.; Schmobert, H. Characterization of designer biochars produced at different temperatures and their effects on a lomay sand. *Ann. Environ. Sci.* **2009**, *3*, 195–206.
13. Atkinson, C.J.; Fitzgerald, J.D.; Hipps, N.A. Potential mechanisms for achieving agricultural benefits from biochar application to temperate soil: A review. *Plant Soil* **2010**, *337*, 1–18. [CrossRef]
14. Haefele, S.M.; Konboon, Y.; Wongboon, W.; Amarante, S.; Maarifat, A.A.; Pfeiffer, E.M.; Knoblauch, C. Effects and fate of biochar from rice residues in rice-based systems. *Field Crop. Res.* **2011**, *121*, 430–440. [CrossRef]
15. Ahmad, M.; Rajapaksha, A.U.; Lim, J.E.; Zhang, M.; Bolan, N.; Mohan, D.; Vithanage, M.; Sang Soo Lee, S.S.; Ok, Y.S. Biochar as a sorbent for contaminant management in soil and water: A review. *Chemosphere* **2014**, *99*, 19–33. [CrossRef] [PubMed]
16. Lehmann, J.; Gaunt, J.; Rondon, M. Bio-char sequestration in terrestrial ecosystems e a review. *Mitig. Adapt. Strateg. Glob. Chang.* **2006**, *11*, 403–427. [CrossRef]

17. Glaser, B.; Haumaier, L.; Guggenberger, G.; Zech, W. The Terra Preta phenomenon: A model for sustainable agriculture in the humic tropics. *Die Nat.* **2001**, *88*, 37–41. [CrossRef]
18. Grossman, J.M.; O'Neill, B.E.; Tsai, S.M.; Liang, B.; Neves, E.; Lehmann, J.; Thies, J.E. Amazonian anthrosols support similar microbial communities that differ distinctly from those extant in adjacent, unmodified soils of the same mineralogy. *Environ. Microbiol.* **2010**, *60*, 192–205. [CrossRef]
19. Mishra, B.K.; Ramakrishnan, P.S. Slash and burn agriculture at higher elevations in North-Eastern India. I. Sediment, water and nutrient losses. *Agric. Ecosyst. Environ.* **1983**, *9*, 69–82. [CrossRef]
20. Brewer, C.E. Biochar Characterization and Engineering. Graduate Theses and Dissertations, Paper 12284. Iowa State University, Ames, IA, USA, 2012. Available online: http://lib.dr.iastate.edu/etd (accessed on 4 February 2019).
21. Glaser, B.; Woods, W.I. *Amazonian Dark Earths: Explorations in Space and Time*; Springer: Berlin, Germany, 2004.
22. Lehmann, J.; Kern, D.C.; Glaser, B.; Woods, W.I. *Amazonian Dark Earths: Origins, Properties, Management*; Kluwer Academic: Dordrecht, The Netherlands, 2003.
23. Keech, O.; Carcaillet, C.; Nilsson, M.C. Adsorption of alleopathic compounds by wood-derived charcoal; the role of wood porosity. *Plant Soil* **2005**, *272*, 291–300. [CrossRef]
24. Glaser, B.; Lehmann, J.; Zech, W. Ameliorating physical and chemical properties of highly weathered soils in the tropics with charcoal—A review. *Biol. Fertil. Soils* **2002**, *35*, 219–230. [CrossRef]
25. Steiner, C.; Teixxeira, W.G.; Lehmann, J.; Nehls, T.; de Macedo, J.L.V.; Blum, W.E.H.; Zech, W. Long-term effects of manure, charcoal and mineral fertilization on crop production and fertility on a highly weathered central Amazonian upland soil. *Plant Soil* **2007**, *292*, 275–290. [CrossRef]
26. Asai, H.; Samson, B.K.; Stephan, H.M.; Songyikhangsuthor, K.; Homma, K.; Kiyono, Y.; Inoue, Y.; Shiraiwa, T.; Horie, T. Biochar amendment techniques for upland rice production in Northern Laos 1. Soil physical properties, leaf SPAD and grain yield. *Field Crops Res.* **2009**, *111*, 81–84. [CrossRef]
27. Hammes, K.; Schmidt, M.W.I. Changes of biochar in soil. In *Biochar for Environmental Management: Science and Technology*; Lehmann, J., Joseph, S., Eds.; Earthscan: London, UK, 2009; pp. 169–178.
28. Hossain, M.K.; Strezov, V.; Yin, C.K.; Nelson, P.F. Agronomic properties of wastewater sludge biochar and bioavailability of metals in production of cherry tomato (*Lycopersicon esculentum*). *Chemosphere* **2010**, *78*, 1167–1171. [CrossRef] [PubMed]
29. Laird, D.A.; Fleming, P.; Davis, D.D.; Horto, R.; Wang, B.; Karlen, D.L. Impact of biochar amendments on the quality of a typical Midwestern agricultural soil. *Geoderma* **2010**, *158*, 443–449. [CrossRef]
30. Elmer, W.H.; Pignatello, J.J. Effect of biochar amendments on mycorrhizal associations and fusarium crown and root rot of asparagus in replant soils. *Plant Dis.* **2011**, *95*, 960–966. [CrossRef] [PubMed]
31. Park, J.H.; Choppala, G.K.; Bolan, N.S.; Chung, J.W.; Chuasavathi, T. Biochar reduces the bioavailability and phytotoxicity of heavy metals. *Plant Soil* **2011**, *348*, 439–451. [CrossRef]
32. Kangoma, E.; Blango, M.M.; Rashid-Noah, A.B.; Sherman-Kamara, J.; Moiwo, J.P.; Kamara, A. Potential of biochar-amended soil to enhance crop productivity under deficit irrigation. *Irrig. Drain.* **2017**. [CrossRef]
33. Schmidt, M.W.I.; Noack, A.G. Black carbon in soils and sediments: Analysis, distribution, implications and current challenges. *Glob. Biogeochem. Cycles* **2000**, *14*, 777–793. [CrossRef]
34. Schmidt, H.P. 55 Uses of Biochar. *Ithaka J.* **2012**, *1*, 286–289.
35. Conteh, A.M.H.; Yan, X.; Moiwo, J.P. The determinants of grain storage technology adoption in Sierra Leone. *Cashier Agric.* **2015**, *24*, 47–55. [CrossRef]
36. Wu, H.; Lai, C.; Zeng, G.; Liang, J.; Chen, J.; Xu, J.; Dai, J.; Li, X.; Liu, J.; Chen, J.; et al. The interactions of composting and biochar and their implications for soil amendment and pollution remediation: a review. *Crit. Rev. Biotech.* **2017**, *37*, 754–764. [CrossRef] [PubMed]
37. Nguyen, B.T.; Lehmann, J.; Hockaday, W.C.; Joseph, S.; Masiello, C.A. Temperature sensitivity of black carbon decomposition and oxidation. *Environ. Sci. Technol.* **2010**, *44*, 3324–3331. [CrossRef] [PubMed]
38. Filiberto, D.M.; Gaunt, J.L. Practicality of biochar additions to enhance soil and crop productivity. *Agriculture* **2013**, *3*, 715–725. [CrossRef]
39. Liu, X.; Zhang, A.; Ji, C.; Joseph, S.; Bian, R.; Li, L.; Pan, G.; Paz-Ferreiro, J. Biochar's effect on crop productivity and the dependence on experimental conditions—A meta-analysis of literature data. *Plant Soil* **2013**. [CrossRef]
40. Biederman, L.B.; Harpole, W.S. Biochar and its effects on plant productivity and nutrient cycling: A meta-analysis. *GCB Bioenergy* **2013**, *5*, 202–214. [CrossRef]

41. Peel, M.C.; Finlayson, B.L.; Mcmahon, T.A. Updated world map of the Köppen-Geiger climate classification. *Hydrol. Earth Syst. Sci. Discuss. Eur. Geosci. Union* **2007**, *11*, 1633–1644. [CrossRef]
42. Denis, M.K.A.; Kamara, A.; Momoh, E. Soil fertility status of three chiefdoms in Pujehun District of southern Sierra Leone. *Res. J. Agric. Sci.* **2013**, *4*, 461–464.
43. Weber, K.; Quicker, P. Properties of biochar. *Fuel* **2018**, *217*, 240–261. [CrossRef]
44. Dixon, W.J. Analysis of extreme values. *Ann. Math. Stat.* **1950**, *21*, 488–506. [CrossRef]
45. Karer, J.; Wimmer, B.; Zehetner, F.; Kloss, S.; Soja, G. Biochar application to temperate soils: Effects on nutrient uptake and crop yield under field conditions. *Agric. Food Sci.* **2013**, *22*, 390–403. [CrossRef]
46. Quintero, D.; Ancel, T.; Cassie, G.; Ceron, R.; Darwish, A.; Felix, G.G.; He, J.J.; Keshavamurthy, B.; Makineedi, S.; Nikalje, G.; et al. *Workload Optimized Systems: Tuning POWER7 for Analytics*; An IBM Redbooks Publication: Atlanta, GA, USA, 2012; p. 200.
47. Augustenborg, C.A.; Hepp, S.; Kammann, C.; Hagan, D.; Schmidt, O.; Muller, C. Biochar and earthworm effects on soil nitrous oxide and carbon dioxide emissions. *J. Environ. Qual.* **2012**, *41*, 1203–1209. [CrossRef]
48. Buss, W.; Kammann, C.; Koyro, H.-W. Biochar reduces copper toxicity in *Chenopodium quinoa* Wild. in a Sandy Soil. *J. Environ. Qual.* **2012**, 1157–1165. Available online: www.crops.org.www.soils.org (accessed on 4 February 2019). [CrossRef] [PubMed]
49. Keuls, M. The use of the "studentized range" in connection with analysis of variance. *Euphytica* **1952**, *1*, 112–122. [CrossRef]
50. Africa Rice Center—WARDA; FAO; SAA. *NERICA®: The New Rice for Africa—A Compendium*; Somado, E.A., Guei, R.G., Keya, S.O., Eds.; Africa Rice Center (WARDA): Cotonou, Benin; FAO: Rome, Italy; Sasakawa Africa Association: Tokyo, Japan, 2008; p. 210.
51. Goyal, S.; Chander, K.; Mundra, M.C.; Kapooret, K.K. Influence of inorganic fertilizers and organic amendments on soil organic matter and soil microbial properties under tropical conditions. *Biol. Fertil. Soils.* **1999**, *29*, 196–200. [CrossRef]
52. Trujillo, L. Fluxos de Nutrientes em solo de Pastagem Abandonada sob Adubacao Organica e Mineral na Amazonia Central. Master's Thesis, INPA and University of Amazonas, Manaus, Brazil, 2002.

© 2019 by the authors. Licensee MDPI, Basel, Switzerland. This article is an open access article distributed under the terms and conditions of the Creative Commons Attribution (CC BY) license (http://creativecommons.org/licenses/by/4.0/).

Article

Optimum Method Uploaded Nutrient Solution for Blended Biochar Pellet with Application of Nutrient Releasing Model as Slow Release Fertilizer

JoungDu Shin [1,*], SangWon Park [2] and SunIl Lee [1]

[1] Department of Climate Change and Agroecology, National Institute of Agricultural Sciences, WanJu Gun 55365, Korea; silee83@rda.go.kr
[2] Chemical Safety Division, National Institute of Agricultural Sciences, WanJu Gun 55365, Korea; swpark@korea.kr
* Correspondence: jdshin1@korea.kr; Tel.: +82-63-238-2494; Fax: +82-63-238-3823

Received: 5 March 2019; Accepted: 29 April 2019; Published: 9 May 2019

Abstract: The nutrient releasing characteristics of a blended biochar pellet comprising a mixture of biochar and pig manure compost ratio (4:6) uploaded with nitrogen (N), phosphorus (P) and potassium (K) nutrient solutions were investigated with the application of a modified Hyperbola model during a 77-day precipitation period. The experiment consisted of five treatments, i.e., the control, as 100% pig manure compost pellet (PMCP), a urea solution made at room temperature (TN), a urea solution heated to 60 °C (HTN), N, P and K solutions made at room temperature (TNPK), and N, P and K solutions heated to 60 °C (HTNPK). The cumulative ammonium nitrogen (NH_4-N) in the blended biochar pellets was slow released over the 77 days of precipitation period, but nitrite nitrogen (NO_3-N) was rapidly released, i.e., within 15 days of precipitation (Phase I), close behind on a slower release rate within the final precipitation (Phase II). Accumulated phosphate phosphorus (PO_4-P) concentrations were not much different, and slowly released until the final precipitation period, while the highest accumulated K amount was 2493.8 mg L^{-1} in the TNPK at 8 days, which then remained at a stage state of K. Accumulated silicon dioxide (SiO_2) concentrations abruptly increased until 20 days of precipitation, regardless of treatments. For the application of the releasing model for nutrient releasing characteristics, the estimations of accumulated NH_4-N, NO_3-N, PO_4-P, K and SiO_2 in all the treatments were significantly ($p < 0.01$) fitted with a modified Hyperbola model. These findings indicate that blended biochar pellets can be used as a slow release fertilizer for agricultural practices.

Keywords: blended biochar pellet; modified Hyperbola; nutrient release; slow release fertilizer

---

1. Introduction

Biomass from the agricultural sector is composed of agricultural wastes such as crop residues and fringing trees. Researchers have shown that agro-biomass derived carbon materials such as biochar are one (1) option for carbon sequestration as well as the reduction of greenhouse gas emissions. It is estimated that Korea produces 50 million tons of organic waste from agriculture every year [1]. Thus, to address concerns about climate change, carbon sequestration using biomass conversion technology should be prioritized, since it has already become feasible in Korea. Biochar consists of multi-porous and carbonaceous material obtained from biomass conversion technology under limited oxygen. It consists of non-degradable carbon with double bonds and an aromatic ring that cannot be broken down by microbial degradation [2]; it could be applied for several purposes [3]. However, 30% of biochar is lost due to wind, while 25% is lost due to runoff water during spreading in cultivation areas [4].

It has been suggested that Biochar's effects come from a plant's adsorption and retention abilities vis-à-vis the available nutrients [5]. Nitrogen fertilizers have been chemically transformed through mineralization, ammonification, nitrification, de-nitrification, and immobilization in soil [6,7]. Biochar from rice hull through biomass conversion technology was reported to have 0.498 mg g^{-1} of maximum absorption capacity of NH_4-N, which can prevent nitrification [8]. Biochar from holm oak tree increased NH_4-N adsorption in sandy Acrisol, but had no effect on NO_3-N adsorption in the column experiment [9]. However, the maximum adsorption of NH_4-N in the biochar pellet is 2.94 mg·g^{-1}, and lettuce yield was enhanced by approximately 13% relative to the control [10]. When biochar pellet is applied to the soil, the application of chemical fertilizers is still needed to enhance crop productivity. Previous studies on the characterization of the NH_4-N adsorption capacity of biochar pellets recommends the mixing of biochar with compost containing nutrients before the soil application [10,11]. Phosphorous (P) is essential for maintaining profitable crop production [12,13]. However, it might cause eutrophication in lakes and ponds, thereby destroying the ecosystem [14]. In recent years, attention for biochar has increased because of its absorption abilities [15,16]. Manure-derived biochar increased the maximum uptake of nitrogen (N), phosphorus (P) and potassium (K) by ryegrass by 66.4%, 161% and 210%, respectively [17]. Biochar from tree prunings enhanced P use efficiency from organic P fertilizer more than the chemical fertilizers for both corn and wheat crops [18]. Silicon (Si), on the other hand, is important to the strength of cell walls and defense against diseases, as well as to enhancing the uptake of other nutrients, and increasing tolerance to drought, salts, and extreme temperature stresses [19–21]. Silicon plays an important role in improving the rice yields. It is a potent stimulator of photosynthesis [22].

Biochar is not suitable for crop cultivation in a practical view because it does not contain sufficient amounts of plant nutrients for crop production. Biochar pellets are an efficient way to decrease field handling costs and to significantly reduce biochar loss during soil application [23]. For soil incorporation, poultry litter was mixed, pelletized and slowly pyrolyzed to produce biochar pellets [24]. However, little information is available on blended biochar pellets as a slow release fertilizer, which is required to allow nutrients to flow slowly from the soil during the cultivation period. It supplies most nutrients into the crop without leaching and denitrification losses [25] to increase profits and minimize environmental destruction [26].

The predictions of the used model confirmed that Michaelis–Meten-type kinetics is probably the most dominant mechanism for the leaching of heavy metals from cement-based waste forms. Furthermore, Michaelis–Menten kinetics have been used to explore nitrogen deposition and climate change in laboratory experiments [27]. For the optimization of the blended biochar pellets, it was described that the release of plant nutrient amounts demonstrated a modified hyperbola model [10].

It is hypothesized that (1) the blended biochar pellet with different nutrient uploaded methods have different means of releasing nutrients, and (2) an optimal nutrient uploaded method can be selected using a modified hyperbola model.

Therefore, this experiment aimed to select an optimum nutrient uploaded method with plant nutrient solutions for blended biochar pellets using a Hyperbola model for agricultural practices.

2. Materials and Methods

2.1. Biochar Pellet Productions

Biochar (Go-Chang, JenBok) from rice hull and pig manure compost (NOUSBO Co., Suwon, Korea) were purchased from a local farming cooperative union. The biochar produced from the pyrolysis system is described in detail in Shin's previous publication [10]. The physicochemical properties of biochar and pig manure compost used in this experiment are presented in Table 1. The pH of biochar and pig manure compost were 9.8 and 8.8, and their total carbon contents were 566.3 and 288.9 g kg^{-1}, respectively.

Table 1. Physicochemical properties of biochar and pig compost used [1)].

	pH	EC (dS m^{-1})	TC	TOC	TIC	TN
			----------g kg^{-1}----------			
Biochar	9.78 (1:20)	16.53	575.5	533.0	42.5	2.0
Pig manure compost	8.77 (1:5)	3.40	288.8	258.6	30.2	29.1

[1)] TC; Total carbon, TOC; Total organic carbon, TIC; Total inorganic carbon, and TN; Total nitrogen. The values are represented mean of triplicates samples.

The chemical properties of different blended biochar pellets uploaded with plant nutrient solutions before use in experiments are presented in Table 2. Biochar pellets uploaded with plant nutrient solutions heated to 60 °C had greater than 1.43–1.80% more total nitrogen (T-N) than those at room temperatures; however, their K concentrations decreased.

Table 2. The chemical properties of different biochar pellets before used experiments.

Treatments *	T-N	T-P	K	SiO$_2$
	------------------------ g kg^{-1} ------------------------			
TN	84.0	35.4	13.5	119.5
HTN	102.0	29.5	11.8	125.5
TNPK	75.2	32.8	57.2	108.6
HTNPK	89.5	35.6	39.0	96.9
PMCP	29.1	79.4	20.8	67.2

*TN, uploaded with urea solution at room temperature, HTN; urea solution heated at 60 °C, TNPK; N, P and K solutions at room temperature, HTNPK; N, P and K solutions heated at 60 °C and PMCP; pig manure compost pellets as control. The values are represented as means of triplicate samples.

Different blended biochar pellets were produced by a combination of biochar and pig manure compost (4: 6 ratio, w/w) uploaded with N, P and K plant nutrient solutions. The blended biochar pellets were differently uploaded using a urea solution at room temperature (TN), a urea solution heated to 60 °C (HTN), N, P and K solutions at room temperature (TNPK), and finally, N, P and K solutions heated to 60 °C (HTNPK) to increase solubility. The combination material (2.5 kg total weight) was thoroughly mixed at a 4:6 ratio of biochar and pig manure compost using an agitator (SungChang Co., KyungGi, Korea) for five minutes. The agitator was run continuously while spraying with 1 L of each nutrient solution for ten minutes. The blended biochar pellet (Ø 0.51 cm × 0.78 cm) were made using a machine (7.5 KW, 10 HP, KumKang Engineering Pellet Mill Co., DaeGu, Korea) and by pouring the combination of materials. A processing diagram of the biochar pellet has already been described by Shin et al. [10].

2.2. Batch Column Experiment of Nutrient Precipitation and Chemical Analysis

The experiments consisted of five treatments, i.e., PMCP as control, TN, HTN, TNPK, and HTNPK. Into each glass column was placed a filter and 5 g of different blended biochar pellets. Then, 50 mL deionized water was added. Successive precipitations were performed for nutrient extraction, and the 50mL of deionized water was replaced over a period of 77 days. The collected water was filtered using Whatman No. 2 filter paper. It was then analyzed for NH_4-N and NO_3-N, and the rest of the water samples were stored in a refrigerator until the analysis of PO_4-P, K, and SiO_2 were undertaken using a UV spectrophotometer (ST-Ammonium, C-Mac, Korea). Total carbon (TC) and total organic carbon (TOC) contents were measured using a TOC analyzer (Elementar Vario EL II, Hanau, Germany). Total P, K, and Si in different types of the blended biochar pellets were analyzed using inductively-coupled plasma atomic emission spectrometry (ICP-AES, IntegraXL, GBC LTd., Braeside, Australia) after digesting the samples with nitric and hydrochloric acids.

2.3. Releasing Nutrient Model

The application of a modified Hyperbola model for selecting an optimum mixing ratio of pig manure compost and biochar to make blended biochar pellets based on a nutrient releasing model [10]. In this study, the estimation of the accumulated releasing nutrient amounts in the treatments fitted well with the modified hyperbola model. Therefore, a modified Hyperbola model from the Michaelis-Menten equation was applied to select an optimal nutrient uploading method for the blended biochar pellets.

$$Y = A_{max} [t]/(t_{1/2(A_{max})} + [t]) \quad (1)$$

where Y is the accumulated concentration (mg L^{-1}); A_{max}: the maximum accumulated concentration (mg L^{-1}); $t_{1/2(A_{max})}$: the required time to reach 1/2 A_{max}; and t the precipitation periods (in days).

2.4. Statistical Analysis

The validity of each parameter for a modified Hyperbola model was assured for a normal distribution using Shapiro–Wilk test ($p < 0.05$). The nutrient releasing model for each treatment was accessed by analysis of data using SigmaPlot 12 (Systat Software, Inc., San Jose, CA, USA). The model based on the correlation coefficient values (R^2) was estimated. The statistical analyses for the total water-soluble amounts of NH_4-N, NO_3-N, PO_4-P, K, and SiO_2 were performed using one-way ANOVA with 5 levels, using SAS version 9.0 (SAS Institute, Carry, NC, USA). After deciding on the significant differences ($p < 0.0001$) among treatment means with analyses of variances (ANOVA), Duncan's multiple range tests were performed for each parameter throughout the whole precipitation period. The means of variables among treatments were compared with parameters using the above equation, according to p-value < 0.0001, after ANOVA analysis.

3. Results

It appeared that the highest accumulated NH_4-N concentration was 371.6 mg L^{-1} in the HNPK, but it was almost the same between the HTN and the TNPK during the final precipitation period (Figure 1). The highest accumulated NO_3-N concentration was 101.0 mg L^{-1} in the HTN throughout the experimental period, but results did not differ significantly ($p > 0.05$) among the other treatments within Phase I. The order of highest NO_3-N release in phase II was HTN>TNPK>HTNPK>TN>PMCP.

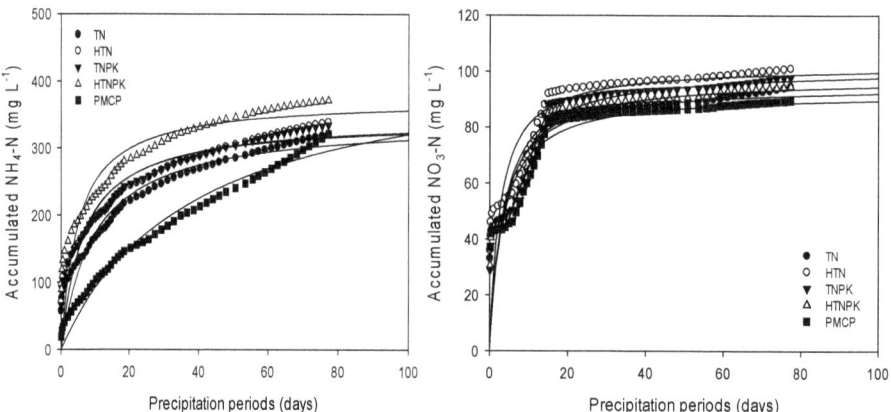

Figure 1. Releasing curves of NH_4-N and NO_3-N from different blended biochar pellets uploaded with plant nutrient solutions. The results are the mean of triplicate samples. Error bars indicate the standard deviation.

The required times of 1/2 maximum accumulated NH_4-N and NO_3-N levels were taken during 6.7 and 3.5 days in the TNPK, respectively (Tables 3 and 4). The model was significantly fitted ($p < 0.01$) and correlated with the R^2 values between the observed and estimated values.

Table 3. Estimation model for NH_4-N release from different types of the blended biochar pellet.

Treatments *	Model Parameters				Analysis of Variance		R^2
	Amax	p-Values	$t_{1/2(Amax)}$	p-Values	F	p-Values	
TN	343.1	<0.0001	10.0	<0.0001	566.7	<0.0001	0.916
HTN	345.2	<0.0001	6.9	<0.0001	348.1	<0.0001	0.870
TNPK	342.7	<0.0001	6.7	<0.0001	405.2	<0.0001	0.886
HTNPK	374.6	<0.0001	5.1	<0.0001	266.0	<0.0001	0.837
PMCP	446.9	<0.0001	39.4	<0.0001	1962.5	<0.0001	0.974

*TN, uploaded with urea solution at room temperature, HTN; urea solution heated to 60 °C, TNPK; N, P and K solutions at room temperature, HTNPK; N, P and K solutions heated to 60 °C and PMCP; pig manure compost pellet as control. Means values indicate significant differences ($p < 0.001$) among treatments (ANOVA).

Table 4. Estimation model for NO_3-N release from different types of the blended biochar pellet.

Treatments *	Model Parameters				Analysis of Variance		R^2
	Amax	p-Values	$t_{1/2(Amax)}$	p-Values	F	p-Values	
TN	95.0	<0.0001	3.2	<0.0001	136.5	<0.0001	0.724
HTN	102.1	<0.0001	2.6	<0.0001	127.5	<0.0001	0.710
TNPK	101.1	<0.0001	3.5	<0.0001	202.0	<0.0001	0.795
HTNPK	97.7	<0.0001	3.5	<0.0001	161.5	<0.0001	0.756
PMCP	92.5	<0.0001	3.3	<0.0001	111.5	<0.0001	0.682

*TN, uploaded with urea solution at room temperature, HTN; urea solution heated to 60 °C, TNPK; N, P and K solutions at room temperature, HTNPK; N, P and K solutions heated to 60 °C and PMCP; control as pig manure compost pellet. Means values indicate significant differences ($p < 0.001$) among treatments (ANOVA).

The accumulated PO_4-P concentration in precipitation water from the different treatments is shown in Figure 2. The lowest accumulated PO_4-P concentration in the PMCP was 423.2 mg L^{-1} from 10 days through 55 days of the precipitation (phase II), with a rapid increase occurring between 55 and 77 days of precipitation.

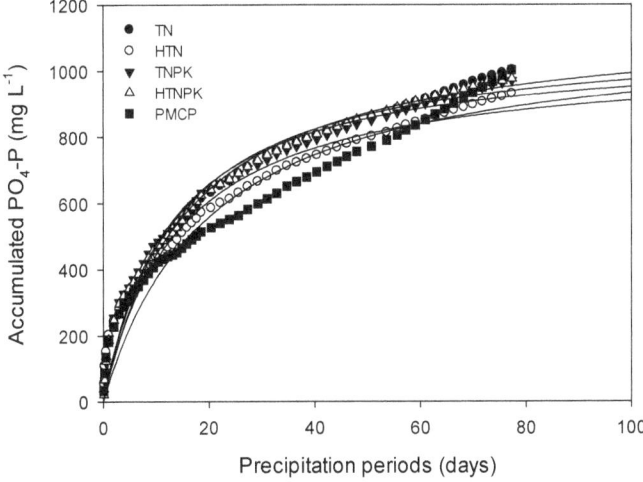

Figure 2. Releasing curves of PO_4-P from the blended biochar pellet uploaded with different plant nutrient solutions. Results are the mean of triplicate samples. Error bars indicate standard deviation.

It was observed that the required times of 1/2 the maximum amount of accumulated PO$_4$-P were taken from 12.6 to 15.9 days, regardless of the blended biochar pellet uploaded with different nutrient solutions, but that they were taken from 20.1 days in the PMCP (Table 5).

Table 5. Estimation model for PO$_4$-P release from different types of the blended biochar pellet.

Treatments *	Model Parameters				Analysis of Variance		R^2
	A$_{max}$	p-Values	t$_{1/2(Amax)}$	p-Values	F	p-Values	
TN	1150.0	<0.0001	15.9	<0.0001	2454.8	<0.0001	0.979
HTN	1040.6	<0.0001	14.3	<0.0001	1436.4	<0.0001	0.965
TNPK	1071.1	<0.0001	12.6	<0.0001	2204.4	<0.0001	0.977
HTNPK	1106.7	<0.0001	13.8	<0.0001	2496.4	<0.0001	0.980
PMCP	1122.0	<0.0001	20.1	<0.0001	750.9	<0.0001	0.935

*TN, uploaded with urea solution at room temperature, HTN; urea solution heated to 60 °C, TNPK; N, P and K solutions at room temperature, HTNPK; N, P and K solutions heated to 60 °C and PMCP; pig manure compost pellet as control. Means values indicate significant differences ($p < 0.001$) among treatments (ANOVA).

The accumulated K concentrations rapidly increased until 7 days of precipitation, and then remained in a steady state through 30 days of precipitation, regardless of the treatment (Figure 3). The study showed that the highest accumulated K concentration was 2680.5 mg L^{-1} in the TNPK, and that the lowest was in the TN and HTN which unloaded with plant nutrient solutions, even though those mixed with 60% of pig manure compost were 1729.0 and 1754.5 mg L^{-1}, respectively.

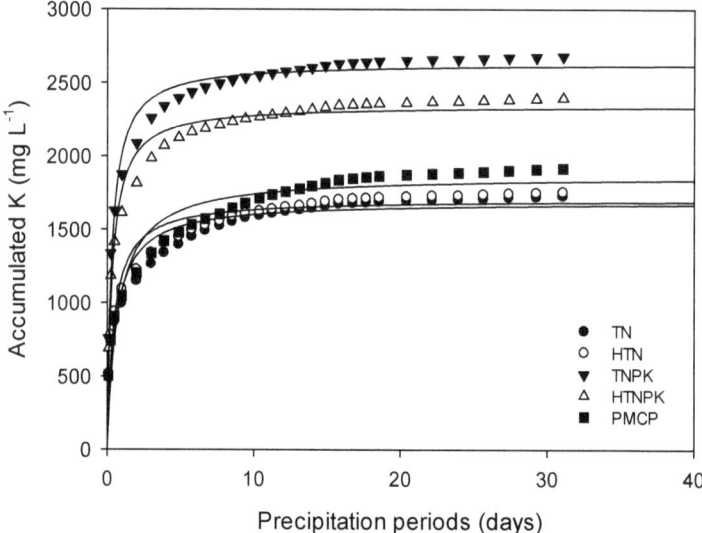

Figure 3. Releasing curves of K from the blended biochar pellet uploaded with plant nutrient solutions. The results are the mean of triplicate samples and error bars indicate standard deviation.

The estimation model for K releases from different blended biochar pellets calculated using a modified hyperbola equation is presented in Table 6. The required times of 1/2 maximum accumulated K amount were taken less than 0.6 days in all the treatments because of a high water-soluble capacity (Table 6).

Table 6. Estimation model for K release from different types of the blended biochar pellet.

Treatments *	Model Parameters				Analysis of Variance		R^2
	A_{max}	p-Values	$t_{1/2(Amax)}$	p-Values	F	p-Values	
TN	1692.3	<0.0001	0.6	<0.0001	266.5	<0.0001	0.905
HTN	1708.6	<0.0001	0.5	<0.0001	323.7	<0.0001	0.920
TNPK	2637.4	<0.0001	0.3	<0.0001	759.0	<0.0001	0.964
HTNPK	2350.4	<0.0001	0.3	<0.0001	576.4	<0.0001	0.954
PMCP	1870.0	<0.0001	0.7	<0.0001	258.1	<0.0001	0.902

*TN, uploaded with urea solution at room temperature, HTN; urea solution heated to 60 °C, TNPK; N, P and K solutions at room temperature, HTNPK; N, P and K solutions heated to 60 °C and PMCP; control as pig manure compost pellet. Means values indicate significant differences (p < 0.001) among treatments (ANOVA).

It appeared that the highest accumulated SiO_2 concentration was 2935.7 mg L^{-1} in the TN, but while lowest in the PMCP was 1599.6 mg L^{-1} in the final precipitation periods. The lowest in the PMCP was continuously observed, compared to the other treatments from 6.7 days through to the end of precipitation period (Figure 4).

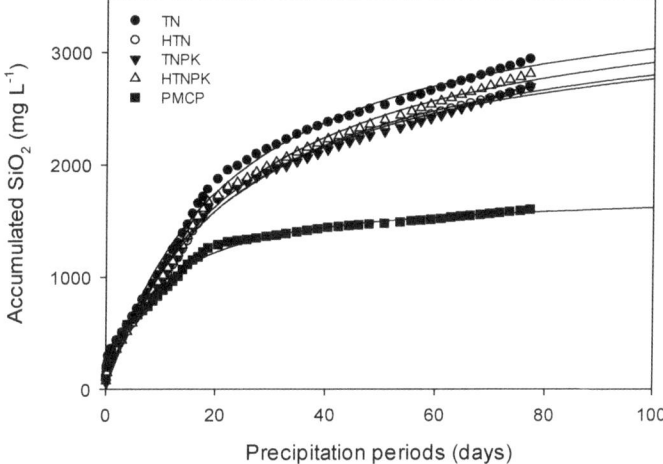

Figure 4. Releasing curves of SiO_2 from the blended biochar pellet uploaded with different plant nutrient solutions. The results are mean of triplicate samples. Error bars indicate standard deviation.

The estimation model for SiO_2 releases from different blended biochar pellets is presented in Table 7. The required times of 1/2 the maximum amount of accumulated SiO_2 were taken from 22.3 to 24.4 days, regardless of the blended biochar pellet uploaded with plant nutrient solutions, but were taken at 8.8 days in the PMCP.

Table 7. Estimation model for SiO_2 release from different types of the blended biochar pellet.

Treatments *	Model Parameters				Analysis of Variance		R^2
	A_{max}	p-Values	$t_{1/2(Amax)}$	p-Values	F	p-Values	
TN	3700.1	<0.0001	22.3	<0.0001	7328.6	<0.0001	0.993
HTN	3448.0	<0.0001	23.4	<0.0001	8354.6	<0.0001	0.994
TNPK	3386.0	<0.0001	22.7	<0.0001	9683.2	<0.0001	0.995
HTNPK	3612.0	<0.0001	24.4	<0.0001	11815.5	<0.0001	0.996
PMCP	1754.9	<0.0001	8.8	<0.0001	3128.4	<0.0001	0.984

*TN, uploaded with urea solution at room temperature, HTN; urea solution heated to 60 °C, TNPK; N, P and K solutions at room temperature, HTNPK; N, P and K solutions heated to 60 °C and PMCP; control as pig manure compost pellet. Means values indicate significant differences ($p < 0.001$) among treatments (ANOVA).

The total water-soluble amounts of NH_4-N, NO_3-N, PO_4-P and SiO_2 were significantly different ($p < 0.01$) among TN, HTN TNPK and HTNPK, but all soluble amounts in the PMCP were significantly lower than those of the other treatment throughout the precipitation period (Table 8).

Table 8. Comparisons of the total water-soluble accumulated amounts of NH_4-N, NO_3-N, PO_4-P, K, and SiO_2 for different treatments during the precipitation periods.

Treatments *	Total Water-Soluble Amounts (mg)				
	NH_4-N	NO_3-N	PO_4-P	K	SiO_2
TN	2935.7 a	94.7 ab	984.8	1729.1 d	2935.7 a
HTN	2689.1 a	101.1 a	931.4	1754.5 d	2689.1 a
TNPK	2693.1 a	97.9 a	920.1	2680.5 a	2693.1 a
HTNPK	2806.4 a	94.5 ab	975.8	2395.5 b	2806.4 a
PMCP	131.0 b	89.5 b	995.6	1971.4 c	1599.6 b
F	53.40	5.64	13.42	142.62	53.40
p-values	0.0003	0.0427	0.0706	<0.0001	0.0003

*TN, uploaded with urea solution at room temperature, HTN; urea solution heated to 60 °C, TNPK; N, P and K solutions at room temperature, HTNPK; N, P and K solutions heated to 60 °C and PMCP; pig manure compost pellet as the control. Means values followed by different letters indicate significant differences ($p < 0.0001$) among treatments with NH_4-N, PO_4-P, K, and SiO_2 (ANOVA and subsequent Duncan Multiple Range Test).

4. Discussions

4.1. Nitrogen Release from the Blended Biochar Pellet

Nitrogen fertilizers in soil usually undergo nitrogen transformation processes such as mineralization, nitrification, de-nitrification, and immobilization [6,7]. Urea is used as a nitrogen fertilizer because of fast N uptake by crop, although only 40% is absorbed by the plant, while 60% is lost in various ways [28], the great part of which, i.e., 26.5% to 29.4%, is lost through evaporation, thereby contributing to greenhouse gases. Therefore, a controlled release N fertilizer is the best way to minimize N_2O emissions from soil [29]. Shin et al. [10] observed that NH_4-N was adsorbed rapidly, with a combination rate (9:1, biochar: pig manure compost) of the blended biochar pellets in both the pseudo first and second order kinetics. It was further observed that the more biochar in the blended biochar pellet, the greater the adsorption of NH_4-N.

With respect to the estimation values from a modified hyperbola model, it appeared that half of the nitrogen was released within 10 days of precipitation, while though the rest was slowly released through end of experiment (Tables 3 and 4). However, it was observed that the cumulative NH_4-N in the blended biochar pellets was slowly released over the course of the 77 days of precipitation, but that the NO_3-N was rapidly released within 15 days of precipitation (Phase I), closely followed by a slower release rate in the final precipitation (Phase II), regardless of the treatment (Figure 1). Thus, this study demonstrated that blended biochar pellets can work as an ideal slow release fertilizer to create an eco-friendly environment through the reduction of non-point source pollution, and that they may be included in a carbon trading mechanism.

Different patterns for the estimation model were observed in the accumulated NH_4-N and NO_3-N releasing model. The estimated releasing model was fit wee ($p < 0.01$) with all treatments (Tables 3 and 4).

4.2. PO_4-P, K, and SiO_2 Releases from the Blended Biochar Pellet

P, K and SiO_2 are essential elements for crop cultivation; these applications are needed to increase crop yields [12,13]. Phosphate fertilizers are applied to soil every year to increase soil fertility. Annually, current consumption of rock phosphorous as a fertilizer is more than one million tons [30]. The excessive flow of phosphorous in lakes and ponds from croplands is a major cause of eutrophication, which disturbs the ecosystem [14]. Busted eutrophication not only affects aquatic ecosystems, but also indirectly interferes with economic development [31]. Kim et al. [32] reported that two apparent phases showed curves, a greater nutrient release rate within one day (Phase I) providing nutrients to crop growth for short times, followed by a slower nutrient release within 18 days (Phase II) providing nutrients to crops for longer periods. However, the accumulated PO_4-P concentrations were rapidly increased up to 77 days of the precipitation period, but K was almost in a steady state after 20 days of precipitation, regardless of treatments. The releasing patterns of PO_4-P and K did not agree with Kim's experimental results. On the other hand, this implies that the blended biochar pellets used in this study released plant nutrients more slowly than those used in Kim's study.

Releasing PO_4-P concentrations in the TNPK and HTNPK were similar compared with those of TN and HTN at 20 days' precipitation (Phase I), even if not uploaded with phosphorous fertilizer. The maximum accumulated concentrations were not significantly different ($p > 0.05$) among all treatments with different uploaded methods of plant nutrient solutions, even if not uploaded with P; this might be attributed to the P content in pig manure compost. The curves were derived from the estimation model calculated by a modified hyperbola equation using the correlation coefficient value (R^2). The estimation models for PO_4-P releases from different blended biochar pellets indicated that estimation values were well fitted ($p < 0.01$) and that they correlated with the observed values for PO_4-P releases, regardless of which blended biochar pellets were used (Table 5).

Furthermore, the K releasing pattern was similar to the results of this research, except regarding the blended biochar pellet mixed with pig manure compost and uploaded with plant nutrient solutions, which had 6 and 12 days' longer releasing periods in Phases I and II, respectively. This might be due to the different mixing materials, i.e., pig manure compost instead of lignin. The accumulated K concentrations in the TN and HTN did not differ significantly ($p > 0.05$) from the PMCP. The estimation values from a modified hyperbola model were significantly correlated ($p < 0.01$) with the observed values for K releases, regardless of which blended biochar pellets were used.

Silicon (Si) in the soil existed in an unavailable form for plant uptake. Si in crop residues is a useful form relative to that from Si fertilizer. This recycling of Si form can occur on crop land after the decomposition of crop residues. However, Si in crop residues is often removed during harvesting [33]. Also, Si is generally available as uncharged monosilicic acid (H_4SiO_4) [20]. Accumulated SiO_2 concentrations in different treatments rapidly increased up to 18 days of precipitation (Phase I), and then gradually increased up to 77 days (Phase II), regardless of the treatment (Figure 4). The accumulated SiO_2 concentration in precipitation water did not significantly differ ($p > 0.05$) among the HTN, TNPK, and HTNPK. The K and SiO_2 releasing patterns showed a similar trend [32], but the PO_4-P in this study was released rapidly at the end of the precipitation period. The estimation values calculated from a modified hyperbola equation were well fitted ($p < 0.01$) and correlated with the observed values for SiO_2 release, regardless of which blended biochar pellets were used.

Regarding the total water soluble amounts of plant nutrients, no differences were observed with the uploaded methods for blended biochar pellets, especially for PO_4-P. However, optimal slow release fertilizer could depend on 1/2 releasing periods, but was still considered to be releasing for the rest of the precipitation periods, because of the additional fertilizer application method for rice cultivation. It was considered that the best uploaded method of plant nutrients might be the TNPK for the blended biochar pellet.

It was also shown that the experimental results were in agreement with Shin and Park's experimental data [10]. Furthermore, Shin's principal law indicated that pig manure compost pellets released nutrients more slowly than pig manure compost. Therefore, this theory was applied for the blended biochar pellets uploaded with plant nutrient solutions as a slow release fertilizer having an adsorption capacity for plant nutrients.

5. Conclusions

This experiment sought to select an optimum nutrient uploading method for blended biochar pellets using a modified hyperbola model based on their nutrient releasing characteristics. The results of this experiment concluded that blended biochar pellets slowly release nutrients over a 77 day precipitation period, and that they create an eco-friendly environment by reducing non-point pollutants. It appeared that the optimal uploaded method was the TNPK based on the nutrient releasing characteristics with a modified hyperbola model. Furthermore, it was observed that a modified hyperbola model fitted well ($p < 0.01$) in all treatments. Furthermore, dose responses for the blended biochar pellet application in the cropland during crop cultivation should be measured in the future.

Overall, these results showed that blended biochar pellets can be used as a slow release fertilizer for agriculture.

Author Contributions: Conceptualization, funding acquisition and writing—original draft preparation, J.S.; visualization and software, S.P.; validation, S.L.

Funding: This study was carried out with the support of "Research Program for Agricultural Science & Technology Development (Project No. PJ014207)", National Institute of Agricultural Sciences, Rural Development Administration, Republic of Korea.

Conflicts of Interest: The author certifies that he has NO affiliations with or involvement in any organization or entity with any financial interest (such as honoraria; educational grants; participation in speakers' bureaus; membership, employment, consultancies, stock ownership, or other equity interest; and expert testimony or patent-licensing arrangements), or non-financial interest (such as personal or professional relationships, affiliations, knowledge or beliefs) in the subject matter or materials discussed in this manuscript.

References

1. MIFAFF. Annual Statistics in Food, Agriculture, Fisheries and Forestry in 2009. Korean Ministry for Food, Agriculture, Fisheries and Forestry. *Environ. Sci. Pollut. Res.* **2010**, *16*, 1–9.
2. Liu, W.J.; Jiang, H.; Yu, H.Q. Development of biochar-based functional materials: Toward a sustainable platform carbon material. *Chem. Rev.* **2015**, *115*, 12251–12285. [CrossRef]
3. Zhang, X.; Zhang, L.; Li, A. Hydrothermal co-carbonization of sewage sludge and pinewood sawdust for nutrient-rich hydrochar production: Synergistic effects and products characterization. *J. Environ. Manag.* **2017**, *201*, 52–62. [CrossRef]
4. Husk, B.; Major, J. *Commercial Scale Agricultural Biochar Field Trial in Quebec, Canada, over Two Years: Effects of Biochar on Soil Fertility, Biology, Crop Productivity and Quality*; Blue Leaf Inc.: DrummondV, QC, Canada, 2010; pp. 1–38.
5. Ding, Y.; Liu, Y.; Liu, S.; Li, Z.; Tan, X.; Huang, X.; Zeng, G.; Zhou, L.; Zheng, B. Biochar to improve soil fertility. A review. *Agron. Sustain. Dev.* **2016**, *36*, 1–18. [CrossRef]
6. Martens, D.A.; Dick, W.A. Recovery of fertilizer nitrogen from continuous corn soils under contrasting tillage management. *Biol. Fertil. Soils* **2003**, *38*, 144–153. [CrossRef]
7. Ortega, R.A.; Westfall, D.G.; Peterson, G.A. Climatic gradient, cropping system, and crop residues impact on carbon and nitrogen mineralization in no-till soils. *Commun. Soil Sci. Plant. Anal.* **2005**, *36*, 2875–2887. [CrossRef]

8. Choi, Y.-S.; Shin, J.-D.; Lee, S.-I.; Kim, S.-C. Adsorption characteristics of aqueous ammonium using rice hull-derived biochar. *Korean J. Environ. Agric.* **2015**, *34*, 1–6. [CrossRef]
9. Teutscherova, N.; Houska, J.; Navas, M.; Masaguer, A.; Benito, M.; Vazquez, E. Leaching ammonium and nitrate from Acrisol and calcisol amended with holm oak biochar: A column study. *Geoderma* **2018**, *323*, 136–145. [CrossRef]
10. Shin, J.; Choi, E.; Jang, E.S.; Hong, S.G.; Lee, S.; Ravindran, B. Adsorption characteristics of ammonium nitrogen and plant responses to biochar pellet. *Sustainability* **2018**, *10*, 1331. [CrossRef]
11. Filiberto, D.M.; Gaunt, J.L. Practicality of biochar additions to enchance soil and crop productivity. *Agriculture* **2013**, *3*, 715–725. [CrossRef]
12. Almeelbi, T.; Bezbaruah, A. Nanoparticle-sorbed phosphate: Iron and phosphate bioavailability studies with spinacia oleracea and selenastrum capricornutum. *ACS Sustain. Chem. Eng.* **2014**, *2*, 1625–1632. [CrossRef]
13. Zhao, L.; Cao, X.; Zheng, W.; Scott, J.W.; Sharma, B.K.; Chen, X. Copyrolysis of biomass with phosphate fertilizers to improve biochar carbon retention, slow nutrient release, and stabilize heavy metals in soil. *ACS Sustain. Chem. Eng.* **2016**, *4*, 1630–1636. [CrossRef]
14. Ramanan, V.; Thiyagarajan, S.K.; Raji, K.; Suresh, R.; Ramamurthy, P. Outright green synthesis of fluorescent carbon dots from eutrophic algal blooms for in vitro imaging. *ACS Sustain. Chem. Eng.* **2016**, *4*, 4724–4731. [CrossRef]
15. Awasthi, M.K.; Wang, Q.; Huang, H.; Li, L.; Shen, F.; Lahori, A.H.; Wang, P.; Guo, D.; Guo, Z.; Jiang, S.; et al. Effect of biochar amendment on greenhouse gas emission and bio-availability of heavy metals during sewage sludge co-composting. *J. Clean. Prod.* **2016**, *135*, 829–835. [CrossRef]
16. Higashikawa, F.S.; Conz, R.F.; Colzato, M.; Cerri, C.E.P.; Alleoni, L.R.F. Effect of feedstock type and slow pyrolysis temperature in the production of biochars on the removal of cadmium and nickel from water. *J. Clean. Prod.* **2016**, *137*, 965–972. [CrossRef]
17. Subedi, R.; Taupe, N.; Ikoyi, I.; Bertora, C.; Zavattaro, L.; Schmalenberger, A.; Leahy, J.J.; Grinani, C. Chemically and biologically–mediated fertilizing value of manure-derived biochar. *Sci. Total Environ.* **2016**, *550*, 924–933. [CrossRef] [PubMed]
18. Arif, M.; Ilyas, M.; Riaz, M.; Ali, K.; Shah, K.; Haq, I.U.; Fahad, S. Biovhar improves phosphorus use efficiency of organic-inorganic fertilizers, maze-wheat productivity and soil quality in a low fertility alkaline soil. *Field Crop. Res.* **2017**, *214*, 25–37. [CrossRef]
19. Ma, J.F.; Yamaji, N. Silicon uptake and accumulation in higher plants. *Trends Plant. Sci.* **2006**, *11*, 392–397. [CrossRef]
20. Epstein, E. Silicon: Its manifold roles in plants. *Ann. Appl. Biol.* **2009**, *155*, 155–160. [CrossRef]
21. Guntzer, F.; Keller, C.; Meunier, J.-D. Benefits of plant silicon for crops: A review. *Agron. Sustain. Dev.* **2012**, *32*, 201–213. [CrossRef]
22. Detmann, K.C.; Araujo, W.L.; Martins, S.C.; Sanglard, L.M.; Reis, J.V.; Detmann, E.; Rodrigues, F.A.; Nunes-Nesi, A.; Fernie, A.R.; DaMatta, F.M. Silicon nutrion increases grain yield, which, in turn, exerts a feed-forward stimulation of photo-synthetic rates via enhanced mesophyll conductance and alters primary metabolism in rice. *New Phytol.* **2012**, *196*, 752–762. [CrossRef]
23. Reza, M.T.; Lynam, L.G.; Vasquez, V.R.; Coronella, C.J. Pelletization of Biochar from Hydrothermally Carbonized Wood. *Environ. Prog. Sustain. Energy* **2012**, *31*, 225–234. [CrossRef]
24. Cantrell, K.B.; Martin, J.H. Poultry litter and switchgrass blending and pelleting characteristics for biochar production. In Proceedings of the ASABE Annual International Meeting, Dallas, TX, USA, 29 July–1 August 2012.
25. Fernandez-Escobar, R.; Benlloch, M.; Herrera, E.; Garcia-Novelo, J.M. Effect of traditional and slow-release N fertilizers on growth of olive nursery plants and N losses by leaching. *Sci. Hortic.* **2004**, *101*, 39–49. [CrossRef]
26. Mortain, L.; Dez, I.; Madec, P.J. Development of new composites materials, carriers of active agents from biodegradable polymers and wood. *C. R. Chim.* **2004**, *7*, 635–640. [CrossRef]
27. Eberwein, J.; Shen, W.; Jenerette, D. Michaelis-Menten kinetics of soil respiration feedbacks to nitrogen deposition and climate change in subtropical forests. *Sci. Rep.* **2017**, *7*, 1752. [CrossRef]
28. Liang, X.Q.; Chen, Y.X.; Li, H.; Tian, G.M.; Ni, W.Z.; He, M.M.; Zhang, Z.J. Modeling transport and fate of nitrogen from urea applied to a near-trench paddy field. *Environ. Pollut.* **2007**, *50*, 313–320. [CrossRef]
29. Chu, H.; Hosen, Y.; Yagi, K. NO, N_2O, CH_4 and CO_2 fluxes in winter barely field of Japanese Andisol as affected by N fertilizer management. *Soil Biol. Biochem.* **2007**, *39*, 330–339. [CrossRef]

30. Rahman, M.M.; Liu, Y.H.; Kwang, J.H.; Ra, C.S. Recovery of struvite from animal wastewater and its nutrient leaching loss in soil. *J. Hazard. Mater.* **2011**, *186*, 2026–2030. [CrossRef] [PubMed]
31. Schindler, D.W.; Carpenter, S.R.; Chpra, S.C.; Hecky, R.E.; Orihel, D.M. Reducing phosphorous to curb lake eutrophication is a success. *Environ. Sci. Technol.* **2016**, *50*, 8923–8929. [CrossRef]
32. Kim, P.; Daniel, H.; Nicole, L. Nutrient release from switchgrass-derived biochar pellets embedded with fertilizers. *Geoderma* **2014**, *232*, 341–351. [CrossRef]
33. Houben, D.; Sonnet, P.; Cornelis, J.T. Biochar from Miscanthus: A potential silicon fertilizer. *Plant Soil* **2013**, *374*, 871–882. [CrossRef]

© 2019 by the authors. Licensee MDPI, Basel, Switzerland. This article is an open access article distributed under the terms and conditions of the Creative Commons Attribution (CC BY) license (http://creativecommons.org/licenses/by/4.0/).

Article

Biochar from Microwave Pyrolysis of Artemisia Slengensis: Characterization and Methylene Blue Adsorption Capacity

Xuhui Li [1], Kunquan Li [1], Chunlei Geng [1], Hamed El Mashad [2], Hua Li [1,*] and Wenqing Yin [1,*]

[1] College of Engineering, Nanjing Agricultural University, Nanjing 210031, China; 2017812101@njau.edu.cn (X.L.); kqlee@njau.edu.cn (K.L.); 2017112027@njau.edu.cn (C.G.)

[2] Department of Biological and Agricultural Engineering, University of California, One Shields Avenue, Davis, CA 95616, USA; heelmashad@UCDAVIS.EDU

* Correspondence: lihua@njau.edu.cn (H.L.); yinwq@njau.edu.cn (W.Y.); Tel.: +86-13951740692 (H.L.); +86-13851938631 (W.Y.)

Received: 23 March 2019; Accepted: 26 April 2019; Published: 1 May 2019

Featured Application: This research provided a new treatment method for artemisia selengensis. Thus, it can produce biochar with high adsorbability instead of just wasted.

Abstract: In this research, artemisia selengensis was used to produce biochar via microwave pyrolysis. The influence of pyrolysis temperature, heating rates, temperature holding time and additive on the biochar yield and adsorbability were all investigated. The results suggest that the biochar yield decreased with the increase of pyrolysis temperature while the adsorbability of the biochar increased with an increase of the pyrolysis temperature; the biochar yield and its adsorbability could achieve the desired value when the heating rate and temperature holding time were in a specific scope; the biochar yield decreased when an additive was added; the adsorbability of the biochar could be increased by adding $ZnCl_2$ (metal chloride) and Na_2CO_3 (metal carbonate). According to the orthogonal experiments, the optimal conditions for biochar production were: pyrolysis temperature 550 °C, heating rate 2 °C/s, temperature holding time 15 min, without additive.

Keywords: artemisia selengensis; microwave pyrolysis; biochar; adsorbability

1. Introduction

Activated carbon is an excellent adsorbent that has been applied in many applications [1]. Due to its stable chemical properties, high mechanical strength, acid and alkali resistance, heat resistance, insolubility in water and organic solvents, and renewability, activated carbon has been widely used in the chemical industry, environmental protection, food processing, refined, metallurgy, medicine, military chemical protection, and other fields [2–4]. Biomass is the main raw material of activated carbon. Several reviews report a great deal of work done on applications for the removal of specific pollutants from an aqueous phase [5]. Under the condition of high temperature and low oxygen, biomass could turn to biochar. However, the biochar often has poor quality, ash and other impurities are in the microporous, thus the adsorption performance is not good. Further improvement of the structure is needed to standardize biochar into activated carbon [6]. In recent years, microwave pyrolysis has gained increased attention from many researchers because of its unique heating mode and higher application prospect. Biochar made by microwave pyrolysis has a better structure than those made by traditional electric pyrolysis [7–11].

Artemisia selengensis, as compositae sagebrush and perennial herbaceous plant, was planted in shoals and moist areas. Due to its developmental nutrition, the cultivated area of artemisia selengensis in Nanjing, Jiangsu province has reached 2.33×10^9 hectares, and the annual production is more than 500,000 tons [12]. After two seasons, the aged stems of artemisia selengensis have no edible value, and they become a big problem that needs to be solved. High humidity and high fiber content made those aging stems difficult to mush. There has been little research focusing on the processing of aged stems. Traditional thermochemical conversion is not suitable for this material because it is highly humid and fibrous. However, microwave pyrolysis can effectively be used for the process because microwaves can affect biomass itself directly [13], and there are not many limitations on its size [14,15]. Water is a good kind of microwave absorber, so high moisture content also brings a high heating rate. Thus, the microwave pyrolysis technique is suitable to convert aged stems into biochar.

In this research, artemisia selengensis was used to produce biochar by using microwave pyrolysis technology. The influence of pyrolysis temperature, heating rates, temperature holding time and additive on the biochar yield and adsorbability were investigated. An orthogonal experiment was designed to find the optimum condition to produce biochar for the maximum adsorbability.

2. Materials and Methods

2.1. Materials

The artemisia selengensis straws used in this study were collected form Baguazhou, Nanjing (30°10′ N, 118°49′ E). And the picture of atemisia selengensis straw is shown in Figure 1. The collected straws had an average moisture content of 92.16% and an average diameter of 6 mm. Then, the straws were cut into 3 cm long pieces and stored for the pyrolysis experiments. The contents of artemisia selengensis straws were measured using the Van Soest. The content of hemicellulose, cellulose, lignin and ash is shown in Table 1. As can be concluded from Table 1, artemisia selengensis straws have more cellulose than rice straws, therefore, they are hard to mush. According to Luo and Wang's researches, the main product of cellulose pyrolysis is volatile, and volatile could make the biochar produced have more potential with more micropores [16,17]. Thus, the biochar produced by artemisia selengensis has porous properties. Furthermore, artemisia selengensis straws have little ash, which means there are little non-organic impurities in the biochar, which is a great condition for producing good adsorptive biochar.

Figure 1. Artemisia selengensis from Baguazhou.

Table 1. Artemisia selengensis cellulose content.

Composition	Artemisia Selengensis	Rice Straw
Hemicellulose	4.056%	31.589%
Cellulose	55.335%	36.242%
Lignin	8.621%	7.128%
Ash	0.240%	1.032%

2.2. Experimental Conditions

The microwave system used in this study was designed by the Biomass and Bioenergy Lab of the Nanjing Agricultural University and Nanjing Jinhaifeng Microwave Technology Ltd. (Nanjing, China).

The experimental system can provide variable power from 1 kW to 3 kW, which means that the heat rate is adjustable by changing the power consumption. The system was controlled by a programmable logic controller (PLC; Model: SIEMENS CPU224XP, Siemens, Erfurt, Germany). The system was equipped with three identical microwave generators that can create a frequency of 2.45 GHz (Samsung OM75P-31, Daegu, Korea). They were installed separately on the rear, left, and right wall. The system was designed capable of controlling the pyrolysis temperature. A thermocouple was used to measure the temperature of the central point of the reaction chamber. A data logger was used to save the temperature data in real time. Figure 2 shows a schematic diagram of the experimental system. The model of the ultraviolet spectrophotometer is UNICO UV-2800 (Unico, Shanghai, China). The model of the scanning electron microscope (SEM) is Phenom ProX (Phenom Scientific, Hillsboro, OR USA).

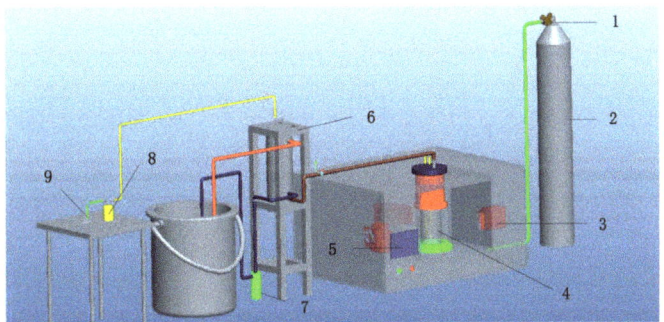

Figure 2. The microwave pyrolysis system: (1: relief valve; 2: nitrogen; 3: microwave generator; 4: reaction chamber; 5: control unit; 6: condenser; 7: circulating pump; 8: gas filter; 9: gas-collection bag).

2.3. Experimental Method

2.3.1. Biochar Production Method

High purity N_2 (99.99%) was purged for over 10 min to completely expel the air in the microwave pyrolysis oven. The pyrolysis temperature was measured by a thermocouple. Related parameters like pyrolysis temperature and microwave power were established according to the experimental design. To eliminate the influence of the quality of material, 45 g ± 0.1 g artemisia selengensis straw was added in every group. The biochar was collected and prepared for acid pickling after cooled.

2.3.2. Acid Pickling

All the biochar samples were mixed with 100 mL hydrochloric acid (pH = 1, 0.1 mol/L). To remove the ash, oxygen-containing functional groups and other impurities, the mixture was boiled for 3 min. After acid pickling, samples were filtered. After the biochar dried, all groups were weighted and prepared for the adsorption experiments.

The biochar yield was calculated using the formula below:

$$\text{Biochar yield} = \frac{m_1}{m_2} \times 100\% \tag{1}$$

where m_1 is the mass of biochar after acid pickling (g) and m_2 is the mass of biomass material (g).

2.3.3. Biochar Adsorption Method

The carbon decolorization ability to methylene blue solution can reflect the adsorption ability of 2 nm mesoporous in carbon. Thus the adsorption value of biochar to methylene blue solution is used as an indicator its adsorption performance in this research. The operations are as follows:

CuSO$_4$ solution (0.025mol/L, blank group) and methylene blue solution (0.025 g/L, 0.05 g/L, 0.2 g/L and 0.5 g/L) were prepared as standard samples. The wavelength of the ultraviolet spectrophotometer was set at 655 nm. The calibration curve was drawn by measuring the absorbance of standard samples.

Take 0.1 g acid pickled biochar from every test group and put it into a 50 mL conical flask. Twenty milliliter of 0.5 g/L methylene blue solution was added in the conical flask, and then put the conical flask onto constant temperature shaker for 20 min, at temperature 20 °C and speed 120 r/min. If the solution decolorizes after shaking, 5 mL more methylene blue solution (0.5 g/L) should be added and then shake the conical flask until the solution stays blue. The adsorption values of biochar are calculated by the calibration curve and absorbance.

2.4. Experimental Design

The effect of pyrolysis temperature, heating rates, temperature holding time and additive (additive amount 5%) on the biochar adsorbability were investigated. The experiment conditions are shown in Tables 2 and 3.

Table 2. The single factor experiment conditions.

No.	Factor		
	Factor	Level	Condition
1–4	A (pyrolysis temperature °C)	450, 500, 550, 600	B = 0.5 °C/s, C = 10 min, D = None
5–8	B (heating rate °C/s)	0.5, 1, 1.5, 2	A = 450 °C, C = 10 min, D = None
9–12	C (temperature holding time min)	10, 15, 20, 25	A = 450 °C, B = 0.5 °C/s, D = None
13–16	D (additive)	None, Char, ZnCl$_2$, Na$_2$CO$_3$	A = 450 °C, B = 0.5 °C/s, C = 10 min

Table 3. The orthogonal experiment schedule and result.

No.	A: Pyrolysis Temperature (°C)	B: Heating Rates (°C/s)	C: Temperature Holding Time	D: Additive	Biochar Yield	Adsorbing Capacity (mg/g)
1	450	0.5	10	None	5.00%	6.989
2	450	1	15	Char	3.66%	7.507
3	450	1.5	20	ZnCl$_2$	3.19%	17.603
4	450	2	25	Na$_2$CO$_3$	2.86%	13.979
5	500	0.5	15	ZnCl$_2$	2.55%	27.958
6	500	1	10	Na$_2$CO$_3$	2.39%	41.160
7	500	1.5	25	None	2.72%	41.936
8	500	2	20	Char	2.68%	15.791
9	550	0.5	20	Na$_2$CO$_3$	2.88%	45.158
10	550	1	25	ZnCl$_2$	2.97%	39.348
11	550	1.5	10	Char	-1.42%	10.096
12	550	2	15	None	2.82%	140.564
13	600	0.5	25	Char	1.72%	16.567
14	600	1	20	None	5.00%	62.128
15	600	1.5	15	Na$_2$CO$_3$	3.66%	9.060
16	600	2	10	ZnCl$_2$	3.19%	67.305

3. Results

3.1. The Effect of Pyrolysis Temperature

The effect of different pyrolysis temperatures (450 °C, 500 °C, 550 °C, 600 °C) on the biochar adsorption is shown in Figure 3. These experiments were conducted at a heating rate of 0.5 °C/s, a temperature holding time of 10 min and without additive.

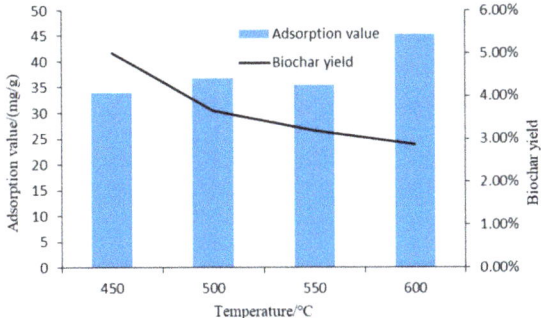

Figure 3. Effect of temperature on the biochar yield and adsorbability.

As shown in Figure 3, with the increasing pyrolysis temperature, the biochar yield decreases, but the adsorbability of the biochar increases. The adsorbability of biochar reached a peak at a pyrolysis temperature of 600 °C, and the highest adsorbing capacity was 45.302 mg/g. It is worth mentioning that the moisture content of the sample was about 98.16%. After pyrolysis, the maximum biochar yield was 5%, and the actual conversion rate was 63.8%, which was relatively high. High temperature can promote the secondary reaction of biochar with other products and vapor in the reactor, thus the biochar yield decreases as the pyrolysis temperature increases. The influence of the pyrolysis temperature on the adsorbability of biochar was not obvious, but from 550 °C to 600 °C, the adsorbability of biochar increased significantly. It may suggest that a high pyrolysis temperature can promote a secondary reaction of biochar with the gas and liquid products [13,18], thus the biochar structure was optimized and the adsorbability of the biochar rose significantly. Combined with the result of carbon yield, the variation trend of the biochar yield reduced after 550 °C, which means that the secondary reaction was roughly completed and this also proved that a secondary reaction could improve the adsorbability of biochar to some extent.

3.2. The Effect of Heating Rate

The effect of different heating rates (0.5 °C/s, 1 °C/s, 1.5 °C/s, 2 °C/s) on the biochar adsorption is shown in Figure 4. These experiments were conducted at a pyrolysis temperature of 450 °C, a holding time of 10 min and without additive.

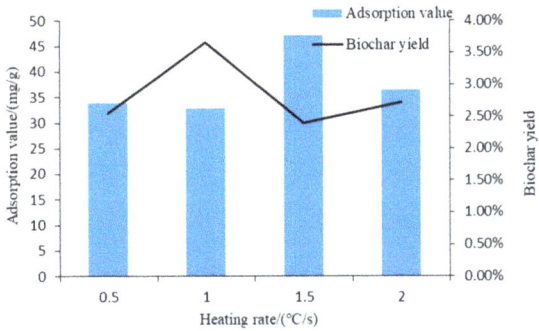

Figure 4. Effect of heating rate on the biochar yield and adsorbability.

As shown in Figure 4, the biochar yield reached a peak of 3.66% when the heating rate was 1 °C/s. The adsorbability of the biochar reached its highest value of 47.114 mg/g when the heating rate was 1.5 °C/s. However, at other heating rates, the biochar yields were close to 2.50%, and adsorbing capacities of biochar were close to 35 mg/g. It may suggest that when the heating rate was low, the reaction speed and the gas production yield were also low. The gas production could not release in time, thus the probability of a secondary reaction increased, and the biochar yield was low. On the contrary, when the heating rate was high, the reaction speed was also high. The gas could not release in time, which increased the probability of the initiation of a secondary reaction. Though the secondary reaction could improve the adsorbability of the biochar, the low heating rate leads to a low gas production yield, thus the quality of the biochar was poor [19]. However, when the heating rate was high, the reaction speed was also high, and the gas products might release in a short time, thus the biochar structure was irregular, and thus resulted in poor adsorption.

3.3. The Effect of Holding Time

The effect of different temperature holding times (10 min, 15 min, 20 min, 25 min) on the biochar adsorption is shown in Figure 5. These experiments were conducted at a pyrolysis temperature of 450 °C, a heating rate of 0.5 °C/s, and without additive.

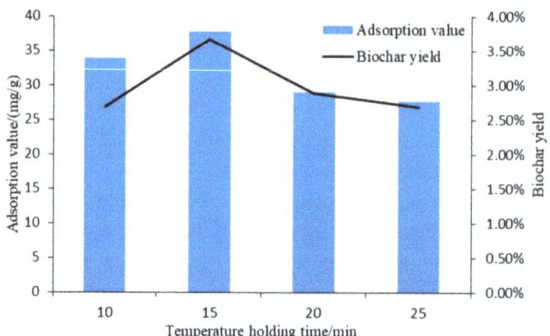

Figure 5. Effect of holding time on the biochar yield and adsorbability.

As shown in Figure 5, with the temperature holding time increasing, the biochar yield and adsorbability of biochar firstly increased at the holding time from 10 to 15 min, then decreased. At a temperature holding time of 15 min, the biochar yield and biochar adsorbability reached a peak value of 3.66% and 37.794 mg/g relatively at the same time. This may suggest during the first 15 min, the remaining material in the reactor continued to react, and the main reaction was coking for the liquid product, thus the biochar yield increased. As the holding time increased, the main reaction became biochar with other products, thus the biochar yield decreased. For the biochar adsorption, because of a short temperature holding time, the biochar structure could not reach an optimum; when the temperature holding time was exceeded, the biochar structure was destroyed by the high temperature [1]. What is more, longer temperature holding times led to more damage to the pore structure. To optimize the adsorbability of biochar, an optimal temperature holding time of approximately 15 min should be selected.

3.4. The Effect of Different Additives

The effect of different additives (none, biochar, metal carbonate-Na_2CO_3, metal chloride-$ZnCl_2$) on the biochar adsorbability is shown in Figure 6. These experiments were conducted at a pyrolysis temperature of 450 °C, a heating rate of 0.5 °C/s, and a temperature holding time of 10 min.

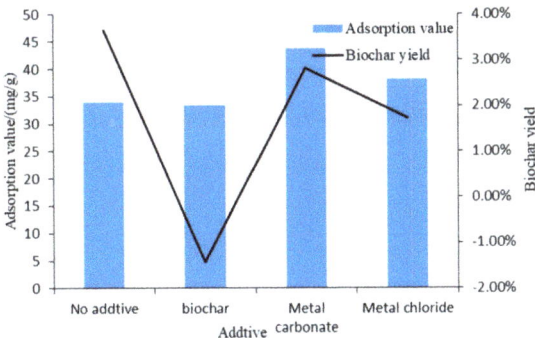

Figure 6. Effect of additive on the biochar yield and adsorbability.

As shown in Figure 6, the additives have a significant influence on biochar yield. The biochar yield reached the highest value without additive. For the biochar adsorbability, the addition of biochar exerted no significant influence on the adsorbability. The values of the adsorption were both around 33.455 mg/g with or without additive. However, the addition of metal carbonate or metal chloride improved the adsorbability of the biochar. The highest adsorbing capacity was 43.748 mg/g when Na_2CO_3 (metal carbonate) was used as an additive. This may suggest that the additive improves the ability of the sample to absorb microwaves, and this could cause hotspots in the samples, thus the adsorbability of the biochar could be improved. Moreover, metal carbonate releases CO_2 when heated, which can improve the biochar pore forming efficiency. It is worth mentioning that when adding biochar, the biochar yield was negative. This is because the additives were not considered as reactants, thus the mass was subtracted during the calculation of the biochar yield. However, in fact, biochar added as an additive indeed reacted, thus the subtraction of reacted additive caused the biochar yield to be negative. For the adsorbability, when the biochar acts as an additive, it could improve the adsorbability of the biochar, but since the moisture content of the sample was 92.16% and the additive content was 5%, the final biochar obtained was mainly the additive biochar. In addition, for the additive biochar, it was re-heated, and the structure was damaged, and this caused the poor adsorbability.

3.5. Orthogonal Experiment Analysis

Due to the high moisture content of the material, all groups of the biochar yield were low, so there is no need to discuss the optimal conditions for biochar yield in this research. Thus, in this part, only biochar adsorbability was analyzed.

ANOVA (analysis of variance) was used in this paper to analyze biochar adsorbability. The results are shown in Tables 4 and 5.

Table 4. The orthogonal experiment results calculation table.

Factor	A	B	C	D
k_1	11.5195	24.1680	31.3875	63.6543
k_2	31.7113	38.2858	46.2723	12.4903
k_3	58.7915	19.6738	35.9200	38.0535
k_4	39.5150	59.4098	27.9575	27.3393

Table 5. The orthogonal experiment results analysis of variance table.

Variance	Sum Of Squares	DOF (Degree Of Freedom)	MS (Mean Square)	F-Value	Significance
A	4591.9187	3	1530.64	1.599	Insignificant
B	3833.0697	3	1277.69	1.335	Insignificant
C	759.8648	3	253.288	0.265	Extremely significant
D	5580.7002	3	1860.23	1.943	Insignificant
Deviation	2871.7577	3	957.253		

Note: $F_{0.01}(3,3) = 29.46$, $F_{0.05}(3.3) = 9.28$

As shown in Table 5, all four factors have an insignificant influence on the adsorbability of the biochar following the order of D > A > B > C, which is: additive > pyrolysis temperature > heating rate > temperature holding time. To improve the biochar adsorbability, the optimal conditions were $A_3B_4C_2D_1$, that means the pyrolysis temperature is 550 °C, the heating rate is 2 °C/s, the temperature holding time is 15 min and without additive.

Two more repeated experiments were done according to the optimal conditions, and the results are shown in Table 6.

Table 6. The verification test results.

Test No.	Adsorbing Capacity mg/g	AVG (Average) mg/g	SD (Standard Deviation) mg/g
1*	140.564		
2	142.172	140.038	2.440
3	137.378		

*: The result was picked from the orthogonal experiment.

For the activated carbon index, the national primary standard (for water purification) requires the adsorbing capacity for methylene blue to be more than 135 mg/g [1], which means that the biochar produced under these conditions only needs acid pickling to reach the national primary standard. This means that the biochar has a high application value.

3.6. SEM Representation

The three SEM tests were from the repeat test in a single factor experiment (pyrolysis temperature 450 °C, heating rate 1 °C/s, temperature holding time 10 min, without additive); metal-carbonate-added test in a single factor experiment (pyrolysis temperature 450 °C, heating rate 1 °C /s, temperature holding time 10 min, Na_2CO_3 additive) and the optimal condition test (pyrolysis temperature 550 °C, heating rate 2 °C/s, temperature holding time 15 min, without additive). For the SEM, the amplification factor is 10,000, the scale length is 8 µm, and images are shown in Figure 7.

Figure 7. SEM (scanning electron microscope) images of biochar: (**a**) repeat test group in single factor experiment; (**b**) metal-carbonate-added test group; (**c**) the optimal condition test.

Figure 7a shows the images of the repeat test group in a single factor experiment with an adsorbing capacity of 33.911 mg/g (pyrolysis temperature 450 °C, heating rate 1 °C/s, temperature holding time 10 min, without additive). As can be seen in Figure 7a, the surface pores of the biochar are quite small, the diameters of those pores are less than 1 μm and have a non-uniform distribution.

Figure 7b represents the result of the metal-carbonate-added test group in a single factor experiment with an adsorbing capacity of 43.748 mg/g (pyrolysis temperature 450 °C, heating rate 1 °C/s, temperature holding time 10 min, Na_2CO_3 additive). As can be seen in Figure 7b, the number of biochar surface pores is larger than that in Figure 7a and the surface pores are more structured and evenly distributed although the surface pores are shallow.

Figure 7c shows an image of the optimal condition test with an adsorbing capacity of 140.564 mg/g (pyrolysis temperature 550 °C, heating rate 2 °C/s, temperature holding time 15 min, without additive). The image shows that in this test, biochar has the largest number of surface pores, the biggest pore diameter, the deepest surface pores, and the most even distribution.

4. Conclusions

As the pyrolysis temperature increased, the biochar yield decreased while the adsorbability of the biochar increased and reached a peak at a pyrolysis temperature of 600 °C.

The biochar yield and adsorbability had a similar trend as the heating rate increased. The biochar yield reached a peak at a heating rate of 1 °C/s, and the adsorbability of the biochar reached a peak at 1.5 °C/s.

The biochar yield and adsorbability both increased when the temperature holding time was increased, reaching the highest value at 15 min. Then they both decreased when the holding time was longer than 15 min.

Additives decreased the biochar yield. There was no significant influence on the adsorbability of the biochar whether carbon (or nothing added) was used as an additive. The addition of Na_2CO_3 (metal carbonate) or $ZnCl_2$ (metal chloride) could improve the adsorbability of the biochar. The highest adsorbing capacity was reached when Na_2CO_3 (metal carbonate) was used as an additive.

For the biochar adsorbability, the order of influence for the four factors is additive > pyrolysis temperature > heating rate > temperature holding time. The optimal conditions were pyrolysis temperature 550 °C, heating rate 2 °C/s, temperature holding time 15 min and without additive. The SEM results also indicate the reasons for the differences of adsorbability from the structures.

Author Contributions: Formal analysis, K.L., H.L. and W.Y.; Investigation, K.L.; Methodology, X.L. and H.E.M.; Project administration, H.L.; Resources, C.G.; Supervision, W.Y.; Writing–original draft, H.L.; Writing–review and editing, H.E.M., H.L. and W.Y.

Funding: This research was funded by Jiangsu Provincial Engineering Laboratory for Biomass Conversion and Process Integration Open Research Funded Projects (JPELBCPI2015002) and Postgraduate Research and Practice Innovation Program of Jiangsu Province (KYCX18-0737).

Conflicts of Interest: The authors declare no conflict of interest.

References

1. Wang, Y.D.; Chen, Z.C.; Roger, R. Research on technology of activated carbon preparation by microwave pyrolysis of shaddock peel. *J. Renew. Energy Resour.* **2014**, *4*, 529–536.
2. Zou, Y.; Han, B.X.; Yan, H.K. Research of high density micropore carbon-base adsorbent for natural gas storage. *J. Carbon Tech.* **1998**, *5*, 23–25.
3. Li, L.T.; Xie, Q.; Hao, L.N. Preparation and Characterization of Metal-Containing Activated Carbon for Electrode. *J. China Univ. Mining Technol.* **2008**, *2*, 225–230.
4. Raffaele, C.; Antonio, P.; Federico, R. An improved method for BTEX extraction from charcoal. *Anal. Methods* **2015**, *7*, 4811–4815.
5. Joana, D.; Maria, A.F.; Manuel, A. Waste materials for activated carbon preparation and its use in aqueous-phase treatment: A review. *J. Environ. Manag.* **2007**, *85*, 833–846.

6. Li, H.H.; Li, H.Y. Effects of different pretreatment methods on structure and adsorption properties of activated carbon. *J. Mater. Sci. Eng. Powder Metall.* **2014**, *4*, 647–653.
7. Huang, Y.; Chiueh, P.; Shih, C. Microwave pyrolysis of rice straw to produce biochar as an adsorbent for CO_2 capture. *J. Energy* **2015**, *84*, 75–82. [CrossRef]
8. Huang, Y.; Chiueh, P.; Lo, S. A review on microwave pyrolysis of lignocellulosic biomass. *J. Sustain. Environ. Res.* **2016**, *3*, 103–109. [CrossRef]
9. Motasemi, F.; Afzal, M.T. A review on the microwave-assisted pyrolysis technique. *J. Renew. Sustain. Energy Rev.* **2013**, *28*, 317–330. [CrossRef]
10. Yunpu, W.; Leilei, D.; Liangliang, F. Review of microwave-assisted lignin conversion for renewable fuels and chemicals. *J. Anal. Appl. Pyrol.* **2016**, *119*, 104–113. [CrossRef]
11. Mašek, O.; Budarin, V.; Gronnow, M. Microwave and slow pyrolysis biochar—Comparison of physical and functional properties. *J. Anal. Appl. Pyrol.* **2013**, *100*, 41–48. [CrossRef]
12. Xiao, M. A new determination of selengensis mineral elements and its development and utilization. *J. Food Ferment. Ind.* **2003**, *8*, 106–107.
13. Li, H.; Li, X.; Liu, L. Experimental study of microwave-assisted pyrolysis of rice straw for hydrogen production. *Int. J. Hydrog. Energy* **2016**, *41*, 2263–2267. [CrossRef]
14. Zhao, X.; Song, Z.; Liu, H. Microwave pyrolysis of corn stalk bale: A promising method for direct utilization of large-sized biomass and syngas production. *J. Anal. Appl. Pyrol.* **2010**, *89*, 87–94. [CrossRef]
15. Huang, Y.; Chiueh, P.; Kuan, W. Microwave pyrolysis of rice straw: Products, mechanism, and kinetics. *J. Bioresour. Technol.* **2013**, *142*, 620–624. [CrossRef] [PubMed]
16. Luo, Z.Y.; Wang, S.R.; Liao, Y.F. Mechanism study of cellulose rapid pyrolysis. *J. Ind. Eng. Chem. Res.* **2004**, *43*, 5605–5610. [CrossRef]
17. Wang, S.R.; Guo, X.J.; Liang, T. Mechanism research on cellulose pyrolysis by Py-GC/MS and subsequent density functional theory studies. *J. Bioresour. Technol.* **2012**, *104*, 722–728. [CrossRef] [PubMed]
18. Li, X.; Li, K.; Li, H. White Poplar Microwave Pyrolysis: Heating Rate and Optimization of Biochar Yield. *J. Bioresour.* **2018**, *13*, 1107–1121. [CrossRef]
19. Li, X.; Li, K.; Geng, C. An economic analysis of rice straw microwave pyrolysis for hydrogen-rich fuel gas. *RSC Adv.* **2017**, *7*, 53396–53400. [CrossRef]

© 2019 by the authors. Licensee MDPI, Basel, Switzerland. This article is an open access article distributed under the terms and conditions of the Creative Commons Attribution (CC BY) license (http://creativecommons.org/licenses/by/4.0/).

Review

Biochar as a Multifunctional Component of the Environment—A Review

Bogdan Saletnik [1,*], Grzegorz Zaguła [1], Marcin Bajcar [1], Maria Tarapatskyy [1], Gabriel Bobula [2] and Czesław Puchalski [1]

1. Department of Bioenergy Technology, Faculty of Biology and Agriculture, Rzeszow University, Ćwiklińskiej 2D, 35-601 Rzeszow, Poland; g_zagula@ur.edu.pl (G.Z.); mbajcar@ur.edu.pl (M.B.); czernicka.maria@gmail.com (M.T.); cpuchal@ur.edu.pl (C.P.)
2. Faculty of Physical Education, Rzeszow University, Towarnickiego 3, 35-959 Rzeszow, Poland; gbobula@ur.edu.pl
* Correspondence: bogdan.saletnik@urz.pl

Received: 4 February 2019; Accepted: 12 March 2019; Published: 18 March 2019

Abstract: The growing demand for electricity, caused by dynamic economic growth, leads to a decrease in the available non-renewable energy resources constituting the foundation of global power generation. A search for alternative sources of energy that can support conventional energy technologies utilizing fossil fuels is not only of key significance for the power industry but is also important from the point of view of environmental conservation and sustainable development. Plant biomass, with its specific chemical structure and high calorific value, is a promising renewable source of energy which can be utilized in numerous conversion processes, enabling the production of solid, liquid, and gaseous fuels. Methods of thermal biomass conversion include pyrolysis, i.e., a process allowing one to obtain a multifunctional product known as biochar. The article presents a review of information related to the broad uses of carbonization products. It also discusses the legal aspects and quality standards applicable to these materials. The paper draws attention to the lack of uniform legal and quality conditions, which would allow for a much better use of biochar. The review also aims to highlight the high potential for a use of biochar in different environments. The presented text attempts to emphasize the importance of biochar as an alternative to classic products used for energy, environmental and agricultural purposes.

Keywords: biochar; pyrolysis; environmental conservation; soil ameliorant

1. Introduction

In recent years, biochar and the wide range of its possible applications have been extensively investigated by researchers worldwide. In accordance with the definition specified by the International Biochar Initiative (IBI), biochar is a fine-grained product of carbonization, characterized by a high content of organic carbon and low susceptibility to degradation, which is obtained through the pyrolysis of biomass and biodegradable waste [1]. It is produced from organic matter as a result of pyrolysis, a process which is carried out in the absence of oxygen. Biochar can be utilized for energy-related purposes associated with environmental conservation and agriculture. The wide range of biochar applications is continuously expanding, mainly in such areas as industry, agriculture and operations related to the natural environment. It can be used as a soil additive, or added to fodder and silage, or applied in water treatment [2,3]. Biochar can also be used for the immobilization of contaminants from soil, and in sewage treatment; it can be applied as a supplementary material in composting and in methane fermentation processes [4–9]. Biochar application can be used as a filter for tar reduction in pyrolysis and gasification, as a fuel when pelletized, and also used as a substrate to produce hydrogen [10–12].

One of the ways to convert biomass is the gasification process, a state-of-the-art method of energetic use of biomass. The advantages of this process in relation to other methods are the possibilities of multi-directional use of the obtained product, namely the gas. Modern gasification methods make it possible to obtain two products: synthesis gas (syngas) being a mixture of gases (H_2, CO, CO_2, CH_4) and residues in the form of ash. When classifying gasification processes, they can be divided using different criteria depending on reactor type, gasification factor, thermal relations, gasification process parameters (pressure, temperature). The first stage of the gasification process is drying the material, which can have up to 50% water content, at a temperature of 100–200 °C. The next step is pyrolysis at increased temperatures (200–600 °C) using an anaerobic atmosphere to release the volatile parts contained in biomass. Products of this stage are solid substances such as charcoal, liquid substances (tars, oils and gas water), flammable gas and aromatic hydrocarbons such as benzene, toluene. The last stage is gasification (temperature above 750 °C) of solid and liquid substances produced at pyrolysis. This is a series of exo- and endothermal reactions that result in the production of flammable gas components [13–15]. Thermal processes of biomass (mainly the ones of lignocellulosic origin) processing also include torrefaction. The process is carried out in an anaerobic atmosphere at a temperature of 200–300 °C and the rate of temperature increase is in the range of 10–100 °C min^{-1}. Given the long duration of the process and relatively low temperatures, this process is also called roasting or mild pyrolysis [16]. Biomass processed as a result of the torrefaction process acquires new physicochemical properties, especially important when used as a fuel for the power industry. Products of torrefaction are characterized by increased milling susceptibility and energy density and their properties resemble low calorific coals. A typical torrefaction process is characterized by loss of mass and chemical energy of the raw material used. When analyzing the mass/energy ratio, an increased concentration of chemical energy of the fuel obtained in relation to the raw material can be observed [17]. When analyzing environmental aspects, biochemical conversions have an important role. The products (fuels) produced by these technologies are biogas, thio-alcohols and biodiesel. The use of biochemical processes seems reasonable when biomass contains large amounts of water. One type of such transformation is alcoholic fermentation, which allows carbohydrates to be broken down under anaerobic conditions with the addition of yeasts. The product of this conversion is bioethanol. Liquid biofuels can also be produced using a biochemical process such as oil esterification, which makes it possible to obtain methyl esters. Methane fermentation, on the other hand, affects the decomposition of multi molecular organic substances under conditions of limited access to oxygen. In the results, we obtain products in the form of alcohols, lower organic acids, as well as methane and carbon dioxide [18].

Thermal biomass conversion methods include pyrolysis, i.e., thermochemical transformation of biomass occurring in anaerobic conditions or in the presence of a small amount of oxygen, insufficient for combustion. Pyrolysis generates carbonization products, i.e., a highly carbonized solid biomass substance, bio-oil, also referred to as pyrolysis oil, as well as gas [19]. Depending on the parameters applied we can distinguish slow (bio-carbonization), fast and moderate pyrolysis as well as gasification. Fast pyrolysis (temperature of 500 °C with the peak (ultimate) temperature maintained for 1 s) produces approximately 12% of biochar. Slightly better results can be achieved by using moderate pyrolysis (temperature of 500 °C, the ultimate temperature maintained for 10–20 s)—approximately 20% of biochar. The highest percentage of biochar, at a level of 35%, may be obtained with the use of slow pyrolysis (at a temperature of 400–500 °C with the ultimate temperature maintained for 5–30 min.). The application of a high temperature, over 800 °C, and a short duration of the process at the ultimate temperature (gasification) leads to a yield of biochar amounting to 10% [20,21]. The low biochar content may also be related to the presence of oxygen and water in the reactor. It should also be pointed out that by using the rate of biomass heating as a classification factor we can distinguish between fast and slow pyrolysis. Reference books state that the heating rate of 1–100 °C min^{-1} is used in slow pyrolysis, whereas reaching the temperature heating rate above 1000 °C min^{-1} is characterized as fast pyrolysis [22].

The properties of biochar closely depend on the temperature of the pyrolysis process [23]. An increase in the temperature of pyrolysis leads to greater carbonification of the feedstock resulting in a higher carbon content and a decrease in the contents of hydrogen and oxygen [24]. The process of pyrolysis optimization towards obtaining a desired product should take into account the temperature of the reactor, rate of temperature increase, and duration of the process at the ultimate temperature. The authors emphasize relevant parameters of pyrolysis optimization in order to increase the utility of biochar as a high carbon material and at the same time cheap in production and generally available for application. This review presents an interdisciplinary approach to the topic of the importance of biochar in environmental, legal aspects of its application and quality standards.

2. Biochar and its Properties

Biochar may be produced from numerous materials of varied origins, e.g., energy crops, forest residues as well as agricultural residues [25–30]. Other materials used for biochar production include sewage sludge, waste from the food processing industries, e.g., oats previously subjected to fermentation, as well as poultry litter and cattle manure (Table 1) [31,32]. Importantly, the choice of feedstock for biochar production depends on e.g., economic and logistic factors as well as the parameters of the pyrolysis process itself and the types and properties of the applied substrates, e.g., water contents [21].

Table 1. Selected feedstocks used in biochar production [25–30].

Origin of Feedstocks	Type
Agriculture	Energy crops, corncob, rice husk, sunflower husk, post-fermentation oats, bamboo, bagasse, waste from olive oil production, straw, wheat husk, cattle manure, poultry litter
Forest	Conifer bark, pellets from sawdust, peat, moss, beech timber,
Waste	Waste from tea factories, paper, sewage sludge, municipal organic waste

The most important properties of biochars include their chemical composition, stability, specific surface and porosity. Importantly, the chemical composition of biochars mainly depends on the chemical composition of the substrates used in biochar production. Biochars contain stable organic carbon, aromatic compounds, aliphatic compounds and ash [33]. Taking into account the type of biomass and parameters of thermal processing applied, the content of carbon in biochar may be in the range of 50–90%, water 1–15%, volatile substances up to 40% and mineral substances up to 5%. Carbonization products have a neutral or alkaline pH and are highly resistant to microbiological degradation and decomposition; applied in the soil they are stable in terms of their chemical composition [21]. Their porous structure on the other hand contributes to improved sorption capacity of soils (Figure 1) [34].

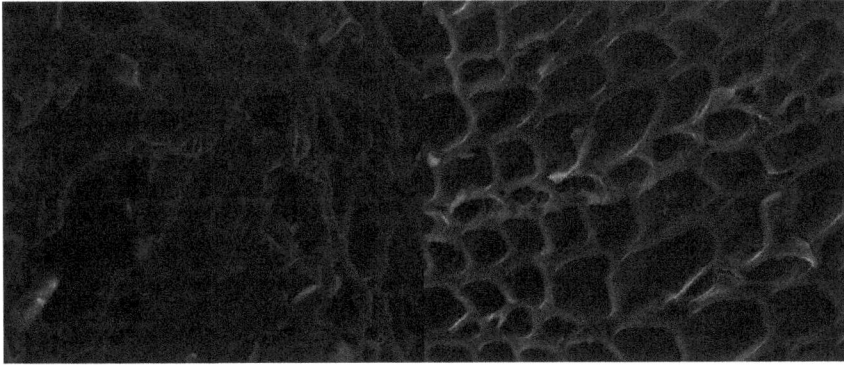

Figure 1. The porous structure of biochar [35].

Biomass pyrolysis conducted at higher temperatures may lead to an increased pH of the biochar. At lower temperatures it is possible to obtain biochar with higher ion-exchange capacity. Because of their physicochemical properties, biochars can be used for such purposes as soil carbon sequestration, the production of soil conditioner, as well as soil rehabilitation [21]. Table 2 presents some selected properties of biochars in relation to the feedstocks used and pyrolysis temperature applied.

Table 2. Selected properties of biochar, depending on the materials applied and the temperature of the pyrolysis process.

Feedstock for Biochar Production	pH	C g kg^{-1}	N g kg^{-1}	C/N	P g kg^{-1}	K g kg^{-1}	Ash %	Temp. of Pyrolysis (°C)	References
Acacia bark	7.4	398	10.4	38	–	–	–	260–360	[36]
Coconut	–	690	9.4	73	–	–	3.38	500	[37]
Corn	–	675	9.3	73	–	10.4	–	350	[38]
Corn	–	790	9.2	86	–	6.7	–	600	[38]
Green waste	6.2	680	1.7	400	0.2	1	–	450	[39]
Peanut shells	–	499	11.0	45	0.6	6.2	–	400	[40]
Pecan shells	7.6	834	3.4	245	–	–	3.8	700	[41]
Pecan shells	–	880	4.0	220	–	–	–	700	[42]
Rice straw	–	490	13.2	37	–	–	9.54	500	[37]
Sewage sludge	–	470	64	7	56	–	35	450	[43]
Sugarcane bagasse	–	710	17.7	40	–	–	4.34	500	[37]
Eucalyptus wood	7.0	824	5.7	144	0.6	–	0.23	350	[44]
Oak wood	–	759	1.0	759	–	1.1	–	350	[38]
Oak wood	–	884	1.2	737	–	2.2	–	600	[38]

Different types of fuel, including waste, can be used for pyrolysis. The types of raw materials for pyrolysis can generally be divided into three groups: mine fuels, e.g., coal, biomass, e.g., wood, sewage sludge and plastics, e.g., tires. Each of these raw materials has different chemical composition and physical properties, which affect the quality of the pyrolysis products obtained. Biomass is a very universal raw material in its morphology and physical characteristics. Biomass material can be partially wet or dry, homogeneous or heterogeneous, of high density or of a loose structure, low or high ash content, high or low fragmentation [45]. This variety of properties of raw materials for biofuel production in gasification reactors is difficult and it is necessary to apply other processing measures focusing on the properties in question. Hence there is a large diversity of bioreactors used in biomass processing as well as a wide range of final products diversified in terms of properties determined by the conditions of the pyrolysis process and the type of bioreactor. Knowledge of the chemical structure and their behavior during the process of pyrolysis is extremely important from the point of view of application of optimal technological solutions, selection of appropriate parameters (temperature, process duration), process efficiency and environmental nuisance.

High heating rates promote cellulose and hemicellulose depolymerization reactions, minimizing the volatiles residence time inside the particle and secondary reactions. It also favors the volatiles cracking. So, the condensable gas release goes on quickly thus achieving high yields of bio-oil and the lowest production of char. Low temperatures and low heating rates promote intra-chain hydrogen bonds of cellulose functional groups, increasing the probability of collision to produce a dehydration reaction. For high heating rates, inter-chain hydrogen bonds are stronger achieving greater separation between the cellulose molecules and thus decreasing the possibility of collisions that facilitate the dehydration reaction. There are many reports in literature pointing out the importance of the effects of the heating rate on the yields of bio-oil and char [22,46,47].

The physical and chemical properties of the raw materials used in the process are the most important. The highest yields of carbonization products are obtained when raw materials with high lignin content are subjected to pyrolysis at moderate temperatures. Basically, biomass containing a significant amount of volatile substances offers a large amount of pyrolytic gas and bio-oil, and the presence of solid coal increases the efficiency of biochar production [19]. The moisture content of biomass has a significant impact on the heat transfer process and product distribution. Numerous experiments in the process of pyrolysis have confirmed that carbonization product's efficiency increases at low heating rates. During fast heating, the efficiency of volatile parts increases and less secondary reactions occur [48]. In the case of reactors with a high biochar production capacity, the fixed bed reactor and fast-moving fluidized bed reactors should be considered. In the case of slow pyrolysis, the use of a fixed bed allows the use of a large working surface of the reactor. Fluidized beds allow the process to be carried out much faster but are limited by volume for technological reasons [49].

3. Legal Aspects and Quality Standards

The term "biochar" does not appear in the legislation of the EU or Poland. The only European country which has adopted regulations relating to it is Switzerland. Considerations linked to biochar production and the wide range of its possible applications should however take into account the legal regulations on waste management, use of fertilizers and product safety [2].

Legal regulations fail to explicitly define the status of biochar, classifying it as a product, by-product, or waste [50–53]. In practice, biochar is perceived as waste, while in the literature it is most frequently described as a by-product of pyrolysis. The relevant EU and Polish regulations [54] define waste as "any substance or object which the holder discards, intends to discard or is required to discard". In accordance with the Waste Act, biochar produced from agricultural biomass wastes, or bio-wastes, may be classified as waste material produced as a result of a thermal conversion process. Accordingly, carbonization products should be treated as waste, however, biochar is not listed in the waste catalogue [55]. Article 14 of the Waste Act provides that "specific types of waste no longer have the status of waste if, following recovery, including recycling, they jointly meet specified conditions:

- the object or substance is commonly used for particular purposes,
- there is a market or demand for such objects or substances,
- the object or substance meets the technical requirements for applications related to specific purposes as well as the requirements set out in the rules and standards applicable to the product,
- use of the object or substance does not lead to negative consequences for human life or well-being or for the environment, as well as the requirements defined by regulations of the European Union" [56]. Given the above, biochar is no longer classified as waste.

In light of the regulations biochar can also be classified as a by-product, if it meets all of the following conditions: "further use of the object or the substance is certain; the object or substance may be used directly with no further processing other than normal industrial practice; the object or substance is produced as an integral part of a production process; the substance or object fulfils all relevant requirements, including legal, product-related, environmental and health protection requirements for the specific use of these substances or objects and their use will not lead to overall adverse environmental or human health impacts" [56]. Biochar can also be treated as a product, if it is the main product of a given process, and it has been produced from biomass obtained specifically for this purpose [53]. In a situation when biochar is treated as a product or by-product, the provisions of the Waste Act are not applicable; on the other hand, it is necessary to take into account other legal requirements (e.g., the regulations of the registration, evaluation, Authorization and restriction of chemicals (REACH) system) [57].

As regards EU laws, it can be concluded that the use of biochar in soils is not explicitly regulated or forbidden [53]. As an exception, Switzerland permits the use of biochar in agriculture provided that the requirements specified by the European Biochar Certificate (EBC) are complied with. The use of biochar as a soil ameliorant in the EU is subject to the provisions set forth by Regulation (EC) No 2003/2003 [58]. In Poland, in order to use biochar as a fertilizer or soil ameliorant it is necessary to complete a registration procedure and obtain approval from the Minister of Agriculture and Rural Development. Notably, these legal requirements are not applicable if biochar is used as a fertilizer in quantities necessary for experimental studies as well as in research and development projects [2].

Given the wide range of substrates used in the production of biochar, as well as the diverse conditions of thermal processing applied, and consequently the varied chemical composition of the final products, attempts have been made to regulate quality requirements related to carbonization products. It was necessary to take adequate steps and develop uniform guidelines and define quality requirements for biochar and substrates used in its production. As a result, global biochar organizations developed their own quality standards (biochar quality certificates):

- Biochar standards defined by the International Biochar Initiative (IBI), USA [1];
- European Biochar Certificate defined by the European Biochar Foundation [59];
- Biochar Quality Mandate developed by the British Biochar Foundation in the United Kingdom [60].

These standards present recommendations related to the substrates applied in the production of biochar, parameters of technological processes, requirements for biochar materials introduced into the soil as well as providing guidelines for the methodology of conducting measurements and analyses. Importantly, the requirements contained therein take the form of guidelines and they are not legally binding in the European Union member states [61]. Developed under the European Union Framework Program, the project entitled REFERTIL (reducing mineral fertilizers and chemicals use in agriculture by recycling treated organic waste as compost and biochar products) was designed to develop quality requirements for biochar to be adopted as recommendations for legal regulations relating to fertilizers [62].

The quality requirements defined for biochar ensure the safety of its soil-related applications, specify permissible contents of heavy metals, furans, polychlorinated biphenyls, dioxins and polycyclic aromatic hydrocarbons [63–65]. Table 3 summarizes and compares permissible levels of contaminants in biochar, as defined by the International Biochar Initiative (IBI), British Biochar Foundation (BQM), European Biochar Foundation (EBC) and under the REFERTIL project.

Table 3. The permissible content of contaminants in biochar, based on the existing quality standards [1,59,60,62].

Parameter (mg kg^{-1} of Dry Matter)	IBI	BQM		EBC		REFERTIL
		Type of Biochar				
		High Grade	Standard	Premium	Basic	
As	13–100	10	100	13	13	10
Cd	1.4–39	3	39	1	1.5	1.5
Cr	93–1200	15	100	80	90	100
Cu	143–6000	40	1500	100	1000	200
Hg	1–17	1	17	1	1	1
Ni	47–420	10	600	30	50	50
Pb	121–300	60	500	120	150	120
Zn	416–7400	150	2800	400	400	600
Se	2–200	5	100	–	–	–
Mo	5–75	10	75	–	–	–
F	–	–	–	–	–	–
WWA	6–300	20	20	4	12	6
PCB	0.2–1	0.5	0.5	0.2	0.2	0.2
Dioxins and furans (ng kg^{-1})	20	20	20	20	20	20

In Poland, the requirements related to the permissible contamination of organic and organic–mineral fertilizers, as well as crop enhancers are defined by the Regulation of the Minister of Agriculture and Rural Development of 18 June 2008 on the implementation of certain provisions of the Act on fertilizers and fertilization [66]. It specifies e.g., permissible contents of cadmium, lead, nickel, mercury, respectively at the levels of 5, 140, 60, 2 mg kg^{-1} of dry matter of the fertilizer or crop enhancer, yet it fails to take into account contaminations taking the form of polycyclic aromatic hydrocarbons or polychlorinated biphenyls.

4. Biochar in Environmental Conservation

The use of biochar in environmental protection falls within the scope of remediation of polluted soils, energy production, climate change aspects, waste management, sustainable development issues. Production of biochar is one of the methods permitting the reduction of the need for the disposal of animal and plant waste. Biodegradable animal waste, agricultural biomass and sewage sludge can be effectively used for the production of energy through pyrolysis. Additional benefits include a reduction in the volume of waste subjected to thermal processing, as well as the elimination of pathogenic microorganisms potentially occurring e.g., in cattle manure and sewage sludge. The use of the above waste material in the production of biochar may also lead indirectly to the reduction of methane emissions from landfills and reduce the necessity of seeking alternative methods of waste management [67].

Increased emissions of CO_2 to the atmosphere in recent years have led to a significant disproportion between the natural emission and absorption of carbon. It is necessary to take action to balance carbon in the atmosphere by its capture and storage e.g., in the soil [68]. One of the solutions to this problem involves the use of biochar obtained from various types of biomass. Introduced into the soil, it enables the long-term sequestration of carbon. According to the literature, by adding biochar to soil at a rate of 13.5 t ha^{-1} it is possible to store the carbon within it for a minimum of two hundred years [69]. Furthermore, it has been shown that biochar may lead to a decrease in emissions of nitrous oxide (N_2O) and methane (CH_4) from the soil, mediated by biotic and abiotic mechanisms [70].

One of the main applications of biochar is its use as a renewable fuel [71–73]. Carbonization products can be incinerated or co-incinerated in combined heat and power plants and power plants. Biochar in comparison with raw biomass contain less of significant amounts of chlorine, and volatile substances influencing the reduction of boiler efficiency and increased emission of inorganic particles. Changes in chlorine content result from the specificity of the pyrolysis process. During pyrolysis, the chlorine contained in the raw biomass is released in gaseous form and goes to the environment. These changes result from the changing structure of the material and the processes of degassing biomass. Inorganic compounds forming fine particles result in increased sludge production in fuel-burning boilers. This problem disappears in the moment of carbonization of a fuel, which is particularly important in the processes of biomass combustion. Biochar is therefore an important element in reducing this disadvantage [74]. Such fuels are an alternative to conventional fossil fuels (Table 4), as well as they offer the possibility of recovering energy from waste deposited in landfills [2]. An additional direction of wide use of biochar in the power industry is the acquisition of many energy products during its production. Additionally, obtained electricity and heat may successfully reduce the costs of the process and creation of cogeneration installations, which may become an important element in the creation of local power grids. Such small installations on the local market quickly and efficiently process e.g., waste from agricultural production, which creates independent power systems essential for the elimination of power and climate risks [33].

Table 4. The calorific value and contents of carbon, ash and volatiles in selected fuels and biochars [2,26,75,76].

Fuel	Calorific Value (MJ kg^{-1})	Carbon	Ash	Volatiles
			%	
Fossil fuels				
Natural gas	48.0	75.0	0.0	100.0
Lignite	25.0	60.0	12.0	25.0
Bituminous coal	7.5–21.0	66.0–73.0	10.0–20.0	40.0–60.0

Table 4. Cont.

Fuel	Calorific Value (MJ kg^{-1})	Carbon	Ash	Volatiles
		%		
Biomass				
Wood	10.5	35.0	1.0	55.0
Straw	15.0	43.0	3.0	73.0
Rapeseed	15.3	44.7	7.3	78.7
Sunflower	15.7	17.2	8.3	74.5
Biochar				
Biochar from rapeseed	23.4	72.7	21.8	13.6
Biochar from sunflower	20.5	63.4	28.9	13.4
Biochar from oil palm (residues)	17.1	53.8	3.1	81.9
Biochar from cherry wood	27.7	59.5	9.1	22.2

Owing to their sorption properties, biochars can effectively immobilize contamination from solid, liquid and gaseous media. Of particular note is that more effective sorption properties are to be found in biochars produced at higher temperatures, since they have larger specific surface and higher microporosity. Research conducted so far shows that biochars can be used as sorbents in processes aimed at immobilizing the residues of pharmaceuticals and bacteriostatic antibiotics, e.g., sulfamethoxazole, from sewage [28,77], as well as heavy metals from aqueous solutions, municipal sewage and industrial wastewater [78–81]. It has also been reported that biochar was used for immobilizing such pesticides as carbaryl, atrazine, simazine and acetochlor from soils [82–84]. Because of their physicochemical and structural properties, carbonization products are an alternative to activated carbon and other treatment technologies applied to various substances, including sewage and wastewater [77]. Table 5 presents possible biochar applications for immobilizing various types of contaminants in soil and water.

Table 5. Possible biochar applications for immobilizing contaminants in soil and water.

Contamination	Type of Biochar (Feedstock/Pyrolysis Temperature)	Type of Environment	References
Agricultural chemicals			
Atrazine	Cattle manure (450 °C)	Soil	[85]
	Cattle manure (200 °C)	Water	[86]
Atrazine and simazine	Green waste (450 °C)	Water	[82]
Pentachlorophenol	Bamboo (600 °C)	Soil	[87]
Antibiotics			
Sulfamethazine	Hardwood (600 °C)	Water	[88]
Sulfamethoxazole	Bamboo (450 and 600 °C)	Water	[28]
Tylosin	Hardwood (850 and 900 °C)	Water	[89]
Tetracycline	Rice husk (450–500 °C)	Water	[90]
Other hydrocarbons			
Pyrene	Corncob (600 °C)	Water	[91]
	Sawdust (400 i 700 °C)		[92]
Trichloroethylene	Peanut shell (300 and 700 °C)	Water	[93]
Naphthalene	Pine needles (100–700 °C)	Water	[94]
Heavy metals			
Cadmium	*Miscanthus sacchariflorus* (300–600 °C)	Water	[95]
Aluminum	Rice straw (100–600 °C)	Water	[29]
Lead	Pine wood (300 °C)	Water	[96]

5. Biochar as a Activated Carbons

In recent years there has been an increase in interest in the use of active coals resulting mainly from their low production cost and favorable physicochemical properties, which include a strongly developed specific surface, very good ion-exchange properties and high mechanical and chemical durability. In general, any material which contains carbon in its composition in organic compounds may be used to produce active carbons. The raw materials used, activation process, and process parameters determine the physical properties and performance characteristics of the resulting carbon. Modifying these activation properties determines the porosity and pore volume distribution in the carbon. Activated carbon is defined as a carbonaceous material with a large internal surface area and highly developed porous structure resulting from the processing of raw materials under high temperature reactions. It is composed of 87–97% carbon but also contains other elements depending on the processing method used and raw material it is derived from. Activated carbon's porous structure allows it to adsorb materials from the liquid and gas phase. On an industrial scale, the precursors of active carbons include mainly fossil coals, wood, peat, and coconut shells. The literature presents information concerning the production of activated carbons by activating a variety of waste materials such as nut shells, sawdust, straw, fruit stones, straw, sewage sludge and many others [97]. These activities are of justified environmental character as they manage a significant amount of waste and their economic aspect is important as well. The studies carried out so far have highlighted the fact that activated carbons produced from waste materials can show a better sorption capacity than products made from traditional precursors. For example, carbons were prepared from combination of the waste tea and K_2CO_3 have high surface area and pore volume [98]. The production of active carbons takes place within two mechanisms, i.e., physical activation and chemical activation. Physical activation, also known as thermal activation, consists of two successive stages. The first step is the pyrolysis of the starting material, carried out at a high temperature (usually 500–1000 °C). During the second step, the carbonization product obtained is activated by exposure to a high temperature (800–1000 °C) using an oxidizing agent such as water vapor, carbon monoxide (IV) or a mixture of these gases. If the above-mentioned stages of physical activation take place at the same time, then it is direct activation process. In the process of chemical activation, the precursor is subjected to high-temperature treatment in an inert gas atmosphere after prior impregnation or mixing with an activating agent. The activators used in chemical activation are mainly potassium and sodium hydroxide, sodium and potassium carbonates, zinc chloride, and phosphoric acid (V). The disadvantage of this method is its high cost caused by the necessity to use expensive activating agents and to introduce a stage aimed at removing the excess of the activating agent and the by-products. The physical and chemical properties of active carbons such as porous structure, highly developed specific surface area or sorbent capacity are strictly dependent on the substrates used, activation methods and conditions of the process. The properties of carbon sorbents can also be significantly altered by means of appropriate chemical modifications both during the manufacturing process and after the activation process. These processes consist mainly of the introduction of functional groups into the structure of the carbon material or on its surface, which significantly change its chemical character. Modification of carbon materials can be carried out using e.g., liquid oxidants, to which we include, primarily HNO_3, H_2O_2 and $(NH_4)_2S_2O_8$. The surface of active carbons can also be modified with various types of organic and inorganic compounds, e.g., pyridine, compounds of platinum, of chromium, of silver, of copper, of potassium, of zinc, of sodium and of cobalt. This type of modification is aimed at obtaining materials with significantly increased and selective sorbent capacity. Table 6 presents the surface area values for physically and chemically activated carbons obtained from different materials [99–102].

Table 6. Surface area values for physically and chemically activated carbons obtained from different material.

Physically Activated Carbons			Chemically Activated Carbons		
Initial Material	Activation Agent	Surface Area $m^2 g^{-1}$	Initial Material	Activation Agent	Surface Area $m^2 g^{-1}$
Rice [103]	Steam	1122	Rice [104]	KOH	3263
Peanut shells [105]	Steam	757	Hazelnut shells [106]	KOH	1700
Cornstarch [107]	Thermal	686	Corncob [108]	KOH	3054

Table 6. Cont.

Physically Activated Carbons			Chemically Activated Carbons		
Grape pomace, grape stalks [109]	Steam	266 300	Grape seeds [110]	KOH	1860
Finish wood [111]	CO_2	590	Eucalyptus wood, Beech wood [112]	KOH	2120, 2460
Olive Stone [113]	CO_2	1355	Olives stones [114]	$ZnCl_2$	1860
Sunflower stem [115]	CO_2	438	Stem of date palm [116]	KOH H_3PO_4	947 1100
Vine shoots [117]	CO_2	1173	Waste tea [98]	K_2CO_3	1722

Activated carbons can be used in many sectors e.g., pharmaceutical, food industrial as additives or the emerging use. A series of studies demonstrated that activated biochar could be available for pharmaceuticals removal, such as acetaminophen, caffeine, atrazine, diclofenac, glyphosate, naproxen, ibuprofen and sulfamethazine [118]. Hoegberg et al. conducted a study to identify the maximum adsorption capacities of amitriptyline and paracetamol, separately and in combination, to activated charcoal [119]. The rapid adsorption tendencies of activated carbon have also extended the biochar applications to biomedical sciences. Activated carbons or activated charcoal have shown their functionalities in preventing gastrointestinal absorption of certain toxins and drugs, thereby enhancing their elimination even after systemic absorption [120]. Ozsoy and van Leeuwen in their study focused on decolorizing solutions of waste fruit candy extract dissolved in deionized water using activated carbon adsorption. The aim of this study was to use activated carbon in treatment methods for candy wastes to facilitate reuse of fruit pulp, sugar and organic acids in the process and minimize the wastes emanating from this industry [121].

Biochar-mediated adsorption of organic contaminants may be based on the principle of electrostatic interactions with polar or non-polar groups (Figure 2) [122–124].

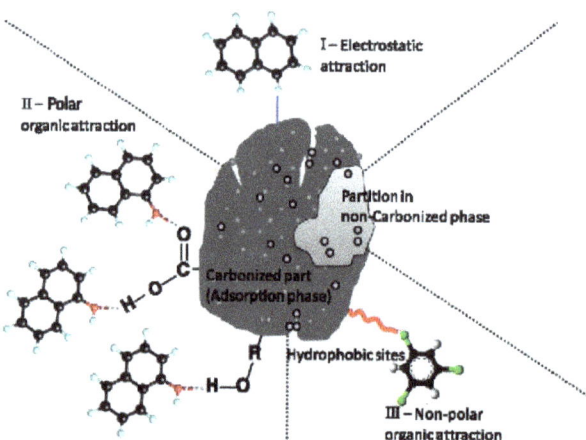

Figure 2. Mechanisms of organic substance adsorption on the surface of carbonization products [124].

Adsorption of inorganic contaminants, including ions of heavy metals, through the use of biochar is characterized by four mechanisms:

- ion exchange (Na^+, K^+ ions are involved),
- precipitation,
- anionic metal attraction,
- cationic metal attraction (Figure 3) [122,124,125].

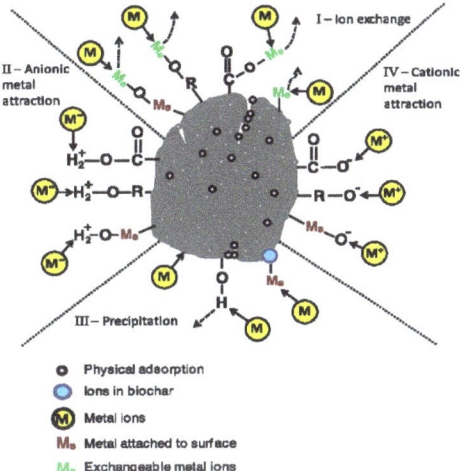

Figure 3. Types (mechanisms) of metal ion adsorption on the surface of carbonization products [124].

Mechanisms I and IV are identical normal cation exchange reactions, albeit with metal cations at I, and H^+ at IV.

In view of the numerous beneficial properties of biochars, in particular the high contents of organic carbon, these materials can be used to enhance the physicochemical and biological properties of soils [126].

6. Biochar as Soil Conditioner

Interest in biochar as a potential soil enhancer began with the discovery of preta de Indio—Indian black earth (in Amazonia), notable for its high content of carbon and nutrients [127]. As shown in previous research, the soils were created a few thousand years ago as a result of the burning of forests and natural fires, as well as soil improvement with the use of charcoal applied by pre-Columbian natives [128].

From the agricultural point of view, the application of carbonization products for soil amelioration seems to be beneficial because the treatment improves the conditions for plant growth, leading to a better yield (Figure 4, Table 7) [129]. Furthermore, due to the rapid effects and relatively low costs of such treatment, biochars are more and more frequently used in processes of soil remediation and conservation [130].

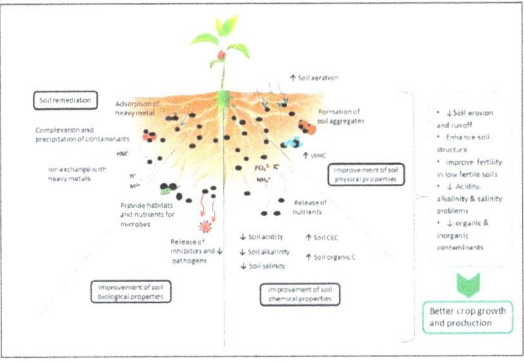

Figure 4. Influence of biochar on soil properties [131].

Table 7. Effect of biochar application on the yield of selected plants.

Type of Biochar (Feedstock)	Dose (t ha^{-1})	Crop	Increase in Yield Compared to the Control %	References
Wood	68	Cowpea	20	[132]
	136.75		100	
	68	Rice	50	
Poultry litter	10	Radish	42	[39,133]
	50.5	Radish	96	
Woodchips from fruit trees	22	Grapes	20	[134]
Cattle manure	15	Maize	150	[135]
Hardwood	19	Maize	10	[136]
	38		17	
	58		48	
Wheat straw	40	Rapeseed	36	[137]
	40	Sweet potatoes	54	

The activity of biochar, after it is introduced into the soil environment, depends predominantly on the feedstocks used in its production and the parameters of the pyrolysis process. These determine the contents of macro- and micro-elements as well as harmful substances, such as e.g., heavy metals. The heterogeneous chemical composition of biochars enables their interactions with a large variety of organic and inorganic compounds present in the soil [138]. The diverse properties of biochar materials enable reactions with mineral and organic fractions of soil and the accumulation of combined mineral and organic complexes [139]. Biochars introduced to the soil are characterized by a high stability and resistance to biological decomposition, therefore they are recognized as a highly effective medium for the sequestration of carbon dioxide in soil [33]. Moreover, the application of biochars to soil leads to increased contents not only of carbon but also of other biogenic compounds, such as phosphorus, potassium, magnesium and nitrogen [39,140]. Owing to their large ion-exchange capacities and specific surface, biochars also produce such effects as the reduced leaching of biogenic elements from soils and a decreased emission of nitrous oxide [141]. Recent study has highlighted that crop biochar also contains biogenic silica (phytoliths) and can quickly release bioavailable silicon to enhance plant biomass and promote the biological silicon cycle in soil [142]. One of the functions of the biocarbon addition is to modify the nitrogen and phosphorus cycle. Biochar as an additive to soils has the ability to store nitrogen by increasing NH_3 and NH_4^+ retention, reducing N_2O emissions and eluting NO_3 ions, as well as inducing the development of nitrogen bacteria which directly affects the increase in soil productivity. Biochar is characterized by very different potassium contents, depending on the type of batch material used in production. Of all macronutrients present in carbonates, potassium is the best available element for plants, and the proportion of biosorptive forms is up to 95% of the total content [33]. Biochar fertilizers also lead to increased soil pH [143]. The related research has shown that carbonization products contain numerous alkaline substances, e.g., calcium carbonate, which may affect soil reaction, and the best effects, i.e., the highest pH increase, may be achieved in strongly acidified soils. Increased pH which is beneficial in acid tropical soils like terra preta but reduces yield in soils of high pH as in many temperate regions [41,144]. Biochar also impacts the physical properties of soil by improving its water retention, capacity to form aggregates, and resistance to erosion [145]. The improvement of the physical characteristics of soils resulting from the use of biochar as a soil ameliorant mainly depends on the properties of the biochar, and these are predominantly determined by the technology applied in its production [146,147]. Owing to their highly porous structure, carbonization products may create favorable conditions for microorganisms, as a consequence improving the fertility and productivity of soils (Table 8). For example, biochar addition to the soil with a high oil organic matter level leads to higher microbial stimulation and consequent higher N mineralization [148].

Table 8. Effects of biochar application on the soil environment of soil microorganisms.

Type of Biochar (Feedstock)	Impact on Soil Microorganisms	References
Willow wood and swine manure: slow pyrolysis at 350 °C slow pyrolysis at 700 °C	Increased microbial biomass in both cases: increased dehydrogenase activity decreased dehydrogenase activity	[149]
Poultry litter and pine woodchips (pyrolysis at 400 and 500 °C)	Increased microbial biomass	[150]
Leaves and fragmented branches	Increased rate of fungal and bacterial growth	[151]
Wood (fast pyrolysis)	Increased microbial count	[152]

7. Conclusions

The research conducted so far focusing on the characteristics and possible applications of biochar in a way attempts to regulate the flow of inorganic matter in the environment and provides significant support for the conservation of the natural environment, in particular taking into account the phytoremediation and utilitarian dimension of sustainable energy management.

Owing to their physicochemical properties, biochars may be used for such purposes as carbon sequestration in soils, reduction of the bioavailability of contaminants affecting living organisms as well as water treatment. They also present significant potential for the immobilization of heavy metals from aqueous solutions and for reducing their mobility in soils. In recent years there has been an increase in interest in the use of activated carbons with elevated and selective sorption capacity. Literature describe important role the waste products as a substrate for the production of such sorbents. Pyrolysis process of biomass allows produce high quality fuel in biochar form. Such fuels are an alternative to conventional fossil fuels, as well as they offer the possibility of recovering energy from waste deposited in landfills.

According to the latest research, the application of carbonization products may effectively enhance the physicochemical properties of soils and improve the fertility of poor soils. Biochar, as a product of thermal biomass conversion carried out with a reduced availability of oxygen, is characterized by a high content of carbon and excellent sorption properties. From the point of view of agriculture, the application of biochar as a soil conditioner produces numerous benefits, such as the enhancement of the physical, chemical and biological properties of soils, and this contributes to an increased crop yield. The application of biochar in soils may be an alternative to traditional forms of mineral amendment and strengthens the ecological aspect of bioenergy engineering.

Continuous progress in research on the use of biochar, its production methods and characterization techniques allow us to conclude that biochar technologies, especially as a means of improving the quality of soil or sorbent of pollution from soil and water, will become much more important in the future. However, it should be stressed that due to the lack of specific regulations for biochar, its commercial use is limited. In order to develop the biochar industry and ensure environmental safety, uniform legal and quality regulations should be ensured.

It should, however, be emphasized that further comprehensive research is needed to investigate the feasibility of the application of biochar and to determine the optimum methods for using this highly productive material.

Funding: This research received no external funding.

Conflicts of Interest: The authors declare no conflict of interest.

References

1. IBI Biochar Standards—Standardized Product Definition and Product Testing Guidelines for Biochar That Is Used in Soil, v.2.1. Available online: http://www.biochar-international.org/sites/default/files/IBI_Biochar_Standards_V2.1_Final.pdf (accessed on 1 February 2019).
2. Malińska, K. Legal and quality aspects of requirements defined for biochar. *Inżynieria i Ochrona Środowiska* **2015**, *18*, 359–371.

3. Pereira, R.C.; Muetzel, S.; Arbestain, M.C.; Bishop, P.; Hina, K.; Hedley, M. Assessment of the influence of biochar on rumen and silage fermentation: A laboratory–scale experiment. *Anim. Feed Sci. Technol.* **2014**, *196*, 22–31. [CrossRef]
4. Tang, J.; Zhy, W.; Kookana, R.; Katayama, A. Characteristics of biochar and its application in remediation of contaminated soil. *J. Biosci. Bioeng.* **2013**, *116*, 653–659. [CrossRef] [PubMed]
5. Mohan, D.; Sarswat, A.; Ok, S.Y.; Pittman, C.U., Jr. Organic and inorganic contaminants removal from water with biochar, a renewable, low cost and sustainable adsorbent—A critical review. *Bioresour. Technol.* **2014**, *160*, 191–202. [CrossRef]
6. Steiner, C.; Das, K.C.; Melear, N.; Lakly, D. Reducing nitrogen loss during poultry litter composting using biochar. *J. Environ. Qual.* **2010**, *39*, 1236–1242. [CrossRef]
7. Steiner, C.; Melear, N.; Harris, K.; Das, K.C. Biochar as bulking agent for poultry litter composting. *Carbon Manag.* **2011**, *2*, 227–230. [CrossRef]
8. Malińska, K.; Zabochnicka-Świątek, M.; Dach, J. Effects of biochar amendment on ammonia emission during composting of sewage sludge. *Ecol. Eng.* **2014**, *71*, 474–478. [CrossRef]
9. Malińska, K.; Dach, J. Biochar as a supplementary material for biogas production. *Ecol. Eng.* **2015**, *4*, 117–124. [CrossRef]
10. Paethanom, A.; Bartocci, P.; D' Alessandro, B.; D' Amico, M.; Testarmata, F.; Moriconi, N.; Slopiecka, K.; Yoshikawa, K.; Fantozzi, F. A low-cost pyrogas cleaning system for power generation: Scaling up from lab to pilot. *Appl. Energy* **2013**, *111*, 1080–1088. [CrossRef]
11. Bartocci, P.; Bidini, G.; Saputo, P.; Fantozzi, F. Biochar pellet carbon footprint. *Chem. Eng. Trans.* **2016**, *50*, 217–222.
12. Bartocci, P.; Zampilli, M.; Bidini, G.; Fantozzi, F. Hydrogen-rich gas production through steam gasification of charcoal pellet. *Appl. Therm. Eng.* **2018**, *132*, 817–823. [CrossRef]
13. Saxena, R.C.; Seal, D.; Kumar, S.; Goyal, H.B. Thermo-chemical routes for hydrogen rich gas from biomass: a review. *Renew. Sustain. Energy Rev.* **2008**, *12*, 1909–1927. [CrossRef]
14. Piskowska-Wasiak, J. Cleaning and conditioning of gas from biomass gasification for production of SNG (Substitute Natural Gas). *Nafta-Gaz* **2011**, *67*, 347–360.
15. Wróblewski, R. The concept of a small cogeneration system integrated with biomass gasification. *Energy Policy J.* **2014**, *17*, 159–170.
16. Chen, W.; Kuo, P. A study on torrefaction of various biomass materials and its impact on lignocellulosic structure simulated by a thermogravimetry. *Energy* **2010**, *35*, 2580–2586. [CrossRef]
17. Kopczński, M.; Zuwała, J. Biomass torrefaction as a way for elimination of technical barriers existing in large-scale co-combustion. *Energy Policy J.* **2013**, *16*, 271–284.
18. Piaskowska-Silarska, M. Analysis of the possibility of obtaining energy from biomass in Poland. *Energy Policy J.* **2014**, *17*, 239–248.
19. Uddin, M.N.; Techato, K.; Taweekun, J.; Rahman, M.M.; Rasul, M.G.; Mahlia, T.M.I.; Ashrafur, S.M. An Overview of Recent Developments in Biomass Pyrolysis Technologies. *Energies* **2018**, *11*, 3115. [CrossRef]
20. Lewandowski, W.M.; Radziemska, E.; Ryms, M.; Ostrowski, P. Modern methods of thermochemical biomass conversion into gas, liquid and solid fuels. *Ecol. Chem. Eng. S* **2011**, *18*, 39–47.
21. Malińska, K. Biochar—A response to current environmental issues. *Eng. Prot. Environ.* **2012**, *15*, 387–403.
22. Montoya, J.I.; Chejne-Janna, F.; Garcia-Pérez, M. Fast pyrolysis of biomass: A review of relevant aspects. Part I: Parametric study. *Dyna* **2015**, *82*, 239–248. [CrossRef]
23. Park, J.; Hung, I.; Gan, Z.; Rojas, O.J.; Lim, K.H.; Park, S. Activated carbon from biochar: influence of its physicochemical properties on the sorption characteristics of phenanthrene. *Bioresour. Technol.* **2013**, *149*, 383–389. [CrossRef] [PubMed]
24. Uchimiya, M.; Wartelle, L.H.; Klasson, K.T.; Fortier, C.A.; Lima, I.M. Influence of pyrolysis temperature on biochar property and function as a heavy metal sorbent in soil. *J. Agric. Food Chem.* **2011**, *59*, 2501–2510. [CrossRef] [PubMed]
25. Kwapinski, W.; Byrne, C.M.P.; Kryachko, E.; Wolfram, P.; Adley, C.; Leahy, J.J.; Novotny, E.H.; Hayes, M.H.B. Biochar from Biomass and Waste. *Waste Biomass Valor.* **2010**, *1*, 177–189. [CrossRef]
26. Sànchez, M.E.; Lindao, E.; Margaleff, D.; Martínez, O.; Morán, A. Pyrolysis of agricultural residues from rape and sunflower: Production and characterization of biofuels and biochar soil management. *J. Anal. Appl. Pyrolysis* **2009**, *85*, 142–144. [CrossRef]

27. Shen, Y.S.; Wang, S.L.; Tzou, Y.M.; Yan, Y.Y.; Kuan, W.H. Removal of hexavalent Cr by coconut coir and derived chars-the effect of surface functionality. *Bioresour. Technol.* **2012**, *104*, 165–172. [CrossRef] [PubMed]
28. Yao, Y.; Gao, B.; Chen, H.; Jiang, L.; Inyang, M.; Zimmerman, A.R.; Cao, X.; Yang, L.; Xue, Y.; Li, H. Adsorption of sulfamethoxazole on biochar and its impact on reclaimed water irrigation. *J. Hazard. Mater.* **2012**, *209*, 408–413. [CrossRef]
29. Qian, L.; Chen, B. Dual role of biochars as adsorbents for aluminum: The effects of oxygen-containing organic components and the scattering of silicate particles. *Environ. Sci. Technol.* **2013**, *47*, 8759–8768. [CrossRef]
30. Xu, X.; Cao, X.; Zhao, L. Comparison of rice husk-and dairy manure–derived biochars for simultaneously removing heavy metals from aqueous solutions: Role of mineral components in biochars. *Chemosphere* **2013**, *92*, 955–961. [CrossRef]
31. Song, W.; Guo, M. Quality variations of poultry litter biochar generated at different pyro lysis temperatures. *J. Anal. Appl. Pyrolysis* **2012**, *94*, 138–145. [CrossRef]
32. Ibarrola, R.; Shackely, S.; Hammond, J. Pyrolysis biochar systems for recovering biodegradable materials: A life cycle carbon assessment. *Waste Manag.* **2012**, *32*, 859–868. [CrossRef] [PubMed]
33. Lehmann, J.; Rilling, M.C.; Thies, J.; Masiello, C.A.; Hockaday, W.C.; Crowley, D. Biochar effects on soil biota—A review. *Soil Biol. Biochem.* **2011**, *43*, 1812–1836. [CrossRef]
34. Atkinson, C.J.; Fitzgerald, J.D.; Hipps, N.A. Potential mechanisms for achieving agricultural benefits from biochar application to temperate soils: Review. *Plant Soil* **2010**, *337*, 1–18. [CrossRef]
35. Burrell, L.D.; Zehetner, F.; Rampazzo, N.; Wimmer, B.; Soja, G. Long-term effects of biochar on soil physical properties. *Geoderma* **2016**, *282*, 96–102. [CrossRef]
36. Yamato, M.; Okimori, Y.; Wibowo, I.F.; Anshori, S.; Ogawa, M. Effects of the application of charred bark in Acacia mangium on the yield of maize, cowpea, peanut and soil chemical properties in south Sumatra, Indonesia. *Soil Sci. Plant Nutr.* **2006**, *52*, 489–495. [CrossRef]
37. Tsai, W.T.; Lee, M.K.; Chang, Y.M. Fast pyrolysis of rice straw, sugarcane bagasse and coconut shell in an induction-heating reactor. *J. Anal. Appl. Pyrolysis* **2006**, *76*, 230–237. [CrossRef]
38. Nguyen, B.T.; Lehmann, J. Black carbon decomposition under varying water regimes. *Org. Geochem.* **2009**, *40*, 846–853. [CrossRef]
39. Chan, K.Y.; Van Zwieten, L.; Meszaros, I.; Downie, A.; Joseph, S. Agronomic values of green waste biochar as a soil amendment. *Aust. J. Soil Res.* **2007**, *45*, 629–634. [CrossRef]
40. Magrini-Bair, K.A.; Czernik, S.; Pilath, H.M.; Evans, R.J.; Maness, P.C.; Leventhal, J. Biomass derived, carbon sequestration, designed fertilizers. *Ann. Environ. Sci.* **2009**, *3*, 217–225.
41. Novak, J.M.; Busscher, W.J.; Laird, D.L.; Ahmedna, M.; Watts, D.W.; Niandou, M.A.S. Impact of biochar amendment on fertility of a Southeastern coastal plain soil. *Soil Sci.* **2009**, *174*, 105–112. [CrossRef]
42. Busscher, W.J.; Novak, J.M.; Evans, D.E.; Watts, D.W.; Niandou, M.A.S.; Ahmedna, M. Influence of pecan biochar on physical properties of Norfolk loamy sand. *Soil Sci.* **2010**, *175*, 10–44. [CrossRef]
43. Bridle, T.R.; Pritchard, D. Energy and nutrient recovery from sewage sludge via pyrolysis. *Water Sci. Technol.* **2004**, *50*, 169–175. [CrossRef] [PubMed]
44. Rondon, M.A.; Lehmann, J.; Ramirez, J.; Hurtado, M. Biological nitrogen fixation by common beans (*Phaseolus vulgaris* L) increases with bio-char additions. *Biol. Fertil. Soils* **2007**, *43*, 699–708. [CrossRef]
45. Popp, J.; Lakner, Z.; Harangi-Rákos, M.; Fári, M. The effect of bioenergy expansion: Food, energy, and environment. *Renew. Sustain. Energy Rev.* **2014**, *32*, 559–578. [CrossRef]
46. Lédé, J. Biomass pyrolysis: Comments on some sources of confusions in the definitions of temperatures and heating rates. *Energies* **2010**, *3*, 886–898. [CrossRef]
47. Onay, O. Influence of pyrolysis temperature and heating rate on the production of bio-oil and char from safflower seed by pyrolysis, using a well-swept fixed-bed reactor. *Fuel Process. Technol.* **2007**, *88*, 523–531. [CrossRef]
48. Sinha, S.; Jhalani, A.; Ravi, M.R.; Ray, A. Modelling of Pyrolysis in Wood: A Review. *Sol. Energy Soc. India J.* **2000**, *10*, 41–62.
49. Jahirul, M.I.; Rasul, M.G.; Chowdhury, A.A.; Ashwath, N. Biofuels Production through Biomass Pyrolysis—A Technological Review. *Energies* **2012**, *5*, 4952–5001. [CrossRef]
50. Van den Bergh, C. Biochar and waste law: A comparative analysis. *Eur. Energy Environ. Law Rev.* **2009**, *18*, 243–253.

51. Montanarella, L.; Lugato, E. The application of biochar in the EU: Challenges and opportunities. *Agron. J.* **2013**, *3*, 462–473. [CrossRef]
52. Vereš, J.; Koloničný, J.; Ochodek, T. Biochar status under international law and regulatory issues for the practical application. *Chem. Eng. Trans.* **2014**, *37*, 799–804.
53. Van Laer, T.; De Smedt, P.; Ronsse, F.; Ruysschaert, G.; Boeckx, P.; Verstraete, W.; Buysse, J.; Lavrysen, L.J. Legal constraints and opportunities for biochar: A case analysis of EU law. *GCB Bioenergy* **2015**, *7*, 14–24. [CrossRef]
54. Directive 2008/98/EC of the European Parliament and of the Council of 19 November 2008 on Waste and Repealing Certain Directives. Available online: http://www.lex.pl/serial-akt/-/akt/dz-u-ue-l-2008-312-3 (accessed on 1 February 2019).
55. Directive 2001/77/EC of the European Parliament and of the Council of 27 September 2001. Available online: http://orka.sejm.gov.pl/Drektywy.nsf/all/32001L0077/%24File/32001L0077.pdf (accessed on 1 February 2019).
56. Ustawa z dnia 14 Grudnia 2012 o Odpadach [Waste Act of 14 December 2012]. Available online: http://isap.sejm.gov.pl/DetailsServlet?id=WDU20130000021 (accessed on 1 February 2019).
57. Regulation (EC) No 1907/2006 of the European Parliament and of the Council of 18 December 2006 Concerning the Registration, Evaluation, Authorisation and Restriction of Chemicals (REACH). Available online: http://eurlex.europa.eu/LexUriServ/LexUriServ.do?uri=OJ:L:2009:036:0084:0084:PL:PDF (accessed on 1 February 2019).
58. Regulation (EC) No 2003/2003 of the European Parliament and of the Council of 13 October 2003 Relating to Fertilisers. Available online: https://eur-lex.europa.eu/legal-content/EN/TXT/?uri=CELEX%3A32003R2003PDF (accessed on 1 February 2019).
59. EBC. *European Biochar Certificate—Guidelines for Sustainable Production of Biochar*; Version 6.2E; European Biochar Foundation (EBC): Arbraz, Switzerland, 2012.
60. Biochar Quality Mandate (BQM) v.1.0. Available online: http://www.britishbiocharfoundation.org/wp-content/uploads/BQM-V1.0.pdf (accessed on 1 February 2019).
61. Malińska, K.; Mełgieś, K. Current quality and legal requirements for biochar as a fertilizers and soil improver. *Prace Inst. Ceram. Mater. Bud.* **2016**, *9*, 82–95.
62. Refertil Biochar EU Policy—Support Abstract Draft 2014. D 2.3. Biochar Policy Supporting Report, Concerning the Absence of Potential Risks for the Different Environmental Compartments, for the Plants and for Human Health through the Food Chain Resulting from the Use of These Materials in Agricultural Soils. Available online: http://www.refertil.info/sites/default/files/REFERTIL_289785_BIOCHAR_POLICY_abstract_draft_2014.pdf (accessed on 1 February 2019).
63. Hibler, I.; Blum, F.; Leifeld, J.; Schmidt, H.P.; Bucheli, T.D. Quantitative determination of PAHs in biochar: A prerequisite to ensure its quality and safe application. *J. Agric. Food Chem.* **2012**, *60*, 3042–3050.
64. Fabbri, D.; Rombolà, A.G.; Torri, C.; Spokas, K.A. Determination of polycyclic aromatic carbons in biochar and biochar amended soil. *J. Anal. Appl. Pyrolysis* **2013**, *103*, 60–67. [CrossRef]
65. Quilliam, R.S.; Rangecroft, S.; Emmett, B.A.; Deluca, T.H.; Jones, D.L. Is biochar a source or sink for polycyclic aromatic hydrocarbons (PAH) compounds in agricultural soils? *GCB Bioenergy* **2013**, *5*, 96–103. [CrossRef]
66. Regulation of the Minister of Agriculture and Rural Development, of 18 June 2008, on Implementation of Certain Provisions of the Act on Fertilizers and Fertilization. Available online: http://isap.sejm.gov.pl/DetailsServlet?id=WDU2008119 (accessed on 1 February 2019).
67. Singh, B.P.; Cowie, A.L.; Smernik, R.J. Biochar carbon stability in a clayey soil as a function of feedstock and pyrolysis temperature. *Environ. Sci. Technol.* **2012**, *46*, 11770–11778. [CrossRef]
68. Bis, Z. Biochar – return to the past, opportunity for the future. Available online: https://www.cire.pl/pliki/2/biowegiel.pdf (accessed on 1 February 2019).
69. Matovic, D. Biochar as a viable carbon sequestration option: Global and Canadian perspective. *Energy* **2011**, *36*, 2011–2016. [CrossRef]
70. Verheijen, F.; Jeffery, S.; Bastos, A.C.; Van der Velde, M.; Diafas, I. *Biochar Application to Soils—A Critical Scientific Review of Effects on Soil Properties, Processes and Functions*; European Commission: Ispra, Italy, 2010.
71. Skreiberg, Ø.; Wang, L.; Khalil, R.; Gjølsjø, S.; Turn, S. Enabling the biocarbon value chains for energy and metallurgical industries. In Proceedings of the European Biomass Conference and Exhibition, Copenhagen, Denmark, 14–18 May 2018; pp. 1221–1228.

72. Olszewski, M.; Kempegowda, R.S.; Skreiberg, Ø.; Wang, L.; Løvås, T. Techno-Economics of Biocarbon Production Processes under Norwegian Conditions. *Energy Fuels* **2017**, *31*, 14338–14356. [CrossRef]
73. Bach, Q.V.; Tran, K.Q.; Skreiberg, Ø. Comparative study on the thermal degradation of dry- and wet-torrefied woods. *Appl. Energy* **2017**, *185*, 1051–1058. [CrossRef]
74. McLaren, D.A. Comparative global assessment of potential negative emissions technologies. *Process Saf. Environ. Prot.* **2012**, *90*, 489–500. [CrossRef]
75. Sukiran, M.A.; Kheang, L.S.; Baker, N.A.; May, C.Y. Production and characterization of biochar from the pyrolysis of empty fruit bunches. *Am. J. Appl. Sci.* **2011**, *8*, 984–988. [CrossRef]
76. Gheorghe, C.; Marculescu, C.; Badea, A.; Dinca, C.; Apostol, T. Effect of pyrolysis conditions on bio-char production from biomass. Conference papers of 3rd WSEAS International Conference on Renewable Energy Sources. 2009. Available online: https://pdfs.semanticscholar.org/c47b/51964c1fc9ba5ca1bdb660b5fe31478532f4.pdf (accessed on 1 February 2019).
77. Sun, K.; Ro, K.; Guo, M.; Novak, J.; Mashayekhi, H.; Xing, B. Sorption of bisphenol A, 17α-ethinyl estradiol and phenanthrene on thermally and hydrothermally produced biochars. *Bioresour. Technol.* **2011**, *102*, 5757–5763. [CrossRef] [PubMed]
78. Tong, X.; Li, J.; Yuan, J.; Xu, R. Adsorption of Cu(II) by biochars generated from three crop straws. *J. Chem. Eng.* **2011**, *172*, 828–834. [CrossRef]
79. Regmi, P.; Moscoso, J.L.G.; Kumar, S.; Cao, X.; Mao, J.; Scharfan, G. Removal of copper and cadmium from aqueous solutions using switchgrass biochar produced via hydrothermal carbonization proces. *J. Environ. Manag.* **2012**, *109*, 61–69. [CrossRef] [PubMed]
80. Inyang, M.; Gao, B.; Yao, Y.; Xue, Y.; Zimmerman, A.R.; Pullammanappallil, P.; Cao, X. Removal of heavy metals from aqueous solution by biochars derived from anaerobically digested biomass. *Bioresour. Technol.* **2012**, *110*, 50–56. [CrossRef] [PubMed]
81. Mohan, D.; Rajput, S.; Singh, V.K.; Steele, P.H.; Pittman, C.U. Modelling and evaluation of chromium remediation from water using low cost bio-char, a green adsorbent. *J. Hazard. Mater.* **2011**, *188*, 319–333. [CrossRef] [PubMed]
82. Zheng, W.; Guo, M.; Chow, T.; Bennet, D.N.; Rajagopalan, N. Sorption properties of greenwaste biochar for two triazine pesticides. *J. Hazard. Mater.* **2010**, *181*, 121–126. [CrossRef]
83. Zhang, P.; Sun, H.; Yu, L.; Sun, T. Adsorption and catalytic hydrolysis of carbaryl and antrazine on pig manure-derived biochars: impact of structural properties of biochars. *J. Hazard. Mater.* **2013**, *244*, 217–224. [CrossRef]
84. Spokas, K.A.; Koskinen, W.C.; Baker, J.M.; Reicosky, D.C. Impacts of woodchip biochar additions on greenhouse gas production and sorption/degradation of two herbicides in a Minnesota soil. *Chemosphere* **2009**, *77*, 574–581. [CrossRef]
85. Cao, X.; Ma, L.; Liang, Y.; Gao, B.; Harris, W. Simultaneous immobilization of lead and atrazine in contaminated soils using dairy-manure biochar. *Environ. Sci. Technol.* **2011**, *45*, 4884–4889. [CrossRef] [PubMed]
86. Cao, X.; Harris, W. Properties of dairy-manure-derived biochar pertinent to its potential use in remediation. *Bioresour. Technol.* **2010**, *101*, 5222–5228. [CrossRef] [PubMed]
87. Xu, T.; Lou, L.; Luo, L.; Cao, R.; Duan, D.; Chen, Y. Effect of bamboo biochar on pentachlorophenol leachability and bioavailability in agricultural soil. *Sci. Total. Environ.* **2012**, *414*, 727–731. [CrossRef] [PubMed]
88. Teixido, M.; Pignatello, J.J.; Beltran, J.L.; Granados, M.; Peccia, J. Speciation of the ionizable antibiotic sulfamethazine on black carbon (biochar). *Environ. Sci. Technol.* **2011**, *45*, 10020–10027. [CrossRef] [PubMed]
89. Jeong, C.Y.; Wang, J.J.; Dodla, S.K.; Eberhardt, T.L.; Groom, L. Effect of biochar amendment on tylosin adsorption-desorption and transport in two different soils. *J. Environ. Qual.* **2012**, *41*, 1185–1192. [CrossRef] [PubMed]
90. Liu, P.; Liu, W.J.; Jiang, H.; Chen, J.J.; Li, W.W.; Yu, H.Q. Modification of biochar derived from fast pyrolysis of biomass and its application in removal of tetracycline from aqueous solution. *Bioresour. Technol.* **2012**, *121*, 235–240. [CrossRef] [PubMed]
91. Hale, S.E.; Hanley, K.; Lehmann, J.; Zimmerman, A.R.; Cornelissen, G. Effects of chemical, biological, and physical aging as well as soil addition on the sorption of pyrene to activated carbon and biochar. *Environ. Sci. Technol.* **2011**, *45*, 10445–10453. [CrossRef]

92. Zhang, W.; Wang, L.; Sun, H. Modifications of black carbons and their influence on pyrene sorption. *Chemosphere* **2011**, *85*, 1306–1311. [CrossRef]
93. Ahmad, M.; Lee, S.S.; Dou, X.; Mohan, D.; Sung, J.K.; Yang, J.E.; Ok, Y.S. Effects of pyrolysis temperature on soybean stover-and peanut shell-derived biochar properties and TCE adsorption in water. *Bioresour. Technol.* **2012**, *118*, 536–544. [CrossRef]
94. Chen, B.; Zhou, D.; Zhu, L. Transitional adsorption and partition on nonpolar and polar aromatic contaminants by biochars of pine needles with different pyrolytic temperatures. *Environ. Sci. Technol.* **2008**, *42*, 5137–5143. [CrossRef]
95. Kim, W.K.; Shim, T.; Kim, Y.S.; Hyun, S.; Ryu, C.; Park, Y.K.; Jung, J. Characterization of cadmium removal from aqueous solution by biochar produced from a giant Miscanthus at different pyrolytic temperatures. *Bioresour. Technol.* **2013**, *138*, 266–270. [CrossRef] [PubMed]
96. Liu, Z.; Zhang, F.S. Removal of lead from water using biochars prepared from hydrothermal liquefaction of biomass. *J. Hazard. Mater.* **2009**, *167*, 933–939. [CrossRef]
97. Purnomo, C.W.; Castello, D.; Fiori, L. Granular Activated Carbon from Grape Seeds Hydrothermal Char. *Appl. Sci.* **2018**, *8*, 331. [CrossRef]
98. Gurten, I.I.; Ozmak, M.; Yagmur, E.; Aktas, Z. Preparation and characterisation of activated carbon from waste tea using K_2CO_3. *Biomass Bioenergy* **2012**, *37*, 73–81. [CrossRef]
99. Tamer, M.A.; Ismail, A.; Mohd, A.A.; Ahmad, A.F. Review: Production of activated carbon from agricultural byproducts via conventional and microwave heating. *J. Chem. Technol. Biotechnol.* **2013**, *88*, 1183–1190.
100. Bedia, J.; Peñas-Garzón, M.; Gómez-Avilés, A.; Rodriguez, J.J.; Belver, C. A Review on the Synthesis and Characterization of Biomass-Derived Carbons for Adsorption of Emerging Contaminants from Water. *J. Carbon Res.* **2018**, *4*, 63. [CrossRef]
101. Hagemann, N.; Spokas, K.; Schmidt, H.-P.; Kägi, R.; Böhler, M.A.; Bucheli, T.D. Activated Carbon, Biochar and Charcoal: Linkages and Synergies across Pyrogenic Carbon's ABCs. *Water* **2018**, *10*, 182. [CrossRef]
102. González-García, P. Activated carbon from lignocellulosics precursors: A review of the synthesis methods, characterization techniques and applications. *Renew. Sustain. Energy Rev.* **2018**, *82*, 1393–1414. [CrossRef]
103. Amaya, A.; Medero, N.; Tancredi, N.; Silva, H.; Deiana, C. Activated carbon briquettes from biomass materials. *Bioresour. Technol.* **2007**, *98*, 1635–1641. [CrossRef]
104. Liu, D.; Zhang, W.; Lin, H.; Li, Y.; Lu, H.; Wang, Y. A green technology for the preparation of high capacitance rice husk-based activated carbon. *J. Clean. Prod.* **2015**, 1–9. [CrossRef]
105. Wilson, K.; Yang, H.; Seo, C.W.; Marshall, W.E. Select metal adsorption by activated carbon made from peanut shells. *Bioresour. Technol.* **2006**, *97*, 2266–2270. [CrossRef]
106. Unur, E. Functional nanoporous carbons from hydrothermally treated biomass for environmental purification. *Microporous Mesoporous Mater.* **2013**, *168*, 92–101. [CrossRef]
107. Lei, H.; Wang, Y.; Huo, J. Porous graphitic carbon materials prepared from cornstarch with the assistance of microwave irradiation. *Microporous Mesoporous Mater.* **2015**, *210*, 39–45. [CrossRef]
108. Wang, D.; Geng, Z.; Li, B.; Zhang, C. High performance electrode materials for electric double-layer capacitors based on biomass-derived activated carbons. *Electrochim. Acta* **2015**, *173*, 377–384. [CrossRef]
109. Sardella, F.; Gimenez, M.; Navas, C.; Morandi, C.; Deiana, C.; Sapag, K. Conversion of viticultural industry wastes into activated carbons for removal of lead and cadmium. *J. Environ. Chem. Eng.* **2014**, *3*, 253–260. [CrossRef]
110. Ruiz, B.; Ruisánchez, E.; Gil, R.R.; Ferrera-Lorenzo, N.; Lozano, M.S.; Fuente, E. Sustainable porous carbons from lignocellulosic wastes obtained from the extraction of tannins. *Microporous Mesoporous Mater.* **2015**, *209*, 23–29. [CrossRef]
111. Kilpimaa, S.; Runtti, H.; Kangas, T.; Lassi, U.; Kuokkanen, T. Physical activation of carbon residue from biomass gasification: Novel sorbent for the removal of phosphates and nitrates from aqueous solution. *J. Ind. Eng. Chem.* **2014**, *21*, 1354–1364. [CrossRef]
112. Marco-Lozar, J.P.; Linares-Solano, A.; Cazorla-Amorós, D. Effect of the porous texture and surface chemistry of activated carbons on the adsorption of a germanium complex from dilute aqueous solutions. *Carbon* **2011**, *49*, 3325–3331. [CrossRef]
113. Tsyntsarski, B.; Stoycheva, I.; Tsoncheva, T.; Genova, I.; Dimitrov, M.; Petrova, B.; Paneva, D.; Zheleva, Z.; Budinova, T.; Kolev, H.; et al. Activated carbons from waste biomass and low rank coals as catalyst supports for hydrogen production by methanol decomposition. *Fuel Process. Technol.* **2015**, *137*, 139–147. [CrossRef]

114. Temdrara, L.; Khelifi, A.; Addoun, A.; Spahis, N. Study of the adsorption properties of lignocellulosic material activated chemically by gas adsorption and immersion calorimetry. *Desalination* **2008**, *223*, 274–282. [CrossRef]
115. Román, S.; Valente-Nabais, J.M.; Ledesma, B.; González, J.F.; Laginhas, C.; Titirici, M.M. Production of low-cost adsorbents with tunable surface chemistry by conjunction of hydrothermal carbonization and activation processes. *Microporous Mesoporous Mater.* **2013**, *165*, 127–133.
116. Jibril, B.; Houache, O.; Al-Maamari, R.; Al-Rashidi, B. Effects of H_3PO_4 and KOH in carbonization of lignocellulosic material. *J. Anal. Appl. Pyrolysis* **2008**, *83*, 151–156. [CrossRef]
117. Nabais, J.M.V.; Laginhas, C.; Carrott, P.J.M.; Carrott, M.M.L.R. Thermal conversion of a novel biomass agricultural residue (vine shoots) into activated carbon using activation with CO_2. *J. Anal. Appl. Pyrolysis* **2010**, *87*, 8–13. [CrossRef]
118. Tan, X.-F.; Liu, S.-B.; Liu, Y.-G.; Gu, Y.-L.; Zeng, G.-M.; Hu, X.-J.; Wang, X.; Liu, S.-H.; Jiang, L.-H. Biochar as potential sustainable precursors for activated carbon production: Multiple applications in environmental protection and energy storage. *Bioresour. Technol.* **2017**, *227*, 359–372. [CrossRef] [PubMed]
119. Hoegberg, L.C.; Groenlykke, T.B.; Abildtrup, U.; Angelo, H.R. Combined paracetamol and amitriptyline adsorption to activated charcoal. *Clin. Toxicol.* **2010**, *48*, 898–903. [CrossRef] [PubMed]
120. Nanda, S.; Dalai, A.K.; Berruti, F.; Kozinski, J.A. Biochar as an experimental bioresource for energy, agronomy, carbon sequestration, activated carbon and specialty materials. *Waste Biomass Valor.* **2016**, *7*, 201–235. [CrossRef]
121. Ozsoy, H.D.; van Leeuwen, J. Removal of color from fruit candy waste by activated carbon adsorption. *J. Food Eng.* **2010**, *101*, 106–112. [CrossRef]
122. Bogusz, A.; Cejner, M. Biochar materials in adsorption of organic and inorganic contaminants. *Inżynieria Środowiska* **2016**, *22*, 9–33.
123. Sun, K.; Jin, J.; Keiluweit, M.; Kleber, M.; Wang, Z.; Pan, Z.; Xing, B. Polar and aliphatic domains regulate sorption of phthalic acid esters (PAEs) to biochars. *Bioresour. Technol.* **2012**, *118*, 120–127. [CrossRef]
124. Ahmad, M.; Rajapaksha, A.U.; Lim, J.E.; Zhang, M.; Bolan, N.; Mohan, D.; Vithanage, M.; Lee, S.S.; Ok, Y.S. Biochar as a sorbent for contaminant management in soil and water: A review. *Chemosphere* **2014**, *99*, 19–33. [CrossRef]
125. Tan, X.; Liu, Y.; Zeng, G.; Wang, X.; Hu, X.; Gu, Y.; Yang, Z. Application of biochar for the removal of pollutants from aqueous solutions. *Chemosphere* **2015**, *125*, 70–85. [CrossRef]
126. Enders, A.; Hanley, K.; Whitman, T.; Joseph, S.; Lehmann, J. Characterization of biochars to evaluate recalcitrance and agronomic performance. *Bioresour. Technol.* **2012**, *114*, 644–653. [CrossRef]
127. Bezerra, J.; Turnhout, E.; Vasquez, I.M.; Rittl, T.F.; Arts, B.; Kuyper, T.W. The promises of the Amazonian soil: Shifts in discourses of Terra Preta and biochar. *J. Environ. Policy Plan.* **2016**, *46*, 1–13. [CrossRef]
128. Medyńska-Juraszek, A. Biochar as a soil amendment. *Soil Sci. Annual.* **2016**, *67*, 151–157. [CrossRef]
129. Macdonald, L.; Farrell, M.; Van Zwieten, L.; Krull, E. Plant growth responses to biochar addition: An Australian soils perspective. *Biol. Fertil. Soils* **2014**, *50*, 1035–1045. [CrossRef]
130. Beesley, L.; Moreno-Jiménez, E.; Gomez-Eyles, J.; Harris, E.; Robinson, B.; Sizmur, T. A review of biochars' potential role in the remediation, revegetation and restoration of contaminated soils. *Environ. Pollut.* **2011**, *159*, 3269–3282. [CrossRef]
131. Palansooriya, K.N.; Ok, Y.S.; Awad, Y.M.; Lee, S.S.; Sung, J.K.; Koutsospyros, A.; Moon, D.H. Impacts of biochar application on upland agriculture: A review. *J. Environ. Manag.* **2019**, *234*, 52–64. [CrossRef]
132. Glaser, B.; Lehmann, J.; Zech, W. Ameliorating physical and chemical properties of highly weathered soils in the tropics with charcoal-a review. *Biol Fertil. Soils* **2002**, *35*, 1719–1730. [CrossRef]
133. Chan, K.Y.; Van Zwieten, L.; Meszaros, I.; Downie, A.; Joseph, S. Using poultry litter biochars as soil amendments. *Aust. J. Soil Res.* **2008**, *46*, 437–444. [CrossRef]
134. Genesio, L.; Miglietta, F.; Baronti, S.; Vaccari, F.P. Biochar increases vineyard productivity without affecting grape quality: Results from a four years field experiment in Tuscany. *Agric. Ecosyst. Environ.* **2015**, *201*, 20–25. [CrossRef]
135. Uzoma, K.C.; Inoue, M.; Andry, H.; Fujimaki, H.; Zahoor, A.; Nishihara, E. Effect of cow manure biochar on maize productivity under sandy soil condition. *Soil Use Manag.* **2011**, *27*, 205–212. [CrossRef]
136. Rogovska, N.; Laird, D.A.; Rathke, S.J.; Karlen, D.L. Biochar impact on Midwestern Mollisols and maize nutrient availability. *Geoderma* **2014**, *230*, 340–347. [CrossRef]

137. Liu, Z.; Chen, X.; Jing, Y.; Li, Q.; Zhang, J.; Huang, Q. Effects of biochar amendment on rapeseed and sweet potato yields and water stable aggregate in upland red soil. *Catena* **2014**, *123*, 45–51. [CrossRef]
138. Cross, A.; Sohi, S. The priming potential of biochar products in relation to labile carbon contents and soil organic matter status. *Soil Biol. Biochem.* **2011**, *43*, 2127–2134. [CrossRef]
139. Zhang, Q.; Du, Z.; Lou, Y.; He, X. A one-year short-term biochar application improved carbon accumulation in large macro aggregate fractions. *Catena* **2015**, *127*, 26–31. [CrossRef]
140. Laird, D.A. The charcoal vision: A win-win-win scenario for simultaneously producing bioenergy, permanently sequestering carbon, while improving soil and water quality. *Agron. J.* **2008**, *100*, 178–181. [CrossRef]
141. Woolf, D.; Amonette, J.E.; Street-Perrott, F.A.; Lehmann, J.; Joseph, S. Sustainable biochar to mitigate global climate change. *Nat. Commun.* **2010**, *1*, 56. [CrossRef]
142. Li, Z.; Delvaux, B.; Yans, J.; Dufour, N.; Houben, D.; Cornelis, J.T. Phytolith-rich biochar increases cotton biomass and silicon-mineralomass in a highly weathered soil. *J. Plant Nutr. Soil Sci.* **2018**, *181*, 537–546. [CrossRef]
143. Yuan, J.; Xu, R.; Zhang, H. The forms of alkalis in the biochar produced from crop residues at different temperatures. *Bioresour. Technol.* **2011**, *102*, 3488–3497. [CrossRef]
144. Zong, Y.; Xiao, Q.; Lu, S. Acidity, water retention, and mechanical physical quality of a strongly acidic Ultisol amended with biochars derived from different feedstocks. *J. Soil Sediments* **2016**, *16*, 177–190. [CrossRef]
145. Jien, S.; Wang, C. Effects of biochar on soil properties and erosion potential in a highly weathered soil. *Catena* **2014**, *110*, 225–233. [CrossRef]
146. Agegnehu, G.; Bass, A.; Nelson, P.; Bird, M. Benefits of biochar, compost and biochar-compost for soil quality, maize yield and greenhouse gas emissions in a tropical agricultural soil. *Sci. Total Environ.* **2016**, *543*, 295–306. [CrossRef]
147. Das, O.; Sarmah, A. The love-hate relationship of pyrolysis biochar and water: A perspective. *Sci. Total Environ.* **2015**, *512/513*, 682–685. [CrossRef]
148. Cayuela, M.; Van Zwieten, L.; Singh, B.; Jeffery, S.; Roig, A.; Sanchez-Monedero, M.A. Biochar's role in mitigating soil nitrous oxide emissions: A review and meta-analysis. *Agric. Ecosyst. Environ.* **2014**, *191*, 5–16. [CrossRef]
149. Ameloot, N.; De Neve, S.; Jegajeevagan, K.; Yildiz, G.; Buchan, D.; Funkuin, Y.N.; Prins, W.; Bouckaert, L.; Sleutel, S. Short-term CO_2 and N_2O emissions and microbial properties of biochar amended sandy loam soils. *Soil Biol. Biochem.* **2013**, *57*, 401–410. [CrossRef]
150. Ameloot, N.; Sleutel, S.; Das, K.C.; Kanagaratnam, J.; De Neve, S. Biochar amendment to soils with contrasting organic matter level: Effects on N mineralization and biological soil properties. *GCB Bioenergy* **2015**, *7*, 135–144. [CrossRef]
151. Jones, D.L.; Rousk, J.; Edwards-Jones, G.; DeLuca, T.H.; Murphy, D.V. Biochar-mediated changes in soil quality and plant growth in a three year field trial. *Soil Biol. Biochem.* **2012**, *45*, 113–124. [CrossRef]
152. Gomez, J.D.; Denef, K.; Stewart, C.E.; Zheng, J.; Cotrufo, M.F. Biochar addition rate influences soil microbial abundance and activity in temperate soils. *Eur. J. Soil Sci.* **2014**, *65*, 28–39. [CrossRef]

© 2019 by the authors. Licensee MDPI, Basel, Switzerland. This article is an open access article distributed under the terms and conditions of the Creative Commons Attribution (CC BY) license (http://creativecommons.org/licenses/by/4.0/).

Article

Process Simulation and Economic Evaluation of Bio-Oil Two-Stage Hydrogenation Production

Xiaoyuechuan Ma [1,2], Shusheng Pang [3], Ruiqin Zhang [1,2] and Qixiang Xu [1,2,*]

1. College of Chemistry and Molecular Engineering, Zhengzhou University, Zhengzhou 450001, China; bushishuziyouxiang@aliyun.com (X.M.); rqzhang@zzu.edu.cn (R.Z.)
2. Environmental Chemistry & Low Carbon Technologies Key Lab of Henan Province, Zhengzhou 450001, China
3. Department of Chemical and Process Engineering, University of Canterbury, Christchurch 8083, New Zealand; shusheng.pang@canterbury.ac.nz
* Correspondence: xuqixiang@zzu.edu.cn

Received: 19 January 2019; Accepted: 13 February 2019; Published: 18 February 2019

Featured Application: The developed process simulation model is used for economical analysis of a two-stage hydrogenation of the bio-oil process, it is also adaptable for simulation and performance evaluation of process with various types of feedstocks. As the catalytic reaction mechanisms and catalysts coking deactivation kinetics are embedded, its specialty is to simulate and evaluate the process with large catalyst usage or significant catalyst decay.

Abstract: Bio-oil hydrogenation upgrading process is a method that can convert crude bio-oil into high-quality bio-fuel oil, which includes two stages of mild and deep hydrogenation. However, coking in the hydrogenation process is the key issue which negatively affects the catalyst activity and consequently the degree of hydrogenation in both stages. In this paper, an Aspen Plus process simulation model was developed for the two-stage bio-oil hydrogenation demonstration plant which was used to evaluate the effect of catalyst coking on the bio-oil upgrading process and the economic performance of the process. The model was also used to investigate the effect of catalyst deactivation caused by coke deposition in the mild stage. Three reaction temperatures in the mild stage (250 °C, 280 °C, and 300 °C) were considered. The simulation results show that 45% yield of final product is obtained at the optimal reaction condition which is 280 °C for the mild stage and 400 °C for the deep stage. Economic analysis shows that the capital cost of industrial production is $15.2 million for a bio-oil upgrading plant at a scale of 107 thousand tons per year. The operating costs are predicted to be $1024.27 per ton of final product.

Keywords: bio-fuel; aspen plus; hydrogenation; simulation; economic analysis

1. Introduction

As a renewable energy source, biomass contributes about 15% to the total global energy consumption. In China, a traditional agricultural country, the use of straw, sugar cane, and rice husk has great potential for future energy and fuels [1]. It is reported that total agricultural and forestry waste in China reaches 1.5 billion tons per year, of which straw has attracted increasing interests due to environmental concerns as most of it is burned on farmland. If straw were used, the bioenergy from available sources would be equivalent to 180 million tons of standard coal [2]. If one takes into account the CO_2 uptake by plants during growth and supposes that the biomass is processed efficiently for energy and fuels, CO_2 emissions can be reduced by about 90% compared with fossil energy [3]. There are various processes for converting biomass to energy and fuels including thermochemical processes and bioprocesses. It is believed that the thermochemical processes, including combustion, gasification pyrolysis, and liquefaction, are most promising in the short and medium terms of 5 to 10 years [4]. In recent years, extensive studies have been reported on the commercialization of these technologies [5]. However, most of these studies were conducted at pilot or demonstration scales due to the difficulties and challenges in constructing full commercial-scale plants.

Aspen Plus, a large-scale chemical process simulation software, has been widely used in the design and optimization of unit operation devices such as distillation, absorption, and gas fractionation in chemical processes. Optimizing and improving the production itself using the software's own analysis tools (sensitivity analysis, design regulations, etc.) can reduce costs for capital and operation, in addition to saving energy and avoiding negative impacts on the environment. Based on the experimental studies carried out on the fast pyrolysis process of biomass, Susanne et al. [6,7] carried out a simulation design for the process of hydrogenation of bio-oil to biodiesel. In these studies, the simulation of the whole process from raw biomass materials through pyrolysis, bio-oil hydrogenation to transportation fuel was established. Atsonios et al. [8] also conducted research on the hydrotreating of pyrolysis oil produced by the co-pyrolysis of coal and biomass. They used specific components of the oil including anisole, guaiacol, acetic acid, ethylene, and furan as model compounds to represent the pyrolysis oils. The effect of the co-pyrolysis, hydrogenation, catalytic reforming hydrogen production, and coke combustion on the overall process was investigated. In a separate study, Wright et al. [9] simulated the hydrogenation process of bio-oil produced by the rapid pyrolysis of cornstalk, and conducted economic and technical assessments of the entire process. It is found that hydrogen sources have significant impacts which include hydrogen production from reforming part of the pyrolysis bio-oil and hydrogen purchased from the commercial process of natural gas reforming. The results show that if the hydrogen is obtained from reforming pyrolysis bio-oil, the final liquid product yield has the greatest impact on the whole process economy. However, if the hydrogen is purchased from the commercial process, the biomass raw material price has the greatest impact on the entire process economy.

In this study, an Aspen Plus-based process model was established to simulate the two stages of bio-oil hydrogenation. The unique feature of the model was that it considered intrinsic reaction kinetics of catalyst coking deactivation in the first mild stage and the influences of the whole production process. The model also included material mass balance and energy balance in the bio-oil hydrogenation process under different reaction conditions. The process was finally optimized based on economic analysis to achieve the lowest operating costs and the lowest capital costs.

2. Development of a Process Model for Two-Stage Bio-Oil Hydrogenation (HDO)

2.1. Specification of Compounds from Bio-Oil

A process model was developed by using Aspen Plus to simulate the process of two-stage bio-oil hydrogenation. According to the properties of raw bio-oil and products shown in Table 1, and for simulation purposes, the hundreds of substances presenting bio-oil and products were therefore divided into 9 model components according to the reported study [10]: (1) BIO-OIL (organic compounds of raw bio-oil, $CH_{1.47}O_{0.56}$); (2) H_2; (3) ODF (oil-phase organic compounds of mild stage, $CH_{1.47}O_{0.11}$); (4) AQO (aqueous phase organic compounds of mild stage, $CH_{3.02}O_{1.09}$); (5) CO_2; (6) H_2O; (7) COKES (soluble coke deposit, Cs); (8) COKEIS (insoluble coke deposit, C_{is}); and (9) PRODUCT (model compound that represents the final product of bio-fuel, with same properties of gasoline, $CH_{1.71}$). Model compounds were selected from the Aspen Plus database to represent the above-mentioned substances in the bio-oil and the upgraded products [11]. The developed HDO process model is adaptable for different types of bio-oil feedstock with given properties and reaction kinetic parameters.

Table 1. Properties of the feedstock and products of the mild and deep stages [9,11,12].

Properties	Bio-Oil	Mild-Stage Production	Deep-Stage Production
Density (kg L^{-1})	1.1–1.2	1.2	0.84
Viscosity (c St at 25 °C)	19–25	7.4	2.8
Water content (wt %)	20–30	11.4	0.02
Higher heating value (MJ kg^{-1})	16–18	29.5	45.6
pH	2.7	4.2	–
Elemental analysis (wt %, a. f)			
C	44–46	78.3	88
H	6–7	9.6	11.5
O	46–48	11.4	0.5

2.2. Process Setup

As discussed above, the whole HDO process was divided into two main stages which included a mild hydrogenation stage and a subsequent deep hydrogenation stage. The system flow diagram for the bio-oil hydrogenation process is shown in Figures 1 and 2. The whole HDO process consists of seven unit operations: (1) hydrogen preheating; (2) bio-oil preheating; (3) mild hydrogenation reactor; (4) mild hydrogenation product solid–liquid separation; (5) mild hydrogenation product liquid–gas separation; (6) deep hydrogenation reaction; and (7) product cooling and pressure separation.

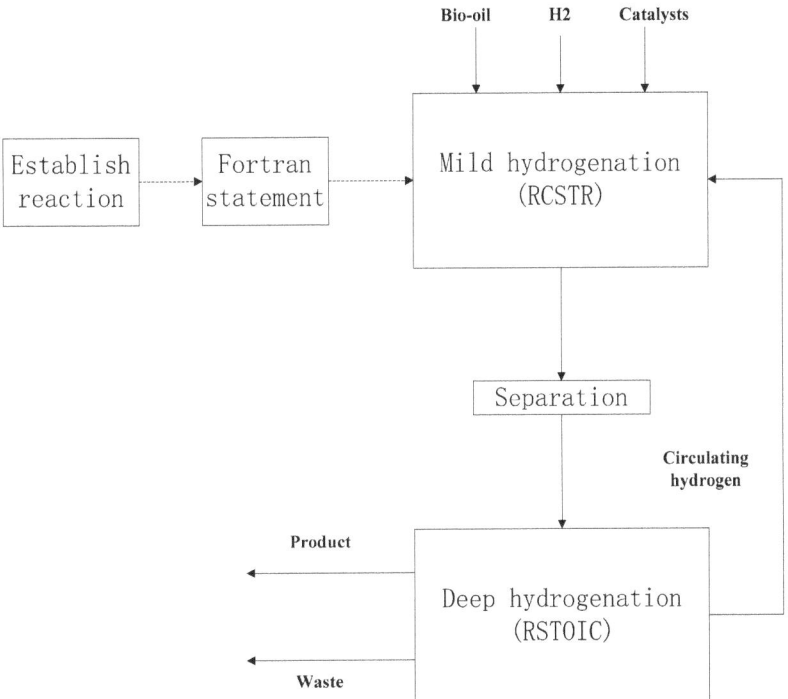

Figure 1. Simulation procedure of the bio-oil hydrogenation (HDO) process.

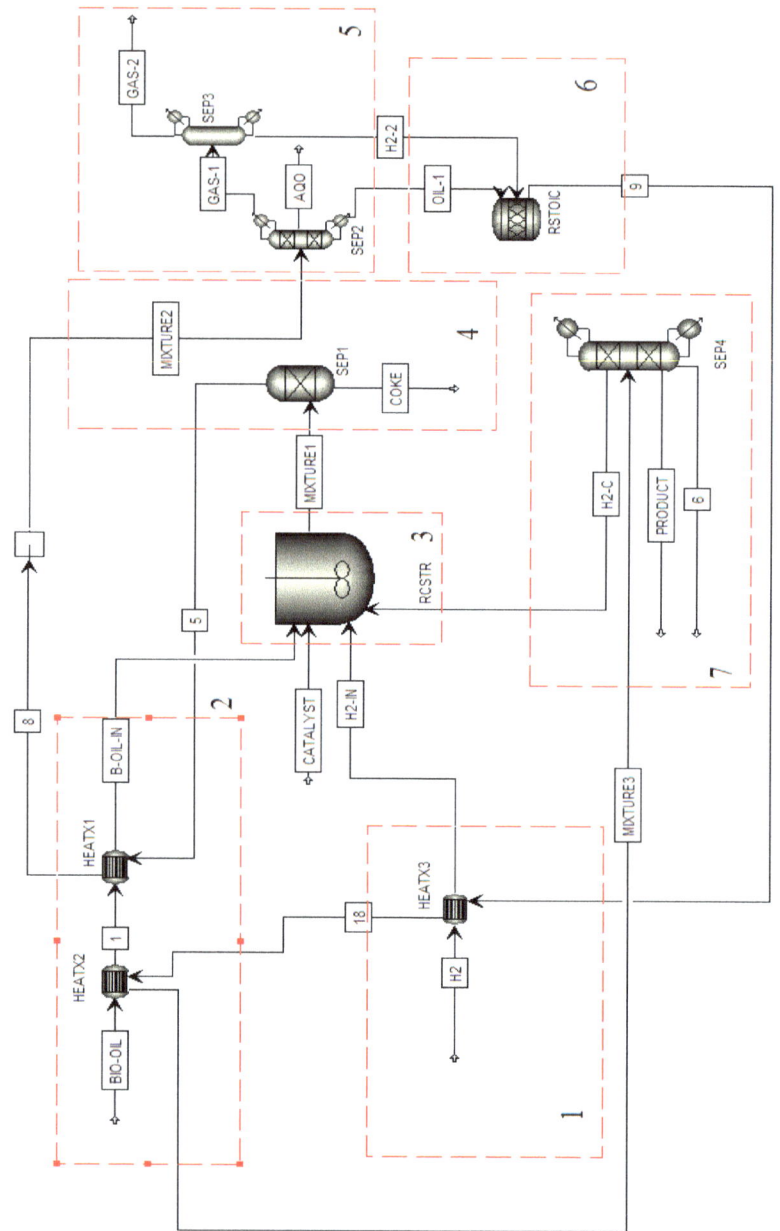

Figure 2. Bio-oil hydrogenation reaction process.

The organic compounds of raw bio-oil (BIO-OIL), hydrogen (H2-1), and the catalyst (CATALYST) are treated as input parameters under normal conditions that are first preheated by the stream (MIXTURE3) and then continuously fed into the first stage HDO, or the mild hydrogenation reactor (RCSTR). According to the reported research and experimental data, operating temperature for the mild hydrogenation reactor of raw bio-oil is set at 250 °C, 280 °C, and 300 °C, respectively. The kinetic reactions added to this RCSTR reactor include the main bio-oil hydrogenation reaction plus side reactions of the catalyst coking [12].

The product stream (MIXTURE1) from mild hydrogenation reactor first enters a solid-fluidized bed separator (SEP-1) to remove the used catalysts (with soluble and insoluble coke deposits), and the liquid product then flows into a three-phase separator (SEP-2) to be separated into: (1) water and aqueous phase organic compounds (AQO) formed during the hydrogenation reaction; (2) the unreacted bio-oil with an oil-phase organic compound (ODF); and (3) a mixture of gaseous products and unreacted hydrogen (GAS-1). The organic compound (ODF) and unreacted hydrogen (H2-2) are separated from stream (GAS-1) and then fed into the secondary hydrogenation reactor (RSTOIC) for the deep hydrogenation. The product stream (MIXTURE3) is separated after the heat exchanger. Excessive hydrogen (H2-C) from the deep hydrogenation is continuously recycled and fed to the mild reactor. The heat recovery of the process is achieved by a series of heat exchangers between process streams and feedstock, and the only unit operation that requires external heat supply within the entire process is the secondary hydrogenation reactor.

2.3. Determination of Properties of Bio-Oil and Upgraded Products

The bio-oil and its upgraded product components in this study are complicated, mostly non-polar mixtures of acids, substances, aldehydes, ethers, and alcohols. Therefore, the PR-BM method was chosen in Aspen Plus to determine the chemical and physical properties for both bio-oil and its upgraded products.

2.4. Input Parameters for Plant Operation and Simulation of Hydrogenation Upgrading of Bio-Oil

The annual throughput of the demonstration plant in this case study was set at 107 thousand tons per year of raw bio-oil. Therefore, the input flow rates of feedstock materials were set as follows: 10,000 kg/h for the organic compounds of raw bio-oil (BIO-OIL), 345 kg/h for the hydrogen (H2), and 10 kg/h for the fresh catalyst feed (CATALYST). Catalyst loading in the reactor (RCSTR) was 200 kg. The ratios among the feed streams were selected from reported values [13]. The Ni-based commercial catalysts were used in the mild stage of hydrogenation (HDO). Three sets of operating conditions for mild HDO step were assessed which were, respectively, (1) 250 °C, 5.6 MPa; (2) 280 °C, 8 MPa; and (3) 300 °C, 10 MPa. The optimum reaction conditions were eventually selected according to the desired product yield.

According to the theory of mild hydrogenation kinetics, the chemical conversion reactions among the components was established [10]. The model for the conversion of the catalytic bio-oil hydrogenation process was obtained by fitting the experimental data of product distribution and catalyst deactivation in the small industrial reaction [14]. The pre-exponential factor A and the activation energy E in the chemical reactions were obtained from the previous study on kinetic modeling of the bio-oil hydrogenation from which the parameters of the mild-stage reactions were fitted. Their values are given in Table 2, with the parameter setting of reaction conditions as described above.

Table 2. Kinetic parameters of bio-oil hydrogenation in the mild hydrogenation stage.

Kinetic Reactions	A	E (kJ/mol)
BIO-OIL + 0.22H_2 →0.742ODF + 0.192AQO + 0.121H_2O + 0.076CO_2	4.31 × 10^6 (m^3 MPa^{-1} kg^{-1} h^{-1})	80
BIO-OIL→1.869C_s	1.67 × 10^4 (m^3 h^{-1} kg^{-1})	9.4
BIO-OIL→1.869C_{is}	5.09 × 10^8 (m^3 h^{-1} kg^{-1})	169.5
C_s→0.535BIO-OIL	0.5003 (h^{-1})	44.3
ODF→1.269C_s	6.58 × 10^4 (m^3 h^{-1} kg^{-1})	16
C_s→0.787ODF	2 (h^{-1})	45.3
AQO→2.705C_s	1.2 × 10^3 (m^3 h^{-1} kg^{-1})	14.6
C_s→0.396AQO	5 (h^{-1})	46.7
C_s →C_{is}	1.54 × 10^3 (h^{-1})	52.6

The deactivation rate of the catalyst is defined as follows:

$$Xc = 1 - \frac{m_{cat,f}}{m_{cat}} \quad (1)$$

where m_{cat} is the total mass of the catalyst at a given time and $m_{cat,f}$ is the mass of fresh catalyst. The relationship between the deactivation rate of the catalyst and catalyst coking has been derived from coking theory [14] and experimental results of the batch hydrogenation reactor [15] and is expressed as follows:

$$\frac{dX_C}{dt} = K_d \left[(1 - X_C) \cdot \left(\alpha \frac{dm_{Cs}}{dt} + (1 - \alpha) \frac{dm_{Cis}}{dt} \right) \right]^\beta \quad (2)$$

where K_d is the deactivation constant of coke deposition, α is the weighing factor of coke type, and β is the deactivation exponent. The mathematical expression of impacts of catalyst coking deactivation (Equation (2)) on the reaction system in the mild HDO stage is embedded into the Aspen model using a FORTRAN subroutine.

The deep hydrogenation step used a Ni/Al_2O_3 catalyst, under the operating conditions of 400 °C and 15 MPa.

The stoichiometry for deep hydrogenation reaction which produces the final upgraded bio-fuel (PRODUCT) is obtained from reported experimental results [16] and shown in Equation (3).

$$ODF + 0.23H_2 \rightarrow PRODUCT + 0.11H_2O \quad (3)$$

3. Results and Discussion

3.1. Analysis on Bio-Oil Conversion under Different Reaction Conditions

In running the process model, the operating temperatures for the mild hydrogenation reactor was set at 250 °C, 280 °C, and 300 °C, respectively, and the corresponding pressures were 5.6 MPa, 8 MPa, and 10 MPa. The deep hydrogenation stage used a Ni/Al_2O_3 catalyst under the operating conditions of 400 °C and 15 MPa. As per the simulation results shown in Figure 3, among the various products from the bio-oil upgrading, the yield of the bio-fuel is the highest (45%) and the total amount of coke is the lowest with the mild hydrogenation reactor temperature of 300 °C. However, the yields of bio-fuel and the total coke deposits at 280 °C and 300 °C are very similar. Moreover, the lower operating temperature is beneficial for the sake of process safety as the corresponding pressure is lower. The selection of optimal reaction temperature for the mild stage will be further discussed in the energy analysis section. The amount of insoluble coke deposit increases with the elevated reaction temperature, which is consistent with the reported trend of insoluble coke deposit with reaction temperature in the literature [17]. According to the Arrhenius reaction formula, the higher temperature promotes endothermic reactions. However, the production of insoluble coke deposits is the main reason for catalyst deactivation. The yield of the oil phase (ODF) in the first stage does not increase linearly with increasing temperature.

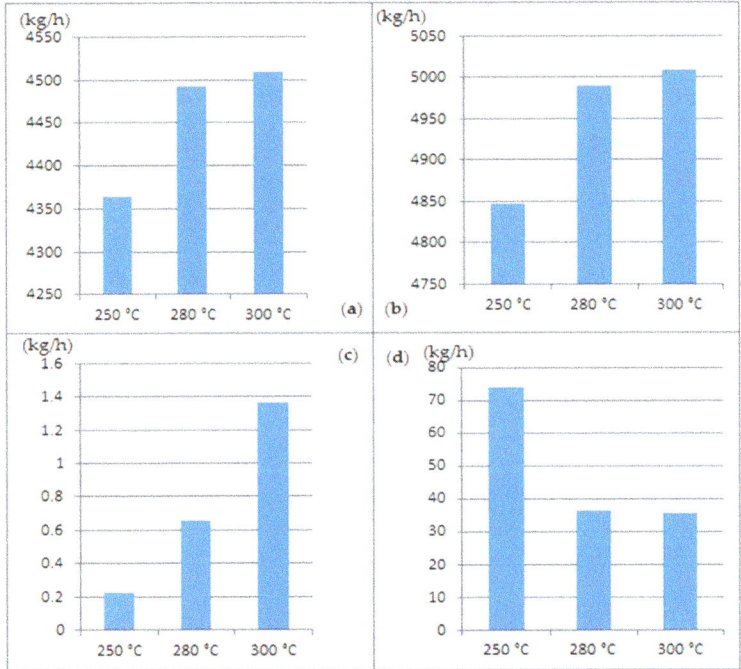

Figure 3. The mass flow at different temperatures of the mild HDO stage. (**a**) PRODUCT (model compound that represents the final product of bio-fuel, with same properties of gasoline, $CH_{1.71}$); (**b**) ODF (oil-phase organic compounds of mild stage, $CH_{1.47}O_{0.11}$); (**c**) COKES (soluble coke deposit, C_s); and (**d**) COKEIS (insoluble coke deposit, C_{is}).

A more detailed material balance of the whole industrial process under different reaction conditions is shown in Tables A1–A3. Since the amount of catalyst is unchanged during the reaction, the flow of catalyst is not shown in the stream summary tables. The slight mass imbalances (0.1–0.6%) in the stream tables were caused by the inherent convergence issues of the Aspen Plus v7.2 software (AspenTech, Boston, Massachusetts, US).

3.2. Analysis on Energy Consumption and Energy Flows

The summary of energy flows in each major unit is shown in Figure 4, in which the negative heat duties of the two reactors indicate that an external heat supply is required for maintaining the required operating temperature. However, the required heat duty of the mild HDO reactor is less than that of the deep stage because the feedstock streams have been partially preheated by the product streams through the heat exchangers.

In total, three heat exchangers are employed for heat recovery. The heat exchange rate in HEATX2 is the highest, because the amount of heat required for bio-oil preheating is large due to high mass flow rate. The optimal reaction temperature for the mild stage was selected as 280 °C, because the product yields at the two operating temperatures are similar but the energy required for 280 °C is less than that for 300 °C.

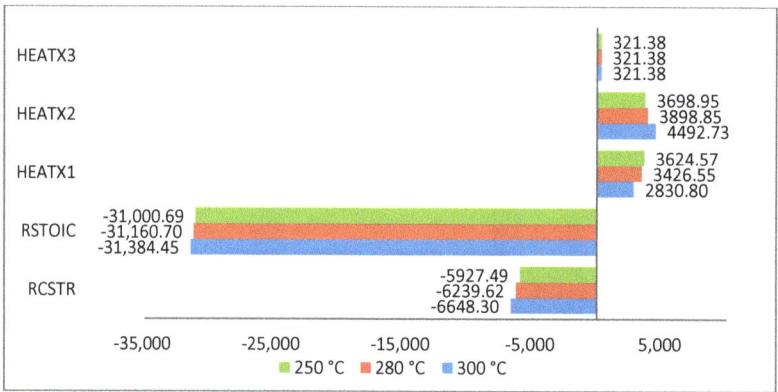

Figure 4. Heat duty of the key blocks in the process.

4. Economic Evaluations

4.1. Fixed Capital Cost

According to the throughput of the process plant defined in this case study, the minimum possible capital investment for the unit product was estimated. The fixed capital cost of the demonstration plant was approximated as the sum of fixed costs (equipment costs and installation costs), because the ancillary costs for a demonstration plant are unpredictable.

The cost for purchasing major equipment (reactors, separators, heat exchangers, etc.) was estimated from the power-law relationship between purchase price, equipment size, and inflation index [18]. The installed costs were estimated using the installation factor method [19]:

$$\text{New Equipment Cost} = \text{Original Equipment Cost} \, (\text{New size}/\text{Original size})^{\exp} \qquad (4)$$

$$\text{Installed Equipment Cost} = \text{Equipment Cost} \times \text{Installation Factor} \qquad (5)$$

The values for the installation factors and exponential factors in Equations (4) and (5) are shown in Table 3, and the equipment purchase and installation costs of the demonstration plant are summarized in Table 4.

Table 3. Installation factor and exponential factor.

Equipment	Installation Factor	Exp
reactor (stainless steel)	1.6	0.5
heat exchanger (stainless steel)	2.2	0.7
separator	3.02	0.7

Table 4. Equipment costs and installation costs.

Equipment	Equipment Cost (millions $)	Installation Cost (millions $)
RCSTR	3.1	5.0
RSTOIC	1.7	2.8
HEATX1	0.2	0.44
HEATX2	0.23	0.5
HEATX3	0.02	0.04
SEP1	0.02	0.06
SEP2	0.18	0.54
SEP3	0.008	0.02
SEP4	0.09	0.28
Total		15.2

4.2. Operating Costs

The operating costs mainly consist of costs for materials and energy consumption, whereas the taxes are not included in this study. The total annual operating time is assumed to be 8000 h. Operating costs include the costs for bio-oil, hydrogen, and catalyst (the average cost of catalyst for the deep and mild stages). The cost of bio-oil is assumed to be the production cost of the pyrolysis process of biomass under ideal conditions [13]. The purchase price of hydrogen considered here is $1.50/gallon of gasoline equivalent (GGE) or nearly $1.50/kg [13]. Prices for the external steam supply to heat the HDO reactors and for cooling water are based on the Chinese industrial standard. The labor costs are estimated as 120 employees in total with an annual salary of $9000. The other costs include the general overheads and the plant maintenance which account for up to 2% of the fixed capital cost.

The overall unit production costs of two scenarios are shown in Table 5 [7,9], which include the practical unit production and the ideal unit production costs. The ideal production assumes that the catalyst activity is not affected by catalyst coking, whereas the practical production takes in account the coking deactivation of the catalyst. By comparing the two cases, one can see that the costs for bio-oil, hydrogen, cooling water, and steam supply are virtually unchanged, because the increases in feed rates for those material and energy streams to maintain the same yield of desired product are not significant. However, the unit cost of the product increases by approximately 10%, because a higher feeding rate of fresh catalyst is required to balance the effect of catalyst coking deactivation.

Table 5. Estimation of raw material costs for the two cases [7,9].

	Unit Costs ($/t)	Practical Production Costs ($/t)	Ideal Production Costs ($/t)
Bio-oil	188.55 [13]	419.75	415.05
Catalyst (Ni-based)	34,444 [7]	76.68	0.0032
Hydrogen	1500 [13]	115.21	113.7
Steam	28.88	346.21	344.58
Cooling water	0.46	0.063	0.062
Labor costs		28.88	28.88
Fixed costs		21.13	21.13
Other costs		16.36	16.36
Total		1024.27	939.76

A sensitivity analysis of the influences on the production costs was performed for both the practical and ideal scenarios, and the resulting minimum and maximum production costs [13,20] that reflect the impact of common variations of the key input parameters from the base case are shown in Figures 5 and 6 The costs of materials and utilities as well as the yield of final products are included. Bio-fuel yield has the strongest positive influence on the production cost; this implies that slight improvements in the overall performance of bio-oil upgrading process could reduce the cost of fuel significantly. In addition, the production costs are sensitive to the purchase price of raw bio-oil, as the cost for acquiring feedstock can vary widely between locations and throughout the year. Moreover, the influence of the catalyst is significant in the practical scenario but is negligible in the ideal case, which is due to the fact that the feeding rate of fresh catalyst in the latter is negligible in comparison with the former because no coking deactivation of the catalyst occurs.

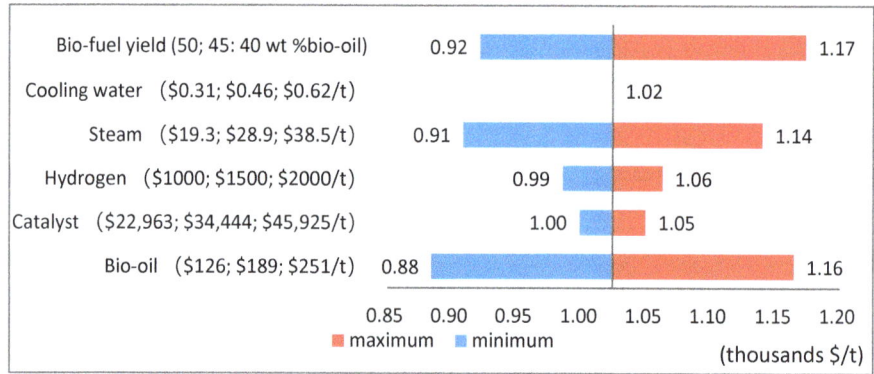

Figure 5. Sensitivity analysis for the practical production costs.

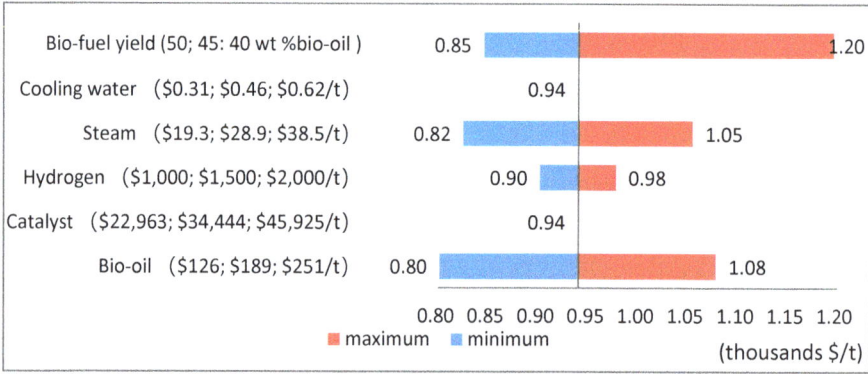

Figure 6. Sensitivity analysis for the ideal production costs.

The comparison of production costs estimated in this study with different research studies [7,13] and of average gasoline prices throughout 2018 is shown in Figure 7. Both practical and ideal (where no catalyst coking occurs) production costs estimated in this study are higher than the results from other research studies. The reason for this is that only the hydrogenation process of bio-oil has been considered in this study, whereas both processes of pyrolysis and hydrogenation were included in Wright et al. [13] and Jones et al. [7]. The overall heat efficiency of the integrated process of pyrolysis and hydrogenation is higher, as the heat surplus of the pyrolysis plant could be utilized for hydrogenation; the cost for external steam could thus be reduced in comparison with the sole hydrogenation process. However, the cost for purchasing bio-oil from biomass pyrolysis can be subsidized according to the renewable energy policy in China. Therefore, the industrial production of upgraded bio-fuel from two-stage hydrogenation of bio-oil is economically feasible as the production cost is $168/ton lower than the average gasoline price in 2018.

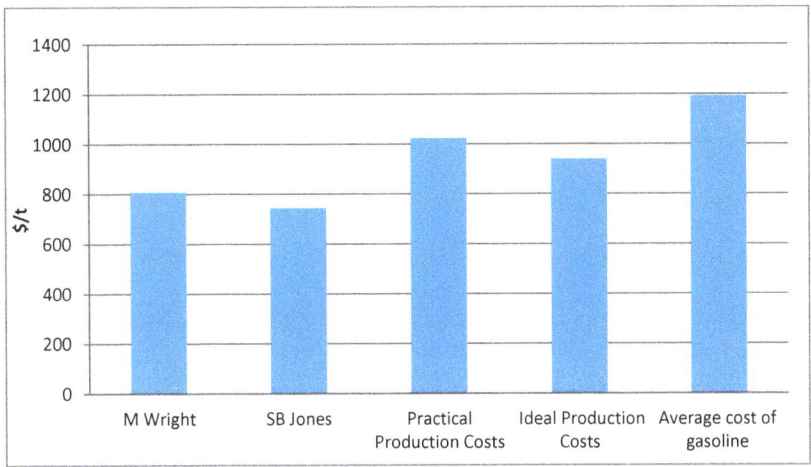

Figure 7. Comparison of simulated production cost and the average cost of Chinese gasoline in 2018 [7,13].

5. Conclusions

In this study, a process simulation and a cost analysis were carried out for a bio-oil hydrogenation demonstrating plant with an annual throughput of 107 thousand tons of raw bio-oil. The following conclusions were drawn from this study.

(1) A process model was developed using Aspen Plus to simulate the bio-oil HDO production process, material and energy balances throughout the plant were carried out, and deactivation of the catalysts due to coking in the mild HDO stage was embedded into the process model using a FORTRAN subroutine. The developed process model was capable of simulating the HDO process with extended ranges of feedstock properties and different reaction kinetics. From the simulation results, the yield of simulated bio-fuel varied between 44.1% and 45% across the selected temperature range of mild HDO reaction (250 °C–300 °C), whereas the optimal reaction temperature for the mild hydrogenation stage was found to be 280 °C, making this trend optimal for industrial design.

(2) The simulation results showed that the catalyst deactivation due to coking could decrease the yield of the desired product, and that a larger feedstock flow for maintaining the productivity was required to compensate for the effect of coking deactivation. All these results showed that the catalyst deactivation affected the performance in the practical production process. Thus, the main reason for the catalyst deactivation was the presence of insoluble coke deposits, which affected the hydrogenation reaction in the mild stage, making it one of the decisive factors of the whole process.

(3) A cost analysis of the bio-oil HDO production was performed and the minimum required costs were estimated. The total fixed capital cost for the demonstration plant with an annual throughput of 107 thousand tons of raw bio-oil was estimated at $15.2 million. The required variable costs for the unit product in the ideal case was found to be $939.76; where the unit production cost for the practical case, which included the catalyst deactivation, was found to be $1024.27/ton. The additional $76.68/ton came mainly from the expenditures related to the extra catalyst. A sensitivity analysis of the key process variables showed that the bio-fuel yield has the most impact on the unit production cost. Variations of ±5% in the bio-fuel yield resulted in production costs rising from 0.92 to 1.2 thousand/ton for the practical scenario. The industrial production of upgraded bio-fuel from the two-stage hydrogenation of bio-oil is economically feasible as the operating costs are $168/ton which is lower than the average gasoline price in 2018.

Author Contributions: Conceptualization, Q.X.; Data curation, X.M.; Formal analysis, X.M.; Investigation, X.M.; Methodology, X.M.; Software, X.M. and Q.X.; Supervision, R.Z.; Validation, Q.X.; Writing—original draft, X.M.; Writing—review and editing, S.P. and Q.X.

Funding: This research was partially funded by National Key Research and Development Program of China, grant number 2017YFC0212400.

Acknowledgments: The authors wish to thank the three anonymous reviewers and the editor for their constructive suggestions for improving the manuscript.

Conflicts of Interest: The authors declare no conflict of interest.

Appendix A. Stream Information and Energy Analysis of the Whole Process

Table A1. Simulated results for main streams of products from mild hydrogenation of bio-oil at 250 °C and 5.6 MPa and deep hydrogenation at 400 °C and 15 MPa.

Streams	B-OIL-IN	H2-IN	MIXTURE1	MIXTURE2	MIXTURE3	PRODUCT	H2-C
Total Flow, kg/h	10,000	345	11,264.79	11,190.29	6147.53	4364.09	849.13
Temperature, °C	192.79	250	250	20	30	30	30
Mass Flow for all components, kg/h							
H_2	0	345	995.55	995.55	849.13	0	849.13
WATER	0	0	934.51	934.51	630.26	0	0
CO_2	0	0	1435.13	1435.13	0	0	0
BIO-OIL	10,000	0	304.05	304.05	304.05	0	0
ODF	0	0	4847.93	4847.93	0	0	0
AQO	0	0	2673.13	2673.13	0	0	0
COKES	0	0	74.28	0	0	0	0
COKEIS	0	0	0.23	0	0	0	0
PRODUCT	0	0	0	0	4364.09	4364.09	0

Table A2. Simulated results for main streams of products from mild hydrogenation of bio-oil at 280 °C and 8 MPa and deep hydrogenation at 400 °C and 15 MPa.

Streams	B-OIL-IN	H2-IN	MIXTURE1	MIXTURE2	MIXTURE3	PRODUCT	H2-C
Total Flow, kg/h	10,000	345	10,862.11	10,824.59	5633.78	4491.99	434.76
Temperature, °C	192.79	250	280	20	3	30	30
Mass Flow for all components, kg/h							
H_2	0	345	585.48	585.48	434.76	0	434.76
WATER	0	0	962.54	962.54	648.73	0	0
CO_2	0	0	1478.17	1478.17	0	0	0
BIO-OIL	10,000	0	58.3	58.3	58.3	0	0
ODF	0	0	4990.01	4990.01	0	0	0
AQO	0	0	2750.11	2750.11	0	0	0
COKES	0	0	36.86	0	0	0	0
COKEIS	0	0	0.66	0	0	0	0
PRODUCT	0	0	0	0	4491.99	4491.99	0

Table A3. Simulated results for main streams of products from mild hydrogenation of bio-oil at 300 °C and 0 MPa and deep hydrogenation at 400 °C and 15 MPa.

Streams	B-OIL-IN	H2-IN	MIXTURE1	MIXTURE2	MIXTURE3	PRODUCT	H2-C
Total Flow, kg/h	10,000	345	10,624.99	10,587.65	5377.13	4508.92	195.85
Temperature, °C	192.79	250.00	300	20	30	30	30
Mass Flow for all components, kg/h							
H_2	0	345	347.13	347.13	195.85	0	195.85
WATER	0	0	967.13	967.13	651.18	0	0
CO_2	0	0	1485.22	1485.22	0	0	0
BIO-OIL	10,000	0	21.19	21.19	21.19	0	0
ODF	0	0	5008.81	5008.81	0	0	0
AQO	0	0	2758.16	2758.16	0	0	0
COKES	0	0	35.97	0	0	0	0
COKEIS	0	0	1.36	0	0	0	0
PRODUCT	0	0	0	0	4508.92	4508.92	0

Table A4. Energy analysis of the overall process at 250 °C and 5.6 MPa and deep hydrogenation at 400 °C and 15 MPa.

	Inlets			Outlets			Operation Temperature (°C)	Heat Medium	Heat Duty(kw)
	Stream	Flow (kg/h)	Temp. (°C)	Stream	Flow (kg/h)	Temp. (°C)			
RCSTR	B-OIL-IN H2-IN H2-C	10,000 345 849.13	192.79 250 30	MIXTURE1	11,264.79	250	250	steam (400 °C)	−5927.49
RSTOIC	OIL-1 H2-2	5151.98 995.55	20	9	6147.53	400	400	steam (600 °C)	−31,000.69
HEATX1	5(Hot) 1(Cold)	11,190.29 10,000	254.07 118.85	8(Hot) B-OIL-IN(Cold)	11,190.29 10,000	192.39 192.79			3624.57
HEATX2	18(Hot) BIO-OIL(Cold)	6147.53 10,000	375.51 20	MIXTURE3(Hot) 1(Cold)	6147.53 10,000	30 118.85			3698.95
HEATX3	9(Hot) H2(Cold)	6147.53 345	400 20	18(Hot) H2-IN(Cold)	6147.53 345	375.51 250			321.38

Table A5. Energy analysis of the overall process at 280 °C and 8 MPa and deep hydrogenation at 400 °C and 15 MPa.

	Inlets			Outlets			Operation Temperature (°C)	Heat Medium	Heat Duty (kw)
	Stream	Flow (kg/h)	Temp. (°C)	Stream	Flow (kg/h)	Temp. (°C)			
RCSTR	B-OIL-IN H2-IN H2-C	10,000 345 434.76	192.79 250 30	MIXTURE1	10,862.11	280	280	steam (400 °C)	−6239.62
RSTOIC	OIL-1 H2-2	5048.30 585.48	20	9	5633.78	400	400	steam (600 °C)	−31,160.70
HEATX1	5(Hot) 1(Cold)	10,824.59 10,000	279.77 118.85	8(Hot) B-OIL-IN(Cold)	10,824.59 10,000	196.85 192.79			3426.55
HEATX2	18(Hot) BIO-OIL(Cold)	5633.78 10,000	371.78 20	MIXTURE3(Hot) 1(Cold)	5633.78 10,000	30 118.85			3898.85
HEATX3	9(Hot) H2(Cold)	5633.78 345	400 20	18(Hot) H2-IN(Cold)	5633.78 345	371.78 250			321.38

Table A6. Energy analysis of the overall process at 300 °C and 10 MPa and deep hydrogenation at 400 °C and 15 MPa.

	Inlets			Outlets			Operation Temperature (°C)	Heat Medium	Heat Duty (kw)
	Stream	Flow (kg/h)	Temp. (°C)	Stream	Flow (kg/h)	Temp. (°C)			
RCSTR	B-OIL-IN	10,000	192.79	MIXTURE1	10,624.99	300	300	steam (400 °C)	−6648.30
	H2-IN	345	250						
	H2-C	195.84	30						
RSTOIC	OIL-1	5030.01	20	9	5377.13	400	400	steam (600 °C)	−31,384.45
	H2-2	347.14							
HEATX1	5(Hot)	10,587.66	284.44	8(Hot)	10,587.66	183.57			2830.80
	1(Cold)	10,000	118.85	B-OIL-IN(Cold)	10,000	192.79			
HEATX2	18(Hot)	5377.15	369.32	MIXTURE3(Hot)	5377.15	30			4492.73
	BIO-OIL(Cold)	10,000	20	1(Cold)	10,000	118.85			
HEATX3	9(Hot)	5377.15	400	18(Hot)	5377.15	369.32			321.38
	H2(Cold)	345	20	H2-IN(Cold)	345	250			

References

1. Werther, J.; Saenger, M.; Hartge, E.U.; Ogada, T.; Siagi, Z. Combustion of agricultural residues. *Prog. Energy Combust. Sci.* **2000**, *26*, 1–27. [CrossRef]
2. He, J.; Liu, Y.; Lin, B. Should China support the development of biomass power generation? *Energy* **2018**, *163*, 416–425. [CrossRef]
3. Demirbas, A. Biofuels sources, biofuel policy, biofuel economy and global biofuel projections. *Energy Convers. Manag.* **2008**, *49*, 2106–2116. [CrossRef]
4. Pang, S. Advances in thermochemical conversion of woody biomass to energy, fuels and chemicals. *Biotechnol. Adv.* **2018**. [CrossRef] [PubMed]
5. Mortensen, P.M.; Grunwaldt, J.D.; Jensen, P.A.; Knudsen, K.G.; Jensen, A.D. A review of catalytic upgrading of bio-oil to engine fuels. *Appl. Catal. A Gen.* **2011**, *407*, 1–19. [CrossRef]
6. Jones, S.B.; Meyer, P.A.; Snowden-Swan, L.J.; Padmaperuma, A.B.; Tan, E.; Dutta, A.; Jacobson, J.; Cafferty, K. *Process Design and Economics for the Conversion of Lignocellulosic Biomass to Hydrocarbon Fuels: Fast Pyrolysis and Hydrotreating Bio-Oil Pathway*; U.S. Department of EnergyOffice of Scientific and Technical Information: Oak Ridge, TN, USA, 1 November 2013.
7. Jones, S.B.; Valkenburt, C.; Walton, C.W.; Elliott, D.C.; Holladay, J.E.; Stevens, D.J.; Kinchin, C.; Czernik, S. *Production of Gasoline and Diesel from Biomass via Fast Pyrolysis, Hydrotreating and Hydrocracking: A Design Case*; Pacific Northwest National Laboratory: Richland, WA, USA, 25 February 2009.
8. Atsonios, K.; Kougioumtzis, M.A.; Grammelis, P.; Kakaras, E. Process Integration of a Polygeneration Plant with Biomass/Coal Co-pyrolysis. *Energy Fuels* **2017**, *31*, 14408–14422. [CrossRef]
9. Wright, M.M.; Daugaard, D.E.; Satrio, J.A.; Brown, R.C. Techno-economic analysis of biomass fast pyrolysis to transportation fuels. *Fuel* **2010**, *89*, S2–S10. [CrossRef]
10. Venderbosch, R.H.; Ardiyanti, A.R.; Wildschut, J.; Oasmaa, A.; Heeres, H.J. Stabilization of biomass-derived pyrolysis oils. *J. Chem. Technol. Biotechnol.* **2010**, *85*, 674–686. [CrossRef]
11. Lv, Q.; Yue, H.; Xu, Q.; Zhang, C.; Zhang, R. Quantifying the exergetic performance of bio-fuel production process including fast pyrolysis and bio-oil hydrodeoxygenation. *J. Renew. Sustain. Energy* **2018**, *10*, 043107. [CrossRef]
12. Xu, X.; Zhang, C.; Zhai, Y.; Liu, Y.; Zhang, R.; Tang, X. Upgrading of Bio-Oil Using Supercritical 1-Butanol over a Ru/C Heterogeneous Catalyst: Role of the Solvent. *Energy Fuels* **2014**, *28*, 4611–4621. [CrossRef]
13. Wright, M.; Satrio, J.A.; Brown, R.C. *Techno-Economic Analysis of Biomass Fast Pyrolysis to Transportation Fuels*; NREL: Golden, CO, USA, 2010.
14. Li, Y.; Zhang, C.; Liu, Y.; Hou, X.; Zhang, R.; Tang, X. Coke Deposition on Ni/HZSM-5 in Bio-oil Hydrodeoxygenation Processing. *Energy Fuels* **2015**, *29*, 1722–1728. [CrossRef]
15. Yamamoto, Y.; Kumata, F.; Massoth, F.E. Hydrotreating catalyst deactivation by coke from SRC-II oil. *Fuel Process. Technol.* **1988**, *19*, 253–263. [CrossRef]
16. Xu, X.; Zhang, C.; Liu, Y.; Zhai, Y.; Zhang, R. Two-step catalytic hydrodeoxygenation of fast pyrolysis oil to hydrocarbon liquid fuels. *Chemosphere* **2013**, *93*, 652–660. [CrossRef] [PubMed]

17. Li, Y.; Zhang, C.; Liu, Y.; Tang, S.; Chen, G.; Zhang, R.; Tang, X. Coke formation on the surface of Ni/HZSM-5 and Ni-Cu/HZSM-5 catalysts during bio-oil hydrodeoxygenation. *Fuel* **2017**, *189*, 23–31. [CrossRef]
18. Yang, Z.; Qian, K.; Zhang, X.; Lei, H.; Xin, C.; Zhang, Y.; Qian, M.; Villota, E. Process design and economics for the conversion of lignocellulosic biomass into jet fuel range cycloalkanes. *Energy* **2018**, *154*, 289–297. [CrossRef]
19. En, W.; Guoyan, Z.; Shandong, T. Technical economic analysis of heat exchangers. *Chem. Ind. Eng. Prog.* **2006**, *25*, 458–461.
20. Wu, W.; Long, M.R.; Zhang, X.; Reed, J.L.; Maravelias, C.T. A framework for the identification of promising bio-based chemicals. *Biotechnol. Bioeng.* **2018**, *115*, 2328–2340. [CrossRef] [PubMed]

© 2019 by the authors. Licensee MDPI, Basel, Switzerland. This article is an open access article distributed under the terms and conditions of the Creative Commons Attribution (CC BY) license (http://creativecommons.org/licenses/by/4.0/).

Review

Preparation and Modification of Biochar Materials and their Application in Soil Remediation

Xue Yang, Shiqiu Zhang, Meiting Ju * and Le Liu *

College of Environmental Science and Engineering, Nankai University, Tianjin 300350, China; 2120170638@mail.nankai.edu.cn (X.Y.); swustzsq@sina.com (S.Z.)
* Correspondence: nkujumeiting@sohu.com (M.J.); tjliule@126.com (L.L.); Tel.: +86-13820988813 (M.J.); +86-13672031215 (L.L.)

Received: 28 February 2019; Accepted: 25 March 2019; Published: 1 April 2019

Abstract: As a new functional material, biochar was usually prepared from biomass and solid wastes such as agricultural and forestry waste, sludge, livestock, and poultry manure. The wide application of biochar is due to its abilities to remove pollutants, remediate contaminated soil, and reduce greenhouse gas emissions. In this paper, the influence of preparation methods, process parameters, and modification methods on the physicochemical properties of biochar were discussed, as well as the mechanisms of biochar in the remediation of soil pollution. The biochar applications in soil remediation in the past years were summarized, such as the removal of heavy metals and persistent organic pollutants (POPs), and the improvement of soil quality. Finally, the potential risks of biochar application and the future research directions were analyzed.

Keywords: biochar preparation; soil pollution; remediation

1. Introduction

With the development of industry and high-intensity human activities in China, soil pollution is becoming more and more serious, mainly due to the reduction of soil area and pollution by chemical compounds such as pesticides, petroleum, heavy metals, persistent organic matter, and acidic substances [1].

Pollutants in soil mainly include heavy metals and organic compounds, such as Cd, Pb, Cr, pesticides, fertilizers, antibiotics, polycyclic aromatic hydrocarbons (PAHs), polychlorinated biphenyls (PCBs), etc. [2,3]. These pollutants not only affect the decline of crop yield and quality, resulting in further deterioration of the atmospheric and water environment quality, but also have carcinogenic, teratogenic, mutagenic effects, and genotoxicity, which endanger human health through the food chain [4].

The remediation methods of contaminated soil are mainly divided into physical, chemical, biological, and plant methods. Physical remediation technologies mainly include soil leaching, thermal desorption, steam extraction, and off-site landfill [5]. But the disadvantages are its high costs and the risk of secondary diffusion. Chemical remediation technologies mainly include immobilization-stabilization techniques, redox, chemical modification, surfactant cleaning, and organic matter improvement [6,7], but the chemicals used may cause secondary pollution to the environment. There is a long repair cycle in bioremediation technology and the repair effect is susceptible to external environmental factors.

Since Lehmann proposed the efficacy of Amazon black soil [8], scholars have found that the biochar produced by the lack of oxygen through pyrolysis of agricultural and forestry wastes is a material with well-developed pore structure, large specific surface area, abundant oxygen-containing functional groups, and excellent adsorption performance [9,10]. Biochar remediation technology is between physical remediation and chemical remediation. On one hand, inorganic pollutants could be removed by physical adsorption, and organic pollutants could be removed by distribution. On the other hand, the application of biochar affects the solubility, valence, and existence of heavy metals in the soil, thus immobilizing the heavy metals in the soil. Finally, toxicity of heavy metals was fixed or reduced [9]. Due to its remarkable effect, low cost, and convenient operation, biochar has advantages in the treatment of heavy metal and organic pollution [11].

Before summarizing the application of biochar in soil remediation, this paper summarized the preparation and modification methods of biochar and analyzed the influence of different processes on the physicochemical properties of biochar to deepen the understanding of biochar. As the adsorbing material, the removal of heavy metals and organic compounds from soil and the main mechanism of biochar were reviewed. As a soil improver, the improvement of soil pH, nutrient, nitrogen, and phosphorus loss by biochar, and the application trend in the future, were summarized. At the same time, the potential risks of biochar were analyzed to effectively avoid the possible harm to the environment.

2. Preparation and Modification of Biochar

2.1. Preparation of Biochar

The preparation methods of biochar are mainly divided into pyrolysis [12], hydrothermal carbonization (HTC) [13], and microwave carbonization [14]. Different preparation methods affect the physical and chemical properties of biochar, such as yield, ash, specific surface area, pore structure, type and number of functional groups, and cation exchange capacity. Compared with the pyrolysis, HTC does not require drying step and has a higher biochar yield [15]. The advantages of microwave carbonization are controllable process, no hysteresis, rapid heating, and energy efficiency [16,17]. However, biochar prepared through HTC and microwave contained high concentrations of organics, which are not actually considered soil remediation material.

2.1.1. Pyrolysis

Pyrolysis, also known as the thermal decomposition under oxygen-free conditions, is the most common method for preparing biochar. In general, pyrolysis involves the heating of organic materials to temperatures greater than 400 °C under inert atmospheres by electric heating or high-temperature medium. There are many parameters influence physicochemical properties of biochar, such as raw material, reaction temperature, heating rate, residence time, and reaction atmosphere.

2.1.2. Factors Affecting the Pyrolysis Process

The raw materials for the preparation of biochar are abundant. Basically, any form of organic materials can be pyrolyzed [18]. Due to the large output of biomass solid waste resources, biomass is a common raw material for biochar, mainly including wheat straw, corn straw, wood chips, melon seed shell, peanut shell, rice husk, livestock and poultry manure, kitchen waste, sludge, fruit skin, etc. [19]. Biochar prepared from different materials contains different proportions of cellulose, hemicellulose, and lignin, so its yield, element composition, and ash content are different [20,21]. Enders et al. [22] found that the ash content of straw biochar is higher than that of other biochar, which is mainly caused by the high Si content of straw. Yuan et al. [23] compared the physicochemical properties of biochar prepared from different feedstocks (the straws of canola, corn, soybean, and peanut). The ash content of biochar from corn straw prepared at 700 °C was the highest (73.30%), compared to canola, soybean, and peanut straw biochar (28.55%, 23.70%, 38.50%, respectively).

The reaction temperature ranges of high temperature anoxic, hydrothermal synthesis, and flash carbonization for the preparation of biochar are 400–900 °C, 180–250 °C, and 300–600 °C, respectively [11]. In general, with the increase of pyrolysis temperature, the yield of biochar and the number of acidic functional groups (-COOH, -OH) decreased, while the alkaline functional groups, ash content, and pH increased. In addition, the effects of pyrolysis temperature on the surface area and pore volume are especially significant. Park et al. [24] showed that the specific surface area and total pore volume of sesame straw biochar increased from 46.9 to 289.2 $m^2 \cdot g^{-1}$, 0.0716 to 0.1433 $cm^3 \cdot g^{-1}$, respectively, with the pyrolysis temperature increased from 500 to 600 °C.

According to the different heating rate, it could be divided into slow pyrolysis (SP) and fast pyrolysis (FP) [25]. SP is characterized by slow heating (minutes to hours) of the organic material in the oxygen-depleted atmosphere and relatively long solids and gas residence times [26,27]. During the SP process, liquid and solid products such as char, bio-oil, and syngas (CO, CO_2, H_2) are produced. The FP involves blowing small particles of organic material into a thermal reactor and exposing it to heat transfer in milliseconds to seconds [18]. Modern FP often takes place in fluidized bed systems, systems using ablative reactors, and systems using pyrolysis centrifuge reactors (PCR) [28]. Slow and fast pyrolysis results in biochars with different physicochemical properties, thus providing different effects on the soil environment upon application. Compared with the FP-biochar contained labile un-pyrolyzed biomass fraction, the SP-biochar can be pyrolyzed completely [26].

At the same pyrolysis temperature, the yield of biochar decreases with the increase of residence time. Chen et al. [29] prepared orange peel biochar with the pyrolysis temperature of 700 °C and residence time of 6 h, and the biochar yield was only 5.93%. The specific surface area and pores of biochar increased with the extension of residence time. But the residence time is not as long as possible. Lu et al. [30] found that the specific surface area and pores decreased from 2 to 3 h. The reason is that the increase of residence time is conducive to the development of biochar pores, but excessive residence time may cause damage to the pore structure [31].

The reaction atmosphere studied by scholars is dominated by inert gas, such as N_2, Ar, which mainly act to isolate oxygen. Besides, the atmosphere of CO_2, H_2O, NH_3, O_3 [32] is also used to prepare biochar, which is known as physical activation, also called gas activation. The gases selectively decompose the non-structural components of the biochar surface, open its internal pores, and increase the specific surface area and pore volume [11]. Table 1 lists the biochar prepared by different process parameters.

2.1.3. Other New Methods

In addition to the pyrolysis, hydrothermal carbonization, and microwave carbonization discussed above, flash carbonization and torrefaction [33] are other methods of biomass transformation. During the flash carbonization process, the flash fire is ignited at a high pressure (1–2 Mpa) on the biomass packed bed to convert the biomass into the gas and solid phase products [11]. It is reported that about 40% of biomass is converted to solid phase products (biochar) at 1 Mpa [34]. In addition to microwave, new pyrolysis technologies such as laser and plasma cracking technologies have also been developed. The sample usage of laser pyrolysis technology is small, and rapid heating and cooling can be carried out, which can effectively avoid the occurrence of secondary reactions [35]. Plasma pyrolysis technology is mainly applied in the preparation of syngas and coke. Compared with the traditional cracking technology, it can greatly increase the syngas and reduce the yield of bio-oil [36,37]. However, it is difficult to popularize the new pyrolysis technology due to its high cost and energy consumption.

Table 1. Physicochemical properties of biochar prepared by different methods and process parameters.

Raw Material	Atmosphere	Temperature (°C)	Heating Rate (°C/min)	Residence Time (h)	Yield (%)	pH	Ash Content (%)	Surface Area (m²·g⁻¹)	Total Pore Volume (cm³·g⁻¹)	Pore Diameter (nm)	Reference
Herb residue	N_2	400	10	3	37.9	10.2	28.3	49.2	0.042	3.39	[38]
		600			31.2	10.1	31.1	51.3	0.051	3.99	
		800			29.1	10.6	37.1	70.3	0.068	3.87	
Sesame straw	oxygen-limited	400			35.6		30.77	37.2	0.0542		[24]
		500	5	2	28.2		28.54	46.9	0.0716		
		600			22.9		21.98	289.2	0.1433		
Corn straw	N_2	600		3		10.0	5.02	61.0	0.036		[39]
Pine cone	N_2	500		1		4.66	2.13	6.6	0.016	23.7	[40]
Rice-husk		450–500				7.0	42.2	34.4	0.028		[41]
Hickory wood	N_2	450	10	2	28.5	7.9	6.47	12.9			[42]
		600			22.7	8.4	4.18	401.0			
Bagasse	N_2	450	10	2	28.0	7.5	13.68	13.6			
		600			26.5	7.5	15.36	388.3			
Bamboo	N_2	450	10	2	26.3	8.5	8.83	10.2			
		600			24.0	9.2	11.86	375.5			
Poplar chips	N_2	550	5	2	23.18		7.56	212.58	0.356	6.70	[43]
Burcucumber plants	oxygen-limited	700	7	2	27.52	12.23	43.72	2.31	0.008	6.780	[44]
Pine wood	N_2	600	10	1			4.02	209.6	0.003		[45]
Orange peel	oxygen-limited	400	5	6	11.3		6.93	28.1	0.0409	2.9	[29]
		700	5	6	5.93		14.9	501	0.390	1.6	
Marine macroalgae	N_2	450	5	2				1.05	0.007	30.41	[46]
Municipal solid waste	N_2	400		0.5		8.0	6.1	20.7	0.027		[47]
		500		0.5		8.5	9.2	29.1	0.039		
		600		0.5		9.0	6.2	29.8	0.038		
Rice straw	oxygen-limited	700		2			58.97	369.26	0.23		[48]
Swine manure	oxygen-limited	700		2			60.73	227.56	0.14		
Auricularia auricula dreg		400		2		10.09	0.55	77.64	0.0612	4.837	[49]
Thalia dealbata	N_2	500		4			22.0	7.1			[50]
Corn straw	N_2	500		1.5			41.0	32.85	0.0148	5.01	[51]

Table 1. *Cont.*

Raw Material	Atmosphere	Temperature (°C)	Heating Rate (°C/min)	Residence Time (h)	Yield (%)	pH	Ash Content (%)	Surface Area ($m^2 \cdot g^{-1}$)	Total Pore Volume ($cm^3 \cdot g^{-1}$)	Pore Diameter (nm)	Reference
Pitch pine	oxygen-free	400		2 s	33.5		7.9	4.8			[52]
		500		2 s	14.4		7.7	175.4			
Wheat straw	N_2	600	10	3			5.65	38.1	0.051	19.9	[53]
Rice straw	N_2	600	10	3			0.03	27.4	0.040	15.8	
Digested sugar beet tailing	N_2	600	10	2	45.5	9.95		336.0			[54]
Raw sugar beet tailing	N_2	600	10	2	36.3	9.45		2.6			
tea waste	oxygen-limited	700	7	2	28.35		10.87	342.22	0.0219	1.756	[55]
	N_2	700	7	2	22.35		11.60	421.31	0.0576	1.904	

2.2. Modification of Biochar

In order to obtain biochar with superior properties, scholars studied the effects of different modification methods on biochar. Modification refers to the activation of the original biochar through physical and chemical methods, so as to achieve the desired purpose. The type of activator, soaking time, activation time, and activation temperature all affect the properties of biochar. Table 2 lists the biochar prepared from different modified.

2.2.1. Chemical Oxidation

Chemical oxidation refers to the oxidation of the biochar surface to increase the oxygen-containing functional groups such as -OH, -COOH, etc., thereby its hydrophilicity is increased. At the same time, the pore size and structure of the biochar would be changed, and, finally, its adsorption capacity for the polar adsorbate would be enhanced. The commonly used oxidants are HCl, HNO_3, H_2O_2, H_3PO_4, etc. [56–60]. Although the specific surface area of biochar modified by HCl, HNO_3, and H_2O_2 has little difference, compared with biochar modified by HCl, the biochar modified by HNO_3 contains more acidic oxygen-containing functional groups [61] and has stronger adsorption capacity for NH_3–N. Compared with other acids, biochar modified by H_3PO_4 has more advantages in removing Pb pollution. The increased specific surface area and pore volume, as well as the role of phosphate precipitation, increase the biochar adsorption capacity of Pb [60].

2.2.2. Chemical Reduction

Chemical reduction is also known as alkali modification method. The reducing agent was used to reduce functional groups on the surface of biochar, so as to improve its non-polarity. Meanwhile, chemical modification also can improve porosity and specific surface area of biochar. Finally, adsorption capacity of biochar for pollutants is enhanced, especially for non-polar adsorbates. The commonly used reducing agents are NaOH [62], KOH [63], NH_4OH [64], etc. [65]. Different reducing agents have different modification effects. In order to determine suitable modified biochar for improving adsorption capacity of volatile organic compounds (VOCs), Li et al. [64] used NH_4OH, NaOH, HNO_3, H_2SO_4, and H_3PO_4 to carry out chemical treatment on coconut shell-based carbon. The results showed that, compared with the poor adsorption capacity of acid-treated carbon, high adsorption capacity was obtained for alkali-treated carbon. The reason is that surface area and pore volume increased and total oxygen containing function groups were diminished when treated by alkalis, while acid treatment was the opposite. Pouretedal et al. [66] found that the process of biochar activation by KOH and NaOH is different. Atomic species, K, formed in situ during KOH activation intercalates between the layers of the carbon crystallite, while there is hardly any evidence for the intercalation of Na with carbon.

2.2.3. Metal Impregnation

Metal impregnation refers to the adsorption of some heteroatoms or metal ions into the surface and pores of the biochar. On one hand, the specific surface area is increased, and on the other hand, metal ions are combined with the adsorbate to improve the adsorption performance. Common metal ions are iron [67], magnesium [68], silver [69], zinc [70], etc. Some scholars have combined the advantages of chemical reagents to achieve better adsorption performance. Lyu et al. [71] prepared a novel biochar material (CMC–FeS@biochar) via combining carboxymethyl cellulose (CMC) and iron sulfide (FeS), and demonstrated the effective sorbent of CMC–FeS@biochar composite for removal Cr(VI).

Table 2. Preparation of different modified biochar.

Raw Material	Reagent	Pollutant	Modification Method	Modification Effects	Reference
Bamboo hardwoods	NaOH, CS_2	Cd	The composition of sulfur modified mixture solution was obtained by stirring NaOH and CS_2. Biochar and sulfur modified mixture solution stirred at 45 °C for 8 h.	Sulfur-modified biochar (S-BC) has more roughness, with a more granular massive structure than that seen on the pristine biochar.	[76]
	$FeSO_4$		S-BC was added to $FeSO_4$ solution and then stirred for 16 h with magnetic stirrer at 40 °C, and cooled slowly to room temperature and filtered through 0.45 μm filters. The feedstock was oven-dried at 40 °C.	Successful impregnation of sulfur and iron onto the SF-BC surface, and it showed various atomic proportions of sulfur and iron, with biochar ranging from 0.48% to 4.66% and 0.44% to 22.25%, respectively.	
Poplar chips	$AlCl_3$	NO_3^-, PO_4^{3-}	The poplar pieces were impregnated into $AlCl_3$ solutions with different concentrations for 6 h. The mixtures were dried at 80 °C for 48 h. The pretreated pristine poplars were pyrolyzed under the N_2 atmosphere at 550 °C with a heating rate of 5 °C/min, and the peak temperature was maintained for 2 h.	The biochar yield increased after modification with Al. The carbon content of the Al-modified biochar significantly decreased compared with the pristine biochar. The BET surface area significantly increased with the Al content of the biochar. NO_3^- and PO_4^{3-} adsorptions significantly improved on the Al-modified biochar.	[43]
Rice straw, swine manure	H_3PO_4	Tetracycline (TC)	Biochars were immersed in H_3PO_4 solution for 24 h at 25 °C. Then, the H_3PO_4 modified biochars were washed by distilled water until the pH of supernatants was stable. Finally, the supernatants were discarded and the biochars were oven-dried overnight at 105 °C.	The H_3PO_4 modification enhanced the surface area of biochars produced from rice straw biochar (RC) and swine manure biochar (SC). Compared with SC, modified SC presented higher total pore, micropore and mesopore volume by 0.25 to 0.14, 0.09 to 0.07, 0.17 to 0.07 $cm^3 \cdot g^{-1}$), but there was no change between RC and RCA modification.	[48]
Wheat straw, cow manure	HNO_3	U(VI)	Biochar powders were treated with 300 mL 25% HNO_3 solution at 90 °C for 4 h. The excess acid was removed by centrifugation. All oxidized biochar samples were washed with deionized distilled water, freeze-dried, and milled to <0.25 mm.	Owing to the higher contents of surface COO groups and more negative surface charge, the modified biochar showed enhanced U(VI) adsorption ability than the unmodified biochar. The maximum adsorption capacity of U(VI) by the oxidized wheat straw biochar showed an improvement of 40 times relative to the untreated biochar.	[77]

Table 2. *Cont.*

Raw Material	Reagent	Pollutant	Modification Method	Modification Effects	Reference
Auricularia auricular dreg (AAD)	cetyltrimethyl ammonium bromide (CTAB)	Cr(VI)	Mixed 5 g of dried AAD biochar with 250 mL of 3.0% CTAB solution in 25 °C for 2 h. Residual CTAB rinsed with deionized water and the material was dried at 70 °C until the weight remained constant.	After modification, the surface area increased 6.1% and the average pore diameter increased 16.5% (77.64 m^2/g and 48.37 Å). Moreover, the number of mesoporous and micropores in unit area increased obviously. The adsorption rate and quantity of modified AAD biochar were 6.4% and 8.0% higher than those of AAD biochar, respectively.	[49]
Thalia dealbata	MgCl$_2$	sulfamethoxazole (SMX), Cd	Thalia dealbata were soaked in 100 mL 1 M MgCl$_2$ solution, after 0.5 h mixing under magnetic stirring, the pre-treated biomass was then separated from the solution and pyrolyzed at 500 °C.	The surface area of MgCl$_2$ modified biochar (BCM, 110.6 m$_2 \cdot$g^{-1}) was higher than untreated biochar (BC, 7.1 m$_2 \cdot$g^{-1}). The addition of BCM increased the sorption of SMX (by 50.8–58.6%) and Cd (by 24.2–25.6%) as compared with BC. In situ remediation with BCM decreased the mobility and bioavailability of SMX and Cd in sediments.	[50]
Corn straw	Na$_2$S and KOH	Hg(II), atrazine	Biochar were mixed with 500 mL of 2 M Na$_2$S or 2 M KOH solution and stirred for 4 h. The suspension was then filtered and washed with deionized water for several times until the pH of the filtrate was nearly 7. The washed biochar was dried overnight in an oven at 105 °C.	Sulfur content significantly increased by 101.29% under Na$_2$S modification. Compared to untreated biochar (BC, 32.85 m$_2 \cdot$g^{-1}), chemical modification increased the BET surface area which was 55.58 and 59.23 m$_2 \cdot$g^{-1} for Na$_2$S modified biochar (BS), KOH modified biochar (BK), respectively. In comparison to BC, the sorption capacity of BS and BK for Hg (II) increased by 76.95%, 32.12%, while that for atrazine increased by 38.66%, 46.39%, respectively.	[51]
Coconut shell	HCl+ultrasonication	Cd, Ni and Zn	5 g of CS biochar and 250 mL of 1 M HCl were mixed in beaker and ultrasonicated for 3 h with interval stirring. Then, the material was filtered, washed, and dried to constant weight.	Modified coconut shell biochar (MCSB) improved surface functional groups and microcosmic pore structure of pristine biochar (CSB).	[78]
Dairy manure	NaOH	Pb and Cd	Biochar and 2 M NaOH were thoroughly mixed with a solid-liquid ratio of 1:5 and then were re-suspended for 12 h with a speed of 30 r min^{-1} at 65 °C. After that, the mixture was filtered, and the precipitate was collected and rinsed with deionized. Finally, material was dried at 105 °C.	The NaOH treatment increased the specific surface area, ion-exchange capacity, and the number of oxygen-containing functional groups of biochar. The adsorption capacities of biochar for Pb and Cd increased after modification. The highest sorption capacities were 175.53 and 68.08 mg·g^{-1}, for Pb and Cd, respectively.	[79]

2.2.4. Other Modification Methods

In addition to the above three modification methods, modification methods such as low-temperature plasma [72,73], organic matter grafting [74], and ozone oxidation [32] have also been studied. Low-temperature plasma modification means that plasmas generated by glow, microwave, and corona were collided with C=C on the surface of biochar, and plasmas were oxidized to the oxygen-containing functional group, and enhanced the polarity of biochar [75]. However, such methods have not been widely used due to high cost and complicated operation.

3. Removal Mechanism of Major Pollutants by Biochar

The remediation mechanisms of soil pollution by biochar include ion exchange, physical adsorption, electrostatic interaction, precipitation, and complexation [9].

3.1. Ion Exchange

Ion exchange means the process that acidic oxygen-containing functional groups on the surface of biochar, such as carboxyl groups, carbonyl groups, and hydroxyl groups, can ionize H^+ or surface base ions such as Na^+, K^+, Ca^{2+}, Mg^{2+}, etc., to exchange with heavy metal ions or cationic organic pollutants [80].

3.2. Physical Adsorption

Physical adsorption means that biochar utilizes its surface characteristics, namely porosity and large specific surface area, so that pollutants such as heavy metals or organic substances could be adsorbed on its surface or diffused into the micropores. The diameter of the heavy metal ions is smaller than the average pore diameter of the biochar. Generally, the smaller the diameter of the heavy metal, the more the pores penetrate into the pores of the biochar, thereby increasing the adsorption capacity [81,82]. The intensity of physical adsorption is closely related to the properties and specific surface area of biochar, the properties and concentration of pollutants, and the temperature during adsorption process. Physical adsorption kinetics is usually fitted by pseudo-first-order and pseudo-second-order kinetic models [83,84]. Physical adsorption can be either single-layer adsorption or multi-layer adsorption, which is usually fitted by Langmuir and Freundlich model [85,86].

3.3. Electrostatic Interaction

Electrostatic interaction refers to the electrostatic adsorption between the surface charge of biochar and heavy metal ions. When the pH value of solution is greater than the charge point of biochar (pH_{pzc}), the negative charge on the surface of biochar and the heavy metal with positive charge causes electrostatic adsorption. Heavy metal ions with positive charge on the surface of biochar combine with oxygen-containing functional groups such as carboxyl, carbonyl, and hydroxyl [87–91].

3.4. Precipitation

Mineral components in biochar, such as CO_3^{2-}, PO_4^{3-}, SiO_3^{4-}, Cl^-, SO_4^{2-}, SO_3^{2-}, and OH^-, combine with heavy metal ions to form water insoluble substances such as metal oxides, metal phosphates, and metal carbonates, which promote the adsorption and immobilization of heavy metals. Xu et al. [92] believed that the adsorption of Cu, Zn, and Cd by fertilizer biochar was mainly attributed to the precipitation of CO_3^{2-} and PO_4^{3-}, while the electron surface complexation via -OH groups or delocalized π was less.

3.5. Complexation

Complexation refers to the interaction between oxygen-containing functional groups on the surface of biochar and heavy metals to form complexes, which could be fixed. Qian et al. [93] studied the aluminum phytotoxicity of cow manure biochar to wheat and concluded that the adsorption of aluminum by biochar was mainly through the complexation of carboxyl group with $[Al(OH)]^{2+}$ and its monomer surface, rather than through the electrostatic attraction of Al^{3+} with negative charge sites. Jia et al. [94] believed that the adsorption of oxytetracycline by biochar was mainly mediated by π–π interaction and metal bridge, with surface complexation as the main factor, and cationic exchange might exist.

In the process of adsorption, it is often not a single mechanism, but a combination of multiple adsorption mechanisms. Table 3 summarizes the adsorption mechanism of biochar for pollution restoration.

Table 3. Adsorption mechanism of biochar for pollution remediation.

Raw Material	Pollutant	Mechanism	Reference
Municipal sewage sludge	Cd	Surface precipitation under alkaline conditions and exchange of exchangeable cations with Cd.	[95]
Fertilizer	Cu, Zn and Cd	Precipitate from CO_3^{2-}, PO_4^{3-} on the surface of the biochar, partially by surface complexing with -OH group or delocalized π electron.	[92]
Rice husk loaded with manganese oxide	Pb	Oxide spherical complexes and biochar surface oxygen complexes; the π-band electron density of graphene-based carbon in the π-electron cloud system reduces vacancies on the surface of biochar, thereby adsorbing Pb^{2+}.	[96]
Wheat straw, pine needles	Zn	The components of -OH, CO_3^{2-}, and Si in biochar can form precipitates with Zn^{2+}.	[97]
Bamboo, eucalyptus	chloramphenicol	Electron-donor-acceptor (EDA) interaction with pH < 2.0, also forms charge-assisted hydrogen bonds (CAHB) and hydrogen bonds at pH 4.0–4.5, and interaction with CAHB and EDA at pH > 7.0.	[98]
Corn straw	Hg(II), atrazine	After Na_2S modification, sulfur impregnated onto the biochar reacted with Hg(II) to form HgS, which greatly facilitated the sorption of Hg(II). Formation of surface complexes between Hg(II) and the functional groups of sorbent, such as phenolic hydroxyl, carboxylic groups. These oxygen-containing functional groups exchanged ion with Hg(II). The electrostatic and EDA interaction also participated in Hg(II) sorption.	[51]
Dairy manure	Pb and Cd	Because of the easy hydrolysis of Pb at low pH, biochar has a higher affinity for Pb than Cd. Besides, precipitation as carbonate minerals ($2PbCO_3·Pb(OH)_2$ and $CdCO_3$) and complexation with functional groups such as carboxyl and hydroxyl, were also important for adsorption of Pb and Cd by biochar.	[79]
Rice straw, swine manure	Tetracycline (TC)	The H-bonding, electrostatic attraction and EDA interaction might be the primary mechanism during adsorption process.	[48]
Sugar beet tailing (SBT)	Cr(VI)	First, SBT biochar reduced Cr(VI) to Cr(III) by electrostatic adsorption. Second, with the participation of hydrogen ions and the electron donors from SBT biochar, Cr(VI) was reduced to Cr(III). Then, the function groups on the SBT biochar complexed with Cr(III).	[99]
Empty fruit bunch, rice husk	As(III), As(V)	Surface complexes were formed between As(III) and As(V) and the functional groups (hydroxyl, carboxyl, and C–O ester of alcohols) of the two biochars.	[100,101]
Bamboo biomass	Sulfathiazole, sulfamethoxazole, sulfamethazine	The sorption of neutral sulfonamide species occurred mainly due to H-bonds followed by EDA, and by Lewis acid-base interaction. EDA was the main mechanism for the sorption of positive sulfonamides species. The sorption of negative species was mainly due to proton exchange with water forming negative CAHB, followed by the neutralization of -OH groups by H^+ released from functionalized biochar surface, and π–π electron-acceptor–acceptor (EAA) interaction.	[102]

233

4. Application of Biochar in Soil Remediation

4.1. Removal of Heavy Metals

The removal of heavy metals by biochar is mainly reflected in two aspects: One is the adsorption of heavy metals in the pores of biochar to reduce the residual amount in the soil; the other is the ion exchange or redox reaction between the effective components in biochar and heavy metal ions to stabilize the formation of heavy metal precipitates or to reduce toxicity by transforming them into low-valent states.

Boostani et al. [103] investigated the effect of sheep and earthworm manure biochars on Pb immobilization in a contaminated calcareous soil. The addition of biochars resulted in a significant increase in the Pb content in the residual state, which reduced the Pb activity in the soil. Chen et al. [79] studied the adsorption mechanisms for removal Pb and Cd with dairy manure biochar. The extractable Pb and Cd contents decreased significantly and were converted to the precipitation as carbonate minerals. However, it may also be due to the lack of selective adsorption capability of biochar, which adsorbs nitrogen in the soil, resulting in a decrease in soil nutrients [104]. When the soil pollution is contaminated by complex heavy metals, although biochar reduces the concentration of extractable heavy metals, biochar has different adsorption effects on different heavy metals due to competitive adsorption. Yang et al. [105] showed that straw and bamboo biochar are more effective than Zn in reducing extractable Cu and Pb. Zhou et al. [106] also reached a similar conclusion. In the single metal adsorption test, the adsorption capacity of sludge biochar to Zn was the largest, while in the polymetallic adsorption test, the adsorption capacity of Mn, Cu, and Zn decreased, but the adsorption capacity of Cr increased. Table 4 shows the research on removing heavy metals in soil by using biochar in the past two years.

Table 4. Study on the application of biochar to the remediation of heavy metal pollution in soil.

Raw Material	Tested Soil	Pollutant	Remediation Effect	Reference
Bamboo, rice straw, and Chinese walnut shell	industrial contaminated soil	Cu	Cu uptake in roots was reduced by 15%, 35%, and 26%, respectively. Rice straw biochar reduced solubility of Cu and Pb.	[107]
Sewage sludge	Brazil soil	Cd, Pb, and Zn	Biochar reduced the concentration and bioavailable levels of Cd, Pb, and Zn of in the leachates.	[108]
Poultry litter	paddy soil near Zn and Pb mines	Cd, Cu, Zn, Pb	Acid-soluble Cd in soils amended with poultry litter biochar was 8% to 10% lower than in the control polluted soil.	[109]
Wheat straw	acid soil	Cd and Cu	Cu concentration in wheat roots was reduced most efficiently to 40.9% by biochar. Available Cd and Cu in soil added biochar decreased 18.8% and 18.6%.	[110]
Rice husk	saturated soil, dryland soil	Cd	The adsorption of Cd on saturated soil increased by 21–41%, and that on dryland soil increased by 38–54%.	[111]
Gliricidia sepium	shooting range soil	Pb, Cu	The addition of biochar to the soil reduced the dissolution rates of Pb and Cu by 10.0–99.5% and 15.6–99.5%, respectively, and was able to fix Pb and Cu released by protons and ligands in the soil.	[112]
Poultry manure, cow manure, and sheep manure	farmland soil	Cr(VI)	Poultry manure decreased 61.54 mg·kg^{-1} Cr(VI) in acidic soil and 73.93 mg·kg^{-1} Cr(VI) in alkaline soil. Cow and Sheep manure decreased by 66.61, 58.67, and 57.81, 68.15 mg·kg^{-1} Cr(VI) in acidic and alkaline soil, respectively.	[113]

In order to achieve better remediation effect, scholars gradually carry out research on the modification of biochar. Modification refers to the activation of the original biochar through physical and chemical methods, so as to achieve the desired purpose. The modification methods of surface structure characteristics are generally divided into physical method, chemical method, and combined method [65]. In the early stage, Monser et al. [114] modified activated carbon with sodium dodecyl sulfonate to reduce the heavy metal content in phosphoric acid and reduce the content of cadmium and chromium. Scholars have modified biochar similarly to activated carbon, mainly by chemical modification, through adding acid, alkali, oxidants, and supporting various metal oxides to aminated, acidify, and alkalinize biochar. Oxidation, etc., increase the surface oxygen-containing functional groups, thereby achieving a good repair effect. Studies on the adsorption effect of modified biochar are dominated by heavy metals, followed by organics, and most of them are adsorption of heavy metals in aqueous solution. Table 5 provides a summary of studies on soil pollution remediation by various types of modified biochar in recent years.

Table 5. The remediation of soil pollution by various types of modified biochar.

Raw Material	Modification	Pollutant	Tested Soil	Remediation Effect	Reference
Bamboo hardwoods	sulfur-iron	Cr	plant farmland	Sulfur-modified biochar (S-BC) and sulfur-iron modified biochar (SF-BC) addition increased the content of soil organic matter, alpha diversity indices, and changed soil bacterial community structure. The exchangeable Cd in soil was decreased by 12.54%, 29.71%, 18.53% under the treatments of BC, S-BC, SF-BC, respectively.	[76]
Poultry, cow, sheep manure	Chitosan, ZVI	Cr	uncontaminated surface soil	Modified sheep manure biochar reduced Cr(VI) by 55%, and poultry manure modified biochar reduced Cr(VI) by 48%.	[113]
corn straw	Fe-Mn	As	paddy soil	Modified biochar decreased the content of available As, increased the residual, amorphous hydrous oxide-bound, and crystalline hydrous oxide-bound As forms.	[115]
Eucalyptus wood and poultry litter	iron	Cd, Cu, Zn, Pb	paddy soil near Zn and Pb mines	Acid-soluble Cd, Zn, Cu in soils amended with poultry litter biochar (PLB) was 8% to 10%, 27% to 29%, 59% to 63%, respectively, lower than in the control polluted soil. Plant biomass increased by 32% in the treatments containing magnetic PLB.	[109]
Coconut shell	HCl + ultrasonication	Cd, Ni, and Zn	topsoil of paddy fields	In groups with 5% MCSB addition, the acid soluble Cd, Ni and Zn decreased by 30.1%, 57.2%, and 12.7%, respectively.	[78]
Rice husk	Sulfur	Hg	Hg contaminated soil	Modification increased the Hg^{2+} adsorptive capacity of biochar by 73%, to 67.11 $mg \cdot g^{-1}$. And freely available Hg in TCLP (toxicity characterization leaching procedure) leachates by 95.4%, 97.4%, and 99.3%, respectively, compared to untreated soil.	[116]
Corn straw	MnO	As	red soil	Modified biochar (MBC) in red soil had a much greater sorption capacity for As(III) than pristine biochar, although both enhanced the sorption of As(III) in red soil.	[117]

4.2. Removal of Persistent Organic Pollutants (POPs)

The persistent organochlorine pesticides in farmland soils in are still seriously polluted, and the polycyclic aromatic hydrocarbon pollution caused by sewage irrigation cannot be ignored. Biochar has a strong adsorption capacity for organic pollutants, and the process can be understood as the accumulation and collection of organic pollutants on biochar. Table 6 shows the research on the removal of POPs from soil by using biochar in the past two years.

Table 6. Study on the application of biochar to remove persistent organic pollutants (POPs) in soil.

Raw Material	Tested Soil	Pollutants	Remediation Effect	Reference
Fir wood chips	rice soil	2,4-dichlorophenol, phenanthrene	Reduced the degradation and mineralization of both pollutants. Increased the accumulation of their metabolites in soil.	[118]
Mixed wood shavings Rice husk	loamy agricultural soil	Pyrene, polychlorinated biphenyl and dichlorodiphenyldichloroethylene (DDE)	At the biochar dose of 10%, bioavailability and accessibility by 37% and 41%, respectively, compared to unamended soil.	[119]
Rice hull	loamy clay, sandy loam, clay loam	oxyfluorfen	Oxyfluorfen degraded faster in biochar amended soil than in unamended soil. Biochar decreased the oxyfluorfen uptake by soybean plants by 18–63%, and the adsorption capacity of oxyfluorfen by soybean decreased.	[120]
Orchard pruning biomass	vineyard	PAHs	During the investigated period, PAH concentrations decreased with time and the change resulted more intense for light PAHs. The soil properties (TOC, pH, CEC, bulk density) were modified after two consecutive applications	[121]
Corn straw and bamboo	soil contaminated with PAHs	PAHs	The bioaccumulation of PAHs in rice roots was reduced, especially high molecular weight PAHs. The total and bioavailable concentration of PAHs in the soil treated with corn straw biochar were both lower than that of the control group.	[100]

Koltowski et al. [122] studied the removal effect of PAHs in soil by microwave, CO_2, and H_2O activation of willow biochar. The results showed that the biochar samples with the best effect after activation reduced the concentration of PAHs dissolved in the coal plant soil (near cooking plant battery) and bitumen plant soil from 153 to 22 ng/L and 174 to 24 ng/L, respectively, and the PAHs concentration decreased by 86%. The concentration of PAHs dissolved in the asphalt soil (from an industrial waste deposit) decreased from 52 to 16 ng/L, and bioacceptable PAHs reduced to almost zero. Zhang et al. [123] applied biochar from corn straw and pig manure to black soil containing thiacloprid, and explored the adsorption and degradation process of thiacloprid. The results showed that the biochar changed the microbial community of soil by changing the physicochemical properties of the soil, thus promoting the biodegradation of thiacloprid.

In general, biochar can enhance the adsorption capacity of soil for organic pollutants, reduce their activities of desorption and flow in the soil, and bioavailability in soil pore water, provide essential nutrients to improve soil microbial activity, and improve soil physical and chemical properties, etc. [124].

4.3. Amelioration of Soil

The improvement of soil by biochar is mainly reflected in the improvement of soil organic matter content, the increase of nitrogen, potassium, and other nutrients contents and utilization rate, and the improvement of soil erosion and acid soil.

The application of biochar can significantly increase the content of soil organic matter, alkali-hydrolyzed nitrogen, ammonium nitrogen, and available potassium, but the more biochar added is not better. Excessive application of biochar can inhibit the content of nutrients. Bayabil et al. [125] mixed acacia, croton, and eucalyptus charcoal into the soil in a basin of the Ethiopian plateau, and found through laboratory and field experiments that it had a good improvement on the water conservancy characteristics of degraded soil, so as to reduce runoff and erosion. Biochar is mostly alkaline, which can improve the utilization and absorption of nutrients in rice by increasing the pH value of acid soil [126]. In addition, the effect of biochar on soil cation exchange capacity was significant. Agegnehu et al. [127] found that biochar, compost, and their compounds significantly improved the availability and use of plant nutrients: Soil organic carbon, moisture content, CEC, and peanut yield all increased, and greenhouse gas emissions decreased.

Nitrogen is an essential nutrient for plant growth. The application of nitrogen fertilizer could replenish soil nitrogen and maintain land productivity. However, over-application will cause a large loss of soil nitrogen, reduce the efficiency and utilization of nitrogen fertilizer, and aggravate the eutrophication pollution degree of surrounding water environment such as rivers and lakes. The inhibition of nitrogen and phosphorus leaching by biochar is considered as follows: Biochar changed the microbial-mediated reactions in soil nitrogen and phosphorus cycles, namely N_2 fixation, nitrogen and phosphorus mineralization, nitrification, ammonia volatilization, and denitrification. At the same time, biochar provided a reactive surface in which nitrogen and phosphorus ions remain in the soil microbial biomass and exchange sites, both of which regulate crop nitrogen and phosphorus availability [128].

4.4. Potential Risk of Biochar

Although biochar has great advantages in remediation of soil pollution, improvement of soil quality, increase of crop yield, and reduction of greenhouse gas emissions, these studies are all short-term and the long-term effects of biochar on soil are still ambiguous. Therefore, in order to make better use of biochar and reduce its possible risks, the long-term effects and risk assessment of biochar on soil should be paid more attention. Studies have shown that, although the application of biochar improved soil quality and crop yield, biochar reduced the efficacy of herbicide and increased weed growth by 200% [129]. The reduction of herbicide efficacy must increase the use of herbicide, which may increase the residual concentration of herbicide in the soil and cause more serious pollution to the soil. In addition, biochar, which is mostly prepared from crop waste, may contain heavy metals on its own and could release pollutants if it gets into the soil. Due to the weathering and aging, biochar would undergo physical, chemical, and biological degradation. Finally, it would form colloids, nanoparticles, and smaller fragments that alter the microbial community in the soil. However, the interaction between these components of biochar and soil, the internal mechanism of microbial transformation and geochemical circulation still need to be further studied [130].

5. Conclusions

In this paper, the effects of preparation, process parameters, and modification of biochar on its physicochemical properties were reviewed. The mechanism of biochar remediation for soil pollution was summarized, the application status of biochar in soil remediation was analyzed, and the research articles on the removal of heavy metals and organic pollutants by biochar in the past two years were listed; lastly, the possible risks in the application of biochar were proposed. The application of biochar in soil remediation can not only reduce the damage of soil wastes to the atmosphere and water environment, but also remove the pollutants in the soil and improve the soil quality. In addition, biochar has advantages in dealing with water pollution and reducing greenhouse gas emissions, so the research on the application of biochar is of great significance to sustainable development.

At present, the following problems still exist in the application of biochar: (i) Although studies are on the same type of biochar to repair the same kind of pollution, the mechanism of action, adsorption kinetics, thermodynamics, etc., are different; (ii) in terms of the characterization of biochar, there is no unified standard, which is difficult to compare; (iii) the number of indoor tests is much more than of field outdoor tests, which results in the incomplete considered factors and difficult practical application; (iv) the research on the mechanism of biochar on compound pollution is not thorough enough; (v) the study on the long-term effects and negative effects of biochar is not well studied; and (vi) there is little research about life cycle assessment of biochar and the overall economic value of biochar applications is not clearly enough.

Author Contributions: Conceptualization, X.Y. and S.Z.; methodology, L.L.; writing—original draft preparation, X.Y.; writing—review and editing, X.Y. and S.Z.; supervision, M.J., L.L.

Funding: This research was funded by National Natural Science Foundation of China, grant number 51708301; 2017 Science and Technology Demonstration Project of Industrial Integration and Development, Tianjin, China, grant number 17ZXYENC00100; Young Elite Scientists Sponsorship Program by Tianjin, grant number TJSQNTJ-2018-06; Natural Science Foundation of Tianjin, China, grant number 17JCZDJC39500.

Acknowledgments: This research was also supported by the Fundamental Research Funds for the Central Universities. The authors appreciate the financial support and thank the editor and reviewers for their very useful suggestions and comments.

Conflicts of Interest: The authors declare no conflict of interest.

References

1. Li, Z.; Ma, Z.; Van Der Kuijp, T.J.; Yuan, Z.; Huang, L. A review of soil heavy metal pollution from mines in China: Pollution and health risk assessment. *Sci. Total Environ.* **2014**, *468*, 843–853. [CrossRef]
2. Renella, G.; Landi, L.; Nannipieri, P. Degradation of low molecular weight organic acids complexed with heavy metals in soil. *Geoderma* **2004**, *122*, 311–315. [CrossRef]
3. Yang, H.; Huang, X.; Thompson, J.R.; Flower, R.J. Soil Pollution: Urban Brownfields. *Science* **2014**, *344*, 691–692. [CrossRef]
4. Tang, X.; Shen, C.; Shi, D.; Cheema, S.A.; Khan, M.I.; Zhang, C.; Chen, Y. Heavy metal and persistent organic compound contamination in soil from Wenling: An emerging e-waste recycling city in Taizhou area, China. *J. Hazard. Mater.* **2010**, *173*, 653–660. [CrossRef]
5. Yang, K.; Zhu, L.; Xing, B. Enhanced Soil Washing of Phenanthrene by Mixed Solutions of TX100 and SDBS. *Environ. Sci. Technol.* **2006**, *40*, 4274–4280. [CrossRef] [PubMed]
6. Ren, L.; Lu, H.; He, L.; Zhang, Y. Enhanced electrokinetic technologies with oxidization–reduction for organically-contaminated soil remediation. *Chem. Eng. J.* **2014**, *247*, 111–124. [CrossRef]

7. Meers, E.; Tack, F.M.G.; Verloo, M. Degradability of ethylenediaminedisuccinic acid (EDDS) in metal contaminated soils: Implications for its use soil remediation. *Chemosphere* **2008**, *70*, 358–363. [CrossRef] [PubMed]
8. Lehmann, J. A handful of carbon. *Nature* **2007**, *447*, 143–144. [CrossRef] [PubMed]
9. Ahmad, M.; Rajapaksha, A.U.; Lim, J.E.; Zhang, M.; Bolan, N.; Mohan, D.; Vithanage, M.; Lee, S.S.; Ok, Y.S. Biochar as a sorbent for contaminant management in soil and water: A review. *Chemosphere* **2014**, *99*, 19–33. [CrossRef] [PubMed]
10. Godlewska, P.; Schmidt, H.P.; Ok, Y.S.; Oleszczuk, P. Biochar for composting improvement and contaminants reduction. A review. *Bioresour. Technol.* **2017**, *246*, 193–202. [CrossRef]
11. Cha, J.S.; Park, S.H.; Jung, S.-C.; Ryu, C.; Jeon, J.-K.; Shin, M.-C.; Park, Y.-K. Production and utilization of biochar: A review. *J. Ind. Eng. Chem.* **2016**, *40*, 1–15. [CrossRef]
12. Zwieten, L.V.; Kimber, S.; Morris, S.; Chan, K.Y.; Downie, A.; Rust, J.; Joseph, S.; Cowie, A. Effects of biochar from slow pyrolysis of papermill waste on agronomic performance and soil fertility. *Plant Soil* **2010**, *327*, 235–246. [CrossRef]
13. Liu, Z.; Quek, A.; Hoekman, S.K.; Balasubramanian, R. Production of solid biochar fuel from waste biomass by hydrothermal carbonization. *Fuel* **2013**, *103*, 943–949. [CrossRef]
14. Yu, F.; Deng, S.; Chen, P.; Liu, Y.; Wan, Y.; Olson, A.; Kittelson, D.; Ruan, R. Physical and Chemical Properties of Bio-Oils from Microwave Pyrolysis of Corn Stover. *Appl. Biochem. Biotecnol.* **2007**, *137*, 957–970.
15. Sabio, E.; Álvarez-Murillo, A.; Román, S.; Ledesma, B. Conversion of tomato-peel waste into solid fuel by hydrothermal carbonization: Influence of the processing variables. *Waste Manag.* **2016**, *47*, 122–132. [CrossRef] [PubMed]
16. Afolabi, O.O.; Sohail, M.; Thomas, C. Characterization of solid fuel chars recovered from microwave hydrothermal carbonization of human biowaste. *Energy* **2017**, *134*, 74–89. [CrossRef]
17. Liu, J.; Zhang, C.; Guo, S.; Xu, L.; Xiao, S.; Shen, Z. Microwave treatment of pre-oxidized fibers for improving their structure and mechanical properties. *Ceram. Int.* **2019**, *45*, 1379–1384. [CrossRef]
18. Laird, D.A.; Brown, R.C.; Amonette, J.E.; Lehmann, J. Review of the pyrolysis platform for coproducing bio-oil and biochar. *Biofuels Bioprod. Biorefining* **2009**, *3*, 547–562. [CrossRef]
19. Tan, Z.; Lin, C.S.; Ji, X.; Rainey, T.J. Returning biochar to fields: A review. *Appl. Soil Ecol.* **2017**, *116*, 1–11. [CrossRef]
20. Williams, P.T.; Besler, S. The pyrolysis of rice husks in a thermogravimetric analyser and static batch reactor. *Fuel* **1993**, *72*, 151–159. [CrossRef]
21. Crombie, K.; Mašek, O.; Sohi, S.P.; Brownsort, P.; Cross, A. The effect of pyrolysis conditions on biochar stability as determined by three methods. *Gcb Bioenergy* **2013**, *5*, 122–131. [CrossRef]
22. Enders, A.; Hanley, K.; Whitman, T.; Joseph, S.; Lehmann, J. Characterization of biochars to evaluate recalcitrance and agronomic performance. *Bioresour. Technol.* **2012**, *114*, 644–653. [CrossRef]
23. Yuan, J.-H.; Xu, R.-K.; Zhang, H. The forms of alkalis in the biochar produced from crop residues at different temperatures. *Bioresour. Technol.* **2011**, *102*, 3488–3497. [CrossRef] [PubMed]
24. Park, J.H.; Ok, Y.S.; Kim, S.H.; Cho, J.S.; Heo, J.S.; Delaune, R.D.; Seo, D.C. Evaluation of phosphorus adsorption capacity of sesame straw biochar on aqueous solution: Influence of activation methods and pyrolysis temperatures. *Environ. Geochem. Health* **2015**, *37*, 969–983. [CrossRef] [PubMed]
25. Meyer, S.; Glaser, B.; Quicker, P. Technical, Economical, and Climate-Related Aspects of Biochar Production Technologies: A Literature Review. *Environ. Sci. Technol.* **2011**, *45*, 9473–9483. [CrossRef] [PubMed]
26. Bruun, E.W.; Ambus, P.; Egsgaard, H.; Hauggaard-Nielsen, H. Effects of slow and fast pyrolysis biochar on soil C and N turnover dynamics. *Soil Boil. Biochem.* **2012**, *46*, 73–79. [CrossRef]
27. Mohan, D.; Pittman, C.U.; Steele, P.H. Pyrolysis of Wood/Biomass for Bio-oil: A Critical Review. *Energy Fuel* **2006**, *20*, 848–889. [CrossRef]
28. Bech, N.; Larsen, M.B.; Jensen, P.A.; Dam-Johansen, K. Modelling solid-convective flash pyrolysis of straw and wood in the Pyrolysis Centrifuge Reactor. *Biomass Bioenergy* **2009**, *33*, 999–1011. [CrossRef]
29. Chen, B.; Chen, Z.; Lv, S. A novel magnetic biochar efficiently sorbs organic pollutants and phosphate. *Bioresour. Technol.* **2011**, *102*, 716–723. [CrossRef] [PubMed]
30. Lu, G.Q.; Low, J.C.F.; Liu, C.Y.; Lua, A.C. Surface area development sludge during pyrolysis of sewage. *Fuel* **1995**, *74*, 344–348. [CrossRef]

31. Tay, J.; Chen, X.; Jeyaseelan, S.; Graham, N. Optimising the preparation of activated carbon from digested sewage sludge and coconut husk. *Chemosphere* **2001**, *44*, 45–51. [CrossRef]
32. Jimenez-Cordero, D.; Heras, F.; Alonso-Morales, N.; Gilarranz, M.A.; Rodriguez, J.J. Ozone as oxidation agent in cyclic activation of biochar. *Fuel Process. Technol.* **2015**, *139*, 42–48. [CrossRef]
33. Chen, D.; Zheng, Z.; Fu, K.; Zeng, Z.; Wang, J.; Lu, M. Torrefaction of biomass stalk and its effect on the yield and quality of pyrolysis products. *Fuel* **2015**, *159*, 27–32. [CrossRef]
34. Mochidzuki, K.; Soutric, F.; Tadokoro, K.; Antal, M.J.; Toth, M.; Zelei, B.; Várhegyi, G. Electrical and Physical Properties of Carbonized Charcoals. *Ind. Eng. Chem. Res.* **2003**, *42*, 5140–5151. [CrossRef]
35. A Metz, L.; Meruva, N.K.; Morgan, S.L.; Goode, S.R. UV laser pyrolysis fast gas chromatography/time-of-flight mass spectrometry for rapid characterization of synthetic polymers: Optimization of instrumental parameters. *J. Anal. Appl.* **2004**, *71*, 327–341. [CrossRef]
36. Tang, L.; Huang, H. Plasma Pyrolysis of Biomass for Production of Syngas and Carbon Adsorbent. *Energy Fuels* **2005**, *19*, 1174–1178. [CrossRef]
37. Yaman, S. Pyrolysis of Biomass to Produce Fuels and Chemical Feedstocks. *ChemInform* **2004**, *35*, 651–671. [CrossRef]
38. Lian, F.; Sun, B.; Song, Z.; Zhu, L.; Qi, X.; Xing, B. Physicochemical properties of herb-residue biochar and its sorption to ionizable antibiotic sulfamethoxazole. *Chem. Eng. J.* **2014**, *248*, 128–134. [CrossRef]
39. Song, Z.; Lian, F.; Yu, Z.; Zhu, L.; Xing, B.; Qiu, W. Synthesis and characterization of a novel MnOx-loaded biochar and its adsorption properties for Cu^{2+} in aqueous solution. *Chem. Eng. J.* **2014**, *242*, 36–42. [CrossRef]
40. Van Vinh, N.; Zafar, M.; Behera, S.K.; Park, H.S. Arsenic(III) removal from aqueous solution by raw and zinc-loaded pine cone biochar: Equilibrium, kinetics, and thermodynamics studies. *Int. J. Environ. Sci. Technol.* **2015**, *12*, 1283–1294. [CrossRef]
41. Liu, P.; Liu, W.-J.; Jiang, H.; Chen, J.-J.; Li, W.-W.; Yu, H.-Q. Modification of bio-char derived from fast pyrolysis of biomass and its application in removal of tetracycline from aqueous solution. *Bioresour. Technol.* **2012**, *121*, 235–240. [CrossRef] [PubMed]
42. Sun, Y.; Gao, B.; Yao, Y.; Fang, J.; Zhang, M.; Zhou, Y.; Chen, H.; Yang, L. Effects of feedstock type, production method, and pyrolysis temperature on biochar and hydrochar properties. *Chem. Eng. J.* **2014**, *240*, 574–578. [CrossRef]
43. Yin, Q.; Ren, H.; Wang, R.; Zhao, Z. Evaluation of nitrate and phosphate adsorption on Al-modified biochar: Influence of Al content. *Sci. Total Environ.* **2018**, *631*, 895–903. [CrossRef] [PubMed]
44. Rajapaksha, A.U.; Vithanage, M.; Ahmad, M.; Seo, D.-C.; Cho, J.-S.; Lee, S.-E.; Lee, S.S.; Ok, Y.S. Enhanced sulfamethazine removal by steam-activated invasive plant-derived biochar. *J. Hazard. Mater.* **2015**, *290*, 43–50. [CrossRef] [PubMed]
45. Wang, S.; Gao, B.; Li, Y.; Mosa, A.; Zimmerman, A.R.; Ma, L.Q.; Harris, W.G.; Migliaccio, K.W. Manganese oxide-modified biochars: Preparation, characterization, and sorption of arsenate and lead. *Bioresour. Technol.* **2015**, *181*, 13–17. [CrossRef] [PubMed]
46. Jung, K.-W.; Hwang, M.-J.; Jeong, T.-U.; Ahn, K.-H. A novel approach for preparation of modified-biochar derived from marine macroalgae: Dual purpose electro-modification for improvement of surface area and metal impregnation. *Bioresour. Technol.* **2015**, *191*, 342–345. [CrossRef] [PubMed]
47. Jin, H.; Capareda, S.; Chang, Z.; Gao, J.; Xu, Y.; Zhang, J. Biochar pyrolytically produced from municipal solid wastes for aqueous As(V) removal: Adsorption property and its improvement with KOH activation. *Bioresour. Technol.* **2014**, *169*, 622–629. [CrossRef]
48. Chen, T.; Luo, L.; Deng, S.; Shi, G.; Zhang, S.; Zhang, Y.; Deng, O.; Wang, L.; Zhang, J.; Wei, L. Sorption of tetracycline on H_3PO_4 modified biochar derived from rice straw and swine manure. *Bioresour. Technol.* **2018**, *267*, 431–437. [CrossRef]
49. Li, Y.; Wei, Y.; Huang, S.; Liu, X.; Jin, Z.; Zhang, M.; Qu, J.; Jin, Y. Biosorption of Cr(VI) onto Auricularia auricula dreg biochar modified by cationic surfactant: Characteristics and mechanism. *J. Mol. Liq.* **2018**, *269*, 824–832. [CrossRef]
50. Tao, Q.; Li, B.; Li, Q.; Han, X.; Jiang, Y.; Jupa, R.; Wang, C.; Li, T. Simultaneous remediation of sediments contaminated with sulfamethoxazole and cadmium using magnesium-modified biochar derived from Thalia dealbata. *Sci. Total Environ.* **2019**, *659*, 1448–1456. [CrossRef]

51. Tan, G.; Sun, W.; Xu, Y.; Wang, H.; Xu, N. Sorption of mercury (II) and atrazine by biochar, modified biochars and biochar based activated carbon in aqueous solution. *Bioresour. Technol.* **2016**, *211*, 727–735. [CrossRef] [PubMed]
52. Kim, K.H.; Kim, J.-Y.; Cho, T.-S.; Choi, J.W. Influence of pyrolysis temperature on physicochemical properties of biochar obtained from the fast pyrolysis of pitch pine (*Pinus rigida*). *Bioresour. Technol.* **2012**, *118*, 158–162. [CrossRef]
53. Sun, B.; Lian, F.; Bao, Q.; Liu, Z.; Song, Z.; Zhu, L. Impact of low molecular weight organic acids (LMWOAs) on biochar micropores and sorption properties for sulfamethoxazole. *Environ. Pollut.* **2016**, *214*, 142–148. [CrossRef] [PubMed]
54. Yao, Y.; Gao, B.; Inyang, M.; Zimmerman, A.R.; Cao, X.; Pullammanappallil, P.; Yang, L. Biochar derived from anaerobically digested sugar beet tailings: Characterization and phosphate removal potential. *Bioresour. Technol.* **2011**, *102*, 6273–6278. [CrossRef]
55. Rajapaksha, A.U.; Vithanage, M.; Zhang, M.; Ahmad, M.; Mohan, D.; Chang, S.X.; Ok, Y.S. Pyrolysis condition affected sulfamethazine sorption by tea waste biochars. *Bioresour. Technol.* **2014**, *166*, 303–308. [CrossRef] [PubMed]
56. Dong, H.; Deng, J.; Xie, Y.; Zhang, C.; Jiang, Z.; Cheng, Y.; Hou, K.; Zeng, G. Stabilization of nanoscale zero-valent iron (nZVI) with modified biochar for Cr(VI) removal from aqueous solution. *J. Hazard. Mater.* **2017**, *332*, 79–86. [CrossRef]
57. Kołodyńska, D.; Krukowska, J.; Thomas, P. Comparison of sorption and desorption studies of heavy metal ions from biochar and commercial active carbon. *Chem. Eng. J.* **2017**, *307*, 353–363. [CrossRef]
58. Fu, D.; Chen, Z.; Xia, D.; Shen, L.; Wang, Y.; Li, Q. A novel solid digestate-derived biochar-Cu NP composite activating H_2O_2 system for simultaneous adsorption and degradation of tetracycline. *Environ. Pollut.* **2017**, *221*, 301–310. [CrossRef]
59. Zuo, X.J.; Liu, Z.; Chen, M.D. Effect of H_2O_2 concentrations on copper removal using the modified hydrothermal biochar. *Bioresour. Technol.* **2016**, *207*, 262–267. [CrossRef]
60. Zhao, L.; Zheng, W.; Mašek, O.; Chen, X.; Gu, B.; Sharma, B.K.; Cao, X. Roles of Phosphoric Acid in Biochar Formation: Synchronously Improving Carbon Retention and Sorption Capacity. *J. Environ. Qual.* **2017**, *46*, 393. [CrossRef]
61. Ho, P.H.; Lee, S.-Y.; Lee, D.; Woo, H.-C. Selective adsorption of tert-butylmercaptan and tetrahydrothiophene on modified activated carbons for fuel processing in fuel cell applications. *Int. J. Hydrogen Energy* **2014**, *39*, 6737–6745. [CrossRef]
62. Zhang, Q.-P.; Liu, Q.-C.; Li, B.; Yang, L.; Wang, C.-Q.; Li, Y.-D.; Xiao, R. Adsorption of Cd(II) from aqueous solutions by rape straw biochar derived from different modification processes. *Chemosphere* **2017**, *175*, 332–340. [CrossRef]
63. Wang, M.; Wang, J.J.; Wang, X. Effect of KOH-enhanced biochar on increasing soil plant-available silicon. *Geoderma* **2018**, *321*, 22–31. [CrossRef]
64. Li, L.; Liu, S.; Liu, J. Surface modification of coconut shell based activated carbon for the improvement of hydrophobic VOC removal. *J. Hazard. Mater.* **2011**, *192*, 683–690. [CrossRef]
65. Sizmur, T.; Fresno, T.; Akgül, G.; Frost, H.; Jiménez, E.M. Biochar modification to enhance sorption of inorganics from water. *Bioresour. Technol.* **2017**, *246*, 34–47. [CrossRef]
66. Pouretedal, H.; Sadegh, N. Effective removal of Amoxicillin, Cephalexin, Tetracycline and Penicillin G from aqueous solutions using activated carbon nanoparticles prepared from vine wood. *J. Process. Eng.* **2014**, *1*, 64–73. [CrossRef]
67. Yin, Z.; Liu, Y.; Liu, S.; Jiang, L.; Tan, X.; Zeng, G.; Li, M.; Liu, S.; Tian, S.; Fang, Y. Activated magnetic biochar by one-step synthesis: Enhanced adsorption and coadsorption for 17β-estradiol and copper. *Sci. Total Environ.* **2018**, *639*, 1530–1542. [CrossRef]
68. Li, R.; Wang, J.J.; Zhou, B.; Awasthi, M.K.; Ali, A.; Zhang, Z.; Gaston, L.A.; Lahori, A.H.; Mahar, A. Enhancing phosphate adsorption by Mg/Al layered double hydroxide functionalized biochar with different Mg/Al ratios. *Sci. Total Environ.* **2016**, *559*, 121–129. [CrossRef]
69. Wu, H.; Feng, Q.; Yang, H.; Alam, E.; Gao, B.; Gu, D. Modified biochar supported Ag/Fe nanoparticles used for removal of cephalexin in solution: Characterization, kinetics and mechanisms. *Colloids Surf. A Physicochem. Eng. Asp.* **2017**, *517*, 63–71. [CrossRef]

70. Angın, D.; Altintig, E.; Köse, T.E. Influence of process parameters on the surface and chemical properties of activated carbon obtained from biochar by chemical activation. *Bioresour. Technol.* **2013**, *148*, 542–549. [CrossRef]
71. Lyu, H.; Tang, J.; Huang, Y.; Gai, L.; Zeng, E.Y.; Liber, K.; Gong, Y. Removal of hexavalent chromium from aqueous solutions by a novel biochar supported nanoscale iron sulfide composite. *Chem. Eng. J.* **2017**, *322*, 516–524. [CrossRef]
72. Bird, M.I.; Charville-Mort, P.D.; Ascough, P.L.; Wood, R.; Higham, T.; Apperley, D. Assessment of oxygen plasma ashing as a pre-treatment for radiocarbon dating. *Quat. Geochronol.* **2010**, *5*, 435–442. [CrossRef]
73. Wu, G.-Q.; Zhang, X.; Hui, H.; Yan, J.; Zhang, Q.-S.; Wan, J.-L.; Dai, Y. Adsorptive removal of aniline from aqueous solution by oxygen plasma irradiated bamboo based activated carbon. *Chem. Eng. J.* **2012**, *185*, 201–210. [CrossRef]
74. Wang, S.; Zhou, Y.; Han, S.; Wang, N.; Yin, W.; Yin, X.; Gao, B.; Wang, X.; Wang, J. Carboxymethyl cellulose stabilized ZnO/biochar nanocomposites: Enhanced adsorption and inhibited photocatalytic degradation of methylene blue. *Chemosphere* **2018**, *197*, 20–25. [CrossRef]
75. Tendero, C.; Tixier, C.; Tristant, P.; Desmaison, J.; Leprince, P. Atmospheric pressure plasmas: A review. *Spectrochim. B At. Spectrosc.* **2006**, *61*, 2–30. [CrossRef]
76. Wu, C.; Shi, L.; Xue, S.; Li, W.; Jiang, X.; Rajendran, M.; Qian, Z. Effect of sulfur-iron modified biochar on the available cadmium and bacterial community structure in contaminated soils. *Sci. Total Environ.* **2019**, *647*, 1158–1168. [CrossRef]
77. Jin, J.; Li, S.; Peng, X.; Liu, W.; Zhang, C.; Yang, Y.; Han, L.; Du, Z.; Sun, K.; Wang, X. HNO_3 modified biochars for uranium (VI) removal from aqueous solution. *Bioresour. Technol.* **2018**, *256*, 247–253. [CrossRef]
78. Liu, H.; Xu, F.; Xie, Y.; Wang, C.; Zhang, A.; Li, L.; Xu, H. Effect of modified coconut shell biochar on availability of heavy metals and biochemical characteristics of soil in multiple heavy metals contaminated soil. *Sci. Total Environ.* **2018**, *645*, 702–709. [CrossRef]
79. Chen, Z.-L.; Zhang, J.-Q.; Huang, L.; Yuan, Z.-H.; Li, Z.-J.; Liu, M.-C. Removal of Cd and Pb with biochar made from dairy manure at low temperature. *J. Integr. Agric.* **2019**, *18*, 201–210. [CrossRef]
80. Hassan, M.M.; Carr, C.M. A critical review on recent advancements of the removal of reactive dyes from dyehouse effluent by ion-exchange adsorbents. *Chemosphere* **2018**, *209*, 201–219. [CrossRef]
81. Ko, D.C.; Cheung, C.W.; Choy, K.K.; Porter, J.F.; McKay, G. Sorption equilibria of metal ions on bone char. *Chemosphere* **2004**, *54*, 273–281. [CrossRef]
82. Ngah, W.W.; Hanafiah, M.; Hanafiah, M.A.K.M. Removal of heavy metal ions from wastewater by chemically modified plant wastes as adsorbents: A review. *Bioresour. Technol.* **2008**, *99*, 3935–3948. [CrossRef]
83. Sarı, A.; Tuzen, M. Kinetic and equilibrium studies of biosorption of Pb(II) and Cd(II) from aqueous solution by macrofungus (Amanita rubescens) biomass. *J. Hazard. Mater.* **2009**, *164*, 1004–1011. [CrossRef]
84. Ho, Y.; McKay, G. Pseudo-second order model for sorption processes. *Process. Biochem.* **1999**, *34*, 451–465. [CrossRef]
85. Qiu, Y.; Xiao, X.; Cheng, H.; Zhou, Z.; Sheng, G.D. Influence of Environmental Factors on Pesticide Adsorption by Black Carbon: pH and Model Dissolved Organic Matter. *Environ. Sci. Technol.* **2009**, *43*, 4973–4978. [CrossRef]
86. Chun, Y.; Sheng, G.; Chiou, C.T.; Xing, B. Compositions and Sorptive Properties of Crop Residue-Derived Chars. *Environ. Sci. Technol.* **2004**, *38*, 4649–4655. [CrossRef]
87. Pan, J.; Jiang, J.; Xu, R. Adsorption of Cr(III) from acidic solutions by crop straw derived biochars. *J. Environ. Sci.* **2013**, *25*, 1957–1965. [CrossRef]
88. Uchimiya, M.; Lima, I.M.; Klasson, K.T.; Chang, S.; Wartelle, L.H.; Rodgers, J.E. Immobilization of Heavy Metal Ions (CuII, CdII, NiII, and PbII) by Broiler Litter-Derived Biochars in Water and Soil. *J. Agric. Chem.* **2010**, *58*, 5538–5544. [CrossRef]
89. Sohi, S.; Krull, E.; Lopez-Capel, E.; Bol, R. A Review of Biochar and Its Use and Function in Soil. *Adv. Agron.* **2010**, *105*, 47–82.
90. Tang, J.; Zhu, W.; Kookana, R.; Katayama, A. Characteristics of biochar and its application in remediation of contaminated soil. *J. Biosci. Bioeng.* **2013**, *116*, 653–659. [CrossRef]
91. Xu, R.-K.; Xiao, S.-C.; Yuan, J.-H.; Zhao, A.-Z. Adsorption of methyl violet from aqueous solutions by the biochars derived from crop residues. *Bioresour. Technol.* **2011**, *102*, 10293–10298. [CrossRef] [PubMed]

92. Xu, X.; Cao, X.; Zhao, L.; Wang, H.; Yu, H.; Gao, B. Removal of Cu, Zn, and Cd from aqueous solutions by the dairy manure-derived biochar. *Environ. Sci. Pollut. Res.* **2013**, *20*, 358–368. [CrossRef]
93. Qian, L.; Chen, B.; Hu, D. Effective Alleviation of Aluminum Phytotoxicity by Manure-Derived Biochar. *Environ. Sci. Technol.* **2013**, *47*, 2737–2745. [CrossRef]
94. Jia, M.; Wang, F.; Bian, Y.; Jin, X.; Song, Y.; Kengara, F.O.; Xu, R.; Jiang, X. Effects of pH and metal ions on oxytetracycline sorption to maize-straw-derived biochar. *Bioresour. Technol.* **2013**, *136*, 87–93. [CrossRef] [PubMed]
95. Chen, T.; Zhou, Z.; Han, R.; Meng, R.; Wang, H.; Lu, W. Adsorption of cadmium by biochar derived from municipal sewage sludge: Impact factors and adsorption mechanism. *Chemosphere* **2015**, *134*, 286–293. [CrossRef]
96. Yu, H.; Liu, J.; Shen, J.; Sun, X.; Li, J.; Wang, L. Preparation of MnOx-loaded biochar for Pb^{2+} removal: Adsorption performance and possible mechanism. *J. Taiwan Inst. Chem. Eng.* **2016**, *66*, 313–320.
97. Qian, T.; Wang, Y.; Fan, T.; Fang, G.; Zhou, D. A new insight into the immobilization mechanism of Zn on biochar: The role of anions dissolved from ash. *Sci. Rep.* **2016**, *6*, 33630. [CrossRef] [PubMed]
98. Ahmed, M.B.; Zhou, J.L.; Ngo, H.H.; Guo, W.; Johir, M.A.H.; Sornalingam, K.; Rahman, M.S. Chloramphenicol interaction with functionalized biochar in water: Sorptive mechanism, molecular imprinting effect and repeatable application. *Sci. Total Environ.* **2017**, *609*, 885–895. [CrossRef]
99. Dong, X.; Ma, L.Q.; Li, Y. Characteristics and mechanisms of hexavalent chromium removal by biochar from sugar beet tailing. *J. Hazard. Mater.* **2011**, *190*, 909–915. [CrossRef]
100. Li, H.; Dong, X.; Da Silva, E.B.; De Oliveira, L.M.; Chen, Y.; Ma, L.Q. Mechanisms of metal sorption by biochars: Biochar characteristics and modifications. *Chemosphere* **2017**, *178*, 466–478. [CrossRef] [PubMed]
101. Samsuri, A.W.; Sadegh-Zadeh, F.; Seh-Bardan, B.J. Adsorption of As(III) and As(V) by Fe coated biochars and biochars produced from empty fruit bunch and rice husk. *J. Environ. Chem. Eng.* **2013**, *1*, 981–988. [CrossRef]
102. Ahmed, M.B.; Zhou, J.L.; Ngo, H.H.; Guo, W.; Johir, M.A.H.; Sornalingam, K. Single and competitive sorption properties and mechanism of functionalized biochar for removing sulfonamide antibiotics from water. *Chem. Eng. J.* **2017**, *311*, 348–358. [CrossRef]
103. Boostani, H.R.; Najafi-Ghiri, M.; Hardie, A.G.; Khalili, D. Comparison of Pb stabilization in a contaminated calcareous soil by application of vermicompost and sheep manure and their biochars produced at two temperatures. *Appl. Geochem.* **2019**, *102*, 121–128. [CrossRef]
104. Kim, H.-S.; Kim, K.-R.; Kim, H.-J.; Yoon, J.-H.; Yang, J.E.; Ok, Y.S.; Owens, G.; Kim, K.-H. Effect of biochar on heavy metal immobilization and uptake by lettuce (*Lactuca sativa* L.) in agricultural soil. *Environ. Earth Sci.* **2015**, *74*, 1249–1259. [CrossRef]
105. Yang, X.; Liu, J.; McGrouther, K.; Huang, H.; Lu, K.; Guo, X.; He, L.; Lin, X.; Che, L.; Ye, Z.; Wang, H. Effect of biochar on the extractability of heavy metals (Cd, Cu, Pb, and Zn) and enzyme activity in soil. *Environ. Sci. Pollut. Res.* **2016**, *23*, 974. [CrossRef]
106. Zhou, D.; Liu, D.; Gao, F.; Li, M.; Luo, X. Effects of Biochar-Derived Sewage Sludge on Heavy Metal Adsorption and Immobilization in Soils. *Int. J. Environ. Res. Public Health* **2017**, *14*, 681. [CrossRef]
107. Wang, Y.; Zhong, B.; Shafi, M.; Ma, J.; Guo, J.; Wu, J.; Ye, Z.; Liu, D.; Jin, H. Effects of biochar on growth, and heavy metals accumulation of moso bamboo (Phyllostachy pubescens), soil physical properties, and heavy metals solubility in soil. *Chemosphere* **2019**, *219*, 510–516. [CrossRef]
108. Penido, E.S.; Martins, G.C.; Mendes, T.B.M.; Melo, L.C.A.; Guimarães, I.D.R.; Guilherme, L.R.G. Combining biochar and sewage sludge for immobilization of heavy metals in mining soils. *Ecotoxicol. Environ. Saf.* **2019**, *172*, 326–333. [CrossRef]
109. Lu, H.; Li, Z.; Gascó, G.; Méndez, A.; Shen, Y.; Paz-Ferreiro, J. Use of magnetic biochars for the immobilization of heavy metals in a multi-contaminated soil. *Sci. Total. Environ.* **2018**, *622*, 892–899. [CrossRef] [PubMed]
110. Jia, W.; Wang, B.; Wang, C.; Sun, H. Tourmaline and biochar for the remediation of acid soil polluted with heavy metals. *J. Environ. Chem. Eng.* **2017**, *5*, 2107–2114. [CrossRef]
111. Khan, M.A.; Khan, S.; Ding, X.; Khan, A.; Alam, M. The effects of biochar and rice husk on adsorption and desorption of cadmium on to soils with different water conditions (upland and saturated). *Chemosphere* **2018**, *193*, 1120–1126. [CrossRef] [PubMed]
112. Kumarathilaka, P.; Ahmad, M.; Herath, I.; Mahatantila, K.; Athapattu, B.; Rinklebe, J.; Ok, Y.S.; Usman, A.; Al-Wabel, M.I.; Abduljabbar, A.; et al. Influence of bioenergy waste biochar on proton- and ligand-promoted release of Pb and Cu in a shooting range soil. *Sci. Total. Environ.* **2018**, *625*, 547–554. [CrossRef]

113. Mandal, S.; Sarkar, B.; Bolan, N.; Ok, Y.S.; Naidu, R. Enhancement of chromate reduction in soils by surface modified biochar. *J. Environ. Manag.* **2017**, *186*, 277–284. [CrossRef]
114. Monser, L.; Ben Amor, M.; Ksibi, M. Purification of wet phosphoric acid using modified activated carbon. *Chem. Eng. Process. Process. Intensif.* **1999**, *38*, 267–271. [CrossRef]
115. Lin, L.; Li, Z.; Liu, X.; Qiu, W.; Song, Z. Effects of Fe-Mn modified biochar composite treatment on the properties of As-polluted paddy soil. *Environ. Pollut.* **2019**, *244*, 600–607. [CrossRef]
116. O'Connor, D.; Peng, T.; Li, G.; Wang, S.; Duan, L.; Mulder, J.; Cornelissen, G.; Cheng, Z.; Yang, S.; Hou, D. Sulfur-modified rice husk biochar: A green method for the remediation of mercury contaminated soil. *Sci. Total. Environ.* **2018**, *621*, 819–826. [CrossRef]
117. Yu, Z.; Zhou, L.; Huang, Y.; Song, Z.; Qiu, W. Effects of a manganese oxide-modified biochar composite on adsorption of arsenic in red soil. *J. Environ. Manag.* **2015**, *163*, 155–162. [CrossRef] [PubMed]
118. Gu, J.; Zhou, W.; Jiang, B.; Wang, L.; Ma, Y.; Guo, H.; Schulin, R.; Ji, R.; Evangelou, M.W. Effects of biochar on the transformation and earthworm bioaccumulation of organic pollutants in soil. *Chemosphere* **2016**, *145*, 431–437. [CrossRef]
119. Bielská, L.; Škulcová, L.; Neuwirthová, N.; Cornelissen, G.; Hale, S.E. Sorption, bioavailability and ecotoxic effects of hydrophobic organic compounds in biochar amended soils. *Sci. Total Environ.* **2018**, *624*, 78–86. [CrossRef] [PubMed]
120. Wu, C.; Liu, X.; Wu, X.; Dong, F.; Xu, J.; Zheng, Y. Sorption, degradation and bioavailability of oxyfluorfen in biochar-amended soils. *Sci. Total Environ.* **2019**, *658*, 87–94. [CrossRef] [PubMed]
121. Rombolà, A.G.; Fabbri, D.; Baronti, S.; Vaccari, F.P.; Genesio, L.; Miglietta, F. Changes in the pattern of polycyclic aromatic hydrocarbons in soil treated with biochar from a multiyear field experiment. *Chemosphere* **2019**, *219*, 662–670. [CrossRef] [PubMed]
122. Kołtowski, M.; Hilber, I.; Bucheli, T.D.; Charmas, B.; Skubiszewska-Zięba, J.; Oleszczuk, P. Activated biochars reduce the exposure of polycyclic aromatic hydrocarbons in industrially contaminated soils. *Chem. Eng. J.* **2017**, *310*, 33–40. [CrossRef]
123. Zhang, P.; Sun, H.; Min, L.; Ren, C. Biochars change the sorption and degradation of thiacloprid in soil: Insights into chemical and biological mechanisms. *Environ. Pollut.* **2018**, *236*, 158–167. [CrossRef] [PubMed]
124. Yu, H.; Zou, W.; Chen, J.; Chen, H.; Yu, Z.; Huang, J.; Tang, H.; Wei, X.; Gao, B. Biochar amendment improves crop production in problem soils: A review. *J. Environ. Manag.* **2019**, *232*, 8–21. [CrossRef] [PubMed]
125. Bayabil, H.K.; Stoof, C.R.; Lehmann, J.C.; Yitaferu, B.; Steenhuis, T.S. Assessing the potential of biochar and charcoal to improve soil hydraulic properties in the humid Ethiopian Highlands: The Anjeni watershed. *Geoderma* **2015**, *243*, 115–123. [CrossRef]
126. Liu, X.; Li, L.; Bian, R.; Chen, D.; Qu, J.; Wanjiru Kibue, G.; Pan, G.; Zhang, X.; Zheng, J.; Zheng, J. Effect of biochar amendment on soil-silicon availability and rice uptake. *J. Plant Nutr. Soil Sci.* **2014**, *177*, 91–96. [CrossRef]
127. Agegnehu, G.; Bass, A.M.; Nelson, P.N.; Muirhead, B.; Wright, G.; Bird, M.I. Biochar and biochar-compost as soil amendments: Effects on peanut yield, soil properties and greenhouse gas emissions in tropical North Queensland, Australia. *Agric. Ecosyst. Environ.* **2015**, *213*, 72–85. [CrossRef]
128. Gul, S.; Whalen, J.K. Biochemical cycling of nitrogen and phosphorus in biochar-amended soils. *Soil Boil. Biochem.* **2016**, *103*, 1–15. [CrossRef]
129. Safaei Khorram, M.; Fatemi, A.; Khan, M.A.; Kiefer, R.; Jafarnia, S. Potential risk of weed outbreak by increasing biochar's application rates in slow-growth legume, lentil (*Lens culinaris* Medik.). *J. Sci. Food Agric.* **2018**, *98*, 2080–2088. [CrossRef]
130. Lian, F.; Xing, B. Black Carbon (Biochar) In Water/Soil Environments: Molecular Structure, Sorption, Stability, and Potential Risk. *Environ. Sci. Technol.* **2017**, *51*, 13517–13532. [CrossRef]

© 2019 by the authors. Licensee MDPI, Basel, Switzerland. This article is an open access article distributed under the terms and conditions of the Creative Commons Attribution (CC BY) license (http://creativecommons.org/licenses/by/4.0/).

Article

Optimization of *Salix* Carbonation Solid Acid Catalysts for One-Step Synthesis by Response Surface Method

Ping Lu [1], Kebing Wang [1,*] and Juhui Gong [2]

[1] College of Science, Inner Mongolia Agricultural University, Huhhot 010018, China; 13643584686@163.com
[2] School of Chemistry and Chemical Engineering, Inner Mongolia University, Huhhot 010018, China; imugongjuhui@163.com
* Correspondence: wkb0803@imau.edu.cn

Received: 16 March 2019; Accepted: 8 April 2019; Published: 12 April 2019

Featured Application: One of the major problems which we faced in this century was energy shortage, *Salix* carbon-based solid acid catalyst is a promising method for preparing biodiesel with low-cost catalysts to alleviate the energy crisis.

Abstract: *Salix* carboniferous solid acid catalysts were successfully obtained via one-step carbonization and sulfonation of *Salix psammophila* in the presence of concentrated sulfuric acid, which was then used in the esterification reaction between oleic acid and methanol to prepare the biodiesel. The esterification rate of the catalyst obtained from the reaction indicated the catalytic performance of the catalyst. Afterwards, the recycling performance of the catalyst was optimized and characterized based on Fourier transform infrared spectrometer. The catalyst performance was examined and optimized through the response surface method, and the catalyst was determined and characterized based on scanning electron microscope (SEM), elemental analysis, thermogravimetric analysis, and infrared analysis. The results suggested that the optimal preparation conditions were as follows: reaction temperature of 125 °C, reaction time of 102 min, solid–liquid ratio of 17 g/100 mL, standing time of 30 min, and the highest conversion level of 94.15%.

Keywords: one-step method; carbonized sulfonation; response surface method; *Salix* carboniferous solid acid catalysts; biodiesel

---

1. Introduction

Nowadays, oil is the most important source of energy worldwide, but the emissions of greenhouse gases threaten the global ecosystem and sustainable resources have become an urgent need in our daily life [1]. The increasing scarcity of fossil fuels and the emissions of greenhouse gases and related climate change are the main driving forces to develop clean energy [2–4]. Biodiesel is a clean fuel energy consisting of C_{12}–C_{22} fatty acids, which are a kind of long-chain fatty acid alkyl. Compared to fossil fuel, biodiesel shows the advantages of being a biodegradable, sulfur-free, and non-toxic sustainable diesel fuel substitute [5]. Biodiesel can be prepared through the transesterification or esterification of the raw materials of various animal and plant fats, as well as waste oils or vegetable oleic acid with low carbon-chain alcohol [6–9]. Concentrated sulfuric acid is the most extensively used catalyst in the preparation of biodiesel, which can hardly be separated from the products and eventually leads to corrosion of equipment and severe environmental pollution. Therefore, it is of great practical significance to investigate and develop the cheap and environmentally friendly catalyst. The solid acid catalyst is mostly inorganic, including the metal oxides, metal sulfates, metal phosphates, zeolites, etc., which are beginning to appear. However, to some extent, the poor applicability of inorganic solid acid catalysts has limited their usage, especially in liquid phase reactions. The carbonation solid acid catalysts that have been intensively studied in recent years have largely solved these problems. Since 2004, the research group of Toda [10] first used polycyclic aromatic compounds as raw materials, and carbonized the material at 400 °C to obtain powdery products between amorphous carbon and incomplete carbonization. Then the products were reacted in a certain amount with concentrated sulfuric acid at 150 °C for a certain period of time to obtain sulfonated carbonation solid acid catalysts, which were applied to the catalytic esterification reaction. The carbonation solid acid catalysts have become promising catalysts. Many countries have studied and prepared different carbon-based solid acid catalysts. There are also more and more examples of esterification reactions. Lou et al. [11] prepared carbonation solid acid catalysts from four different raw materials, among which starch had the highest catalytic activity, which is higher than SO_4^{2-}/ZrO_2 and niobic acid. Zong et al. [12] prepared carbonation solid acid catalysts of type $CH_{1.14}S_{0.03}O_{0.39}$ with D-glucose powder. The acid strength of the catalysts corresponds to the acid strength of concentrated sulfuric acid. XRD analysis showed that the glucose carbonation solid acid catalysts had an amorphous structure. Kastner et al. [13] used two kinds of biomass, namely peanut shell and wood chips, as raw materials, which were cracked, carbonized, and mixed with concentrated sulfuric acid for sulfonation to obtain biomass carbonation solid acid catalysts. The catalysts have high catalytic activity in the catalytic esterification of palmitic acid and stearic acid, as well as remarkable stability.

As a highly abundant, natural carbon source, biomass is considered as a promising renewable source alternative to fossil fuels [14–17]. The utilization of waste biomass to prepare the biomass carboniferous solid acid catalyst exerts a win-win effect on both the economy and environment [18–23]. *Salix psammophila* is a characteristic plant in Inner Mongolia, and every year the even stubble rejuvenation products become the agricultural and forest residues [24–26]. Using cheap *Salix psammophila* as the raw material, the one-step carbonation and sulfonation of concentrated sulfuric acid was used to prepare the *Salix* carboniferous solid acid catalyst, with the biodiesel prepared by oleic acid and methanol. The catalyst preparation conditions were optimized using the response surface method. Based on the advantages of special resources in Inner Mongolia, we not only realize the new energy utilization of sandy shrubs, but also strive to obtain new catalytic materials of *Salix* carboniferous solid acid catalysts with excellent catalytic performance to improve biodiesel production.

2. Materials and Methods

2.1. Materials

Salix was collected from a local supermarket in Ordos. (Inner Mongolia Autonomous Region, China). Concentrated Sulfuric Acid (98%, Analytical Pure) was provided by Sinopharm Chemical Reagent Co., Ltd. (Shanghai, China). Ethanol (95%, Analytical Pure) was purchased from Tianjin Fengchuan Chemical Reagent Science and Technology Co., Ltd. (Tianjin, China). Cis-9-octadecenoic acid (99%, Analytical Pure) was provided by Tianjin Fuchen Chemical Reagent Factory (Tianjin, China). Absolute methanol (99%, Analytical Pure) was purchased from Tianjin Fengchuan Chemical Reagent Science and Technology Co., Ltd. (Tianjin, China).

2.2. Methods

The *Salix psammophila* was smashed and filtered with a 50–80 mesh sieve, then a certain mass of *Salix psammophila* powder was weighed and put into the polytetrafluoroethylene (PTFE) reactor, and different volumes of concentrated sulfuric acid were added at different solid–liquid ratios. After standing for a certain period of time, the reactor body was put into the homogeneous reactor, a certain temperature was adjusted to react under rotation for a certain period of time, and the reactor was then taken out and put into ice water for sudden cooling. The cooled solid–liquid mixture was added into 300 mL distilled water, stirred, filtered, and washed repeatedly with hot distilled water until it had become neutral. Afterwards, the mixture was dried for 24 h in the oven at 100 °C, then ground, packaged into a seal bag, labeled, and stored in the dryer for use.

A certain amount of oleic acid, methanol, and *Salix* carboniferous solid acid catalyst were added into a 3-neck flask (the quality of catalyst: the quality of oleic acid = 7%, the amount of substance of methanol: the amount of substance of oleic acid = 10:1), which was then put into a water bath at a certain temperature (68 ± 1 °C), and a reflux condenser was placed on the top of the 3-neck flask. After reaching the reaction temperature, the reaction time was counted when liquid reflux was observed in the reflux condensing tube above. After reaching the designated reaction time, the 3-neck flask was taken out and cooled rapidly. The solid acid catalyst was washed with ethanol, and the aqueous phase in the filtrate was evaporated using the rotatory evaporator to obtain the mixture of methyl oleate and oleic acid, which was then titrated.

According to the determination of animal and plant fat acid value and acidity through the third method of hot ethanol method in GB 5009.229-2016, the mixture was dissolved into 50 mL of 95% ethanol, which was heated through a water bath and titrated using the 0.1 mol/L KOH standard solution, with phenolphthalein as the indicator. The titration end-point was reached when the solution became light red and did not fade within 1 min, and the oleic acid conversion rate, namely, the conversion level (X), was calculated.

$$X = (A_0 - A_T)/A_0 \times 100\% \tag{1}$$

where A_0 is the initial acid value of the reaction(mmol/g), and A_T is the instantaneous acid value of the reaction(mmol/g).

The Box–Behnken four-factor three-level response surface method (RSM) design method was utilized, among which, reaction temperature, reaction time, solid–liquid ratio, and standing time were treated as the independent variables, while conversion level was considered as the response value. The factors and level coding are presented in Table 1. In total, 9 experiments (including 4 analytical experiments and 5 central experiments) were designed to estimate the experimental error. The response curve experimental design and results are displayed in Table 2. The Design-Expert software was used for analysis, and the following response surface model was constructed.

$$X = 88.24 + 1.17A + 0.49B - 2.53C - 1.01D - 3.02AB + 4.60AC + 2.47AD + 2.28C - 2.27B + 4.41CD - 5.65A^2 - 2.92B^2 - 1.68C^2 - 1.47D^2 \tag{2}$$

Table 1. Factor levels for the experiments.

Levels	A: Reaction Temperature/°C	B: Reaction Time/min	C: Solid-Liquid Ratio	D: Standing Time/min
−1	120	60	1:4	30
0	135	90	1:5	60
1	150	120	1:6	90

Table 2. Design of RSM and its experimental values.

Number	A: Reaction Temperature/°C	B: Reaction Time/min	C: Solid-Liquid Ratio	D: Standing Time/min	Conversion Level/%
1	135	120	1:5	30	83.19
2	135	90	1:6	30	89.98
3	135	90	1:5	60	86.69
4	135	60	1:5	30	84.01
5	135	60	1:6	60	86.04
6	150	90	1:5	90	78.93
7	150	120	1:5	60	81.71
8	120	90	1:4	60	74.49
9	135	120	1:4	60	85.89
10	135	90	1:6	90	84.67
11	135	120	1:5	90	79.61
12	150	60	1:5	60	84.15
13	150	90	1:5	30	84.51
14	120	90	1:6	60	88.97
15	135	120	1:6	60	86.20
16	120	90	1:5	30	88.33
17	135	90	1:5	60	87.33
18	150	90	1:4	60	82.53
19	120	120	1:5	60	80.70
20	150	90	1:6	60	78.60
21	135	90	1:5	60	85.98
22	135	90	1:5	60	90.27
23	135	60	1:4	60	76.60
24	120	90	1:5	90	72.86
25	120	60	1:5	60	71.05
26	135	60	1:5	90	89.51
27	135	90	1:4	90	88.49
28	135	90	1:5	60	90.93
29	135	90	1:4	30	76.14

3. Results and Discussion

The regression model analysis of variance is shown in Table 3. The regression parameters of the response surface were analyzed through Analysis of Variance. The F-test was conducted to evaluate the significance of the influence of variables in the regression equation on the response value. Typically, a smaller probability P (P-Value) is associated with a higher significance of the corresponding variable. In the proposed model, $P = 0.0406 < 0.05$, and the response surface regression model reached a significant level; while the lack of fit was $P = 0.0873 > 0.05$, which was not significant, suggesting that all experimental points were expressed using the model, high degree of fitting was achieved, and the regression equation model was successfully established. These findings demonstrate that, the *Salix* carbon-based solid acid catalyst can well catalyze the preparation of biodiesel. Among the linear terms, the solid–liquid ratio C had the greatest influence $P = 0.0486 < 0.05$, ranking a significant level. Among the quadratic terms, reaction temperature A^2 ($P = 0.0032 < 0.01$) reached an extremely significant level and the interaction terms $P_{AC} = 0.0394$ (significant level of model in reaction temperature and solid-liquid ratio factors) and $P_{CD} = 0.0469$ (significant level of model in solid-liquid ratio and standing time factors) exerted significant influence on the catalyst performance ($P < 0.05$). Within the range of each factor selected in the present study, the solid–liquid ratio had the greatest influence on the catalytic performance of the *Salix* carboniferous solid acid catalyst.

Table 3. Results of variance analysis of regression models.

Source Model	Sum of Squares	Degrees Freedom	Mean Square	F-Value	P-Value > F-Value	Significant Level
model	603.943	14	43.139	2.6279	0.041	significant
A	16.4034	1	16.403	0.9993	0.335	
B	2.9403	1	2.940	0.1791	0.679	
C	76.6085	1	76.609	4.6669	0.049	significant
D	12.1807	1	12.181	0.7420	0.404	
AB	36.5420	1	36.542	2.2261	0.158	
AC	84.7320	1	84.732	5.1617	0.039	significant
AD	24.4530	1	24.453	1.4896	0.242	
BC	20.8392	1	20.839	1.2695	0.279	
BD	20.6116	1	20.612	1.2556	0.281	
CD	77.9689	1	77.969	4.7497	0.047	significant
A^2	206.912	1	206.91	12.605	0.003	significant
B^2	55.2748	1	55.275	3.3672	0.088	
C^2	18.2349	1	18.235	1.1108	0.310	
D^2	14.0723	1	14.072	0.8573	0.370	
Residual	229.816	14	16.415			
Lack of Fit	210.121	10	21.012	4.2675	0.087	not significant
Pure Error	19.6952	4	4.924			
Cor Total		28				

A 3D diagram can represent not only the characters of the response surface function, but also intuitively illustrate the effect of interaction between factors on the response value.

As can be observed from Figure 1, the interactions between reaction temperature and the solid–liquid ratio, as well as the between standing time and solid–liquid ratio, are extremely significant. They are manifested by a relatively steep curve, verifying that the interactive items of reaction temperature, standing time, and solid–liquid ratio exert significant influence on the catalytic activity of the catalyst.

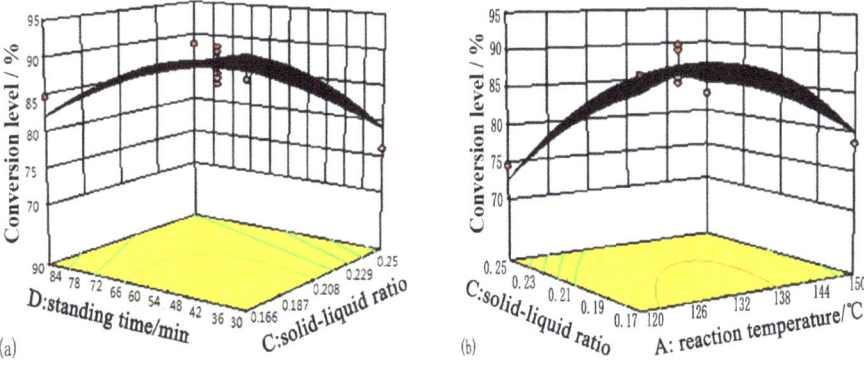

Figure 1. (**a**) Response surface diagram of the effect of solid–liquid ratio and standing time interaction on conversion level; (**b**) response surface diagram of the effects of reaction temperature and solid–liquid ratio interaction on conversion level.

According to Figure 2a, when the reaction temperature remains unchanged, the conversion level shows an increasing trend with an extension in reaction time, and the catalyst activity is enhanced. When the reaction time remains unchanged, the conversion level also displays an increasing trend with the increase in reaction temperature. In linear terms, reaction time and reaction temperature make basically the same influence on the catalyst. Figure 2b shows that with the reaction temperature remaining unchanged, the conversion level presents a decreasing trend with an increase in standing time, and the catalyst activity is weakened; when the standing time remains unchanged, the conversion level is first increased and then decreased with the increase in reaction temperature.

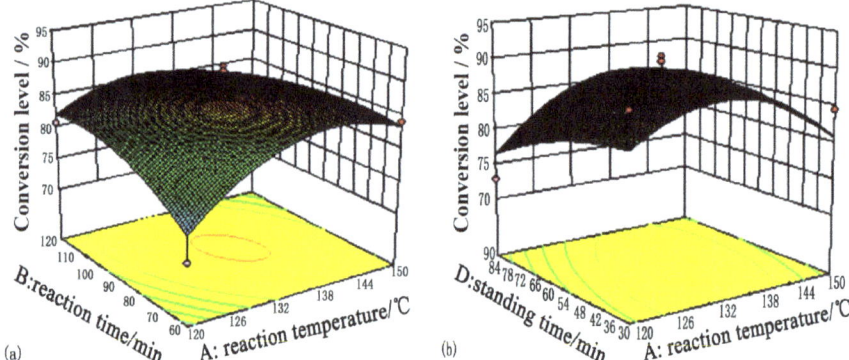

Figure 2. (a) Response surface diagram of the effect of reaction temperature, reaction time interaction on conversion level; (b) response surface diagram of the effect of reaction temperature, standing time interaction on conversion level.

As can be observed from Figure 3a, with a constant reaction time, the conversion level shows an increasing trend with the increase of concentrated sulfuric acid in solid–liquid ratio, and the catalyst activity is also enhanced. When the solid–liquid ratio remains unchanged, the conversion level is relatively gentle with the increase in reaction time. Among the linear terms, solid–liquid ratio has more significant influence than reaction temperature on the catalyst. As presented in Figure 3b, under the condition that the standing time remains unchanged, the conversion level shows an increasing trend with the extension in reaction time, and the catalyst activity is promoted. When the reaction time is unchanged, the conversion level also displays an increasing trend with the increase in standing time.

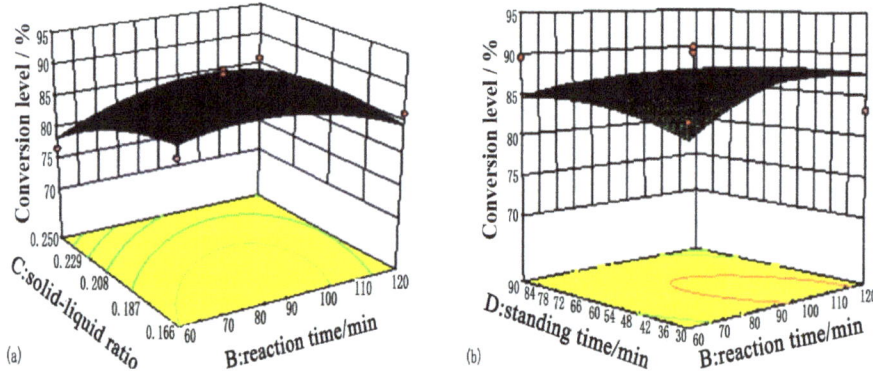

Figure 3. (a) Response surface diagram of the effect of solid–liquid ratio and reaction time interaction on conversion level; (b) response surface diagram of the effect of standing time and reaction time interaction on conversion level.

The maximum response value maximum was determined using the numerical function of Design-Expert software, and the optimal reaction conditions are obtained as follows: reaction temperature of 125 °C, reaction time of 102 min, solid–liquid ratio of 17 g/100 mL, standing time of 30 min, and highest conversion level of 95%.

After optimization with the above-mentioned software, the above conditions were employed to verify the experiment. The conversion level of esterification reaction for the preparation of biodiesel catalyzed by the *Salix* carbon-based solid acid catalyst was 94.15%. The relative deviation between the software predicted value and the experimental value was 0.89%. Therefore, it is feasible to employ this model to analyze and predict the catalytic performance of the *Salix* carboniferous solid acid catalyst.

When comparing (Table 4), the contents of C, O, and H in *Salix psammophila* raw materials were 51.886%, 41.348%, and 6.316%, respectively. The contents of N and S were lower, which were 0.31% and 0.14%, respectively. After the one-step method, the content of N was not obvious, while the content of O and H were significantly decreased, which was caused by the precipitation of O and H on a part of the aromatic ring in the reaction substituted by the sulfonic acid group. The content of S in the catalyst was obviously increased to 26.378% and the measured surface acid amount was 1.28 mmol/g, indicating that the S was introduced into the raw materials of *Salix*, and the sulfonic acid group was formed.

Table 4. Elemental analysis.

	C	H	N	O	S
Salix psammophila	51.886%	6.316%	0.31%	41.348%	0.14%
Catalyst	50.122%	7.583%	0.293%	15.624%	26.378%

As shown, Figure 4 presents that in the infrared spectrum peak of the catalyst, the solid acid catalysts prepared by the one-step method still had a peak of 3400 cm^{-1} containing the associated hydroxyl group and the stretching vibration of the hydroxyl group contained in the carboxyl group; the carbonyl peak contained in the carboxyl group at 1600 cm^{-1}; the stretching vibration peaks of SO$_3$H and O=S=O bonds occur at about 1000 cm^{-1} and 1100 cm^{-1}, respectively, demonstrating that concentrated sulfuric acid has played a role in sulfonation and the sulfonyl has been successfully introduced after the one-step method treatment of *Salix psammophila*. In the one-step process of concentrated sulfuric acid, the positively charged sulfonic acid group electrophilically replaced C on the fused ring structure to form a sulfonic acid complex intermediate. After deprotonation, a carbon-based solid acid catalyst with three acidic groups, SO$_3$H, OH, and COOH were formed, in which a sulfonic acid group was grafted on a carbon skeleton of a polycyclic aromatic carbon ring arranged in a disorderly manner.

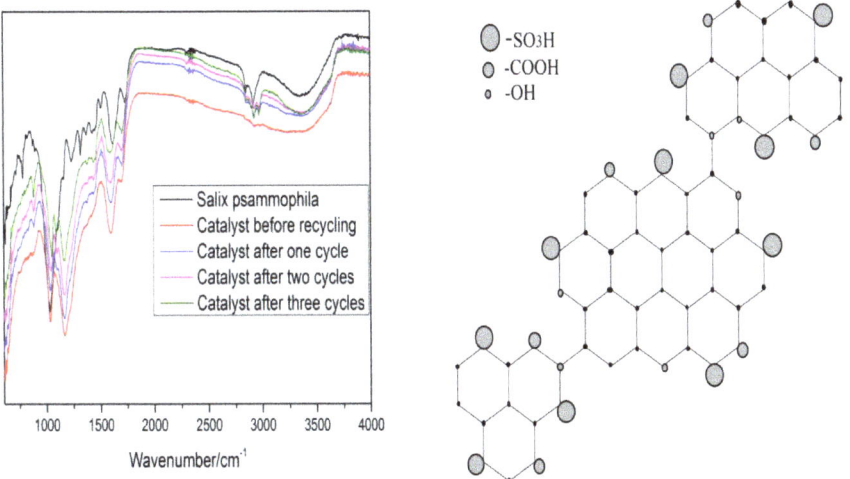

Figure 4. Infrared spectra and syntactic model of catalysts.

Simultaneously, Figure 4 shows that the infrared spectrum peak image of the *Salix* carboniferous solid acid catalyst prepared under the optimal conditions were reused after washing with hot distilled water. The change of conversion level after four cycles is as follows: 92.35%, 76.23%, 74.22%, 75.73%. The methyl oleate absorption peak could be observed at 3000 cm^{-1}, while the catalytic activity of catalyst had decreased, to a large extent, caused by the effect of coating of the catalyst with methyl oleate. The conversion level of catalytic biodiesel reaction was obviously improved by changing the washing liquid of the catalyst, soaking in ethanol solution, and repeatedly washing with hot ethanol solution. The change of conversion level after four cycles is as follows: 92.35%, 87.28%, 84.74%, 85.36%.

Figure 5 shows the thermogravimetry (TG)–derivative thermogravimetry (DTG) curve of the catalyst under N_2 atmosphere at the heating rate of 20 °C/min and an ending temperature of 800 °C. The catalyst dehydration phase occurred at about 105 °C, and the first weight loss rate peak occurred in the DTG curve, and the TG curve showed that the weight loss ratio was 3.73%. The catalyst thermolysis phase occurred at about 180–280 °C. The second weight loss rate peak could be observed in the DTG curve, and the TG curve showed that the weight loss ratio was 5.12%. The weight loss rate peak of the DTG curve at 290 °C was consistent with the peak shape of the experimental initial current instability, which proves that the peak appears to be related to the instrument current instability. The figure presented that the catalyst had been obviously decomposed at 238 °C, which indicated excellent thermal stability.

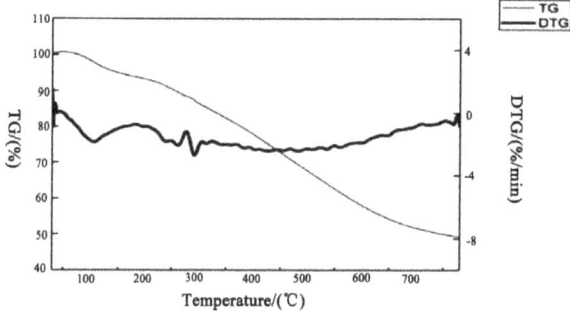

Figure 5. Thermogravimetry –derivative thermogravimetry curve of catalyst.

In the scanning electron micrograph of Figure 6, the left picture shows the *Salix psammophila* and the right picture shows the catalyst prepared under the optimal conditions. As can be observed, under the same scanning electron microscope conditions, when the scale is 5 microns, the structure of *Salix psammophila* is greatly damaged compared with the catalyst. The water molecules and organic ingredients contained in the *Salix psammophila* at the one-step method were evaporated, and the catalyst surface showed the porous frame structure. Meanwhile, the space was mainly filled with the amorphous carbon particles with irregular shapes, and these carbon particles were arranged in a spherical structure. In the field of view, the surface of the sand willow is relatively smooth, while the surface of the catalyst is rough. The particles have showed obvious brightness and darkness, which may be caused by the introduction of sulfonic acid groups or the different particle positions.

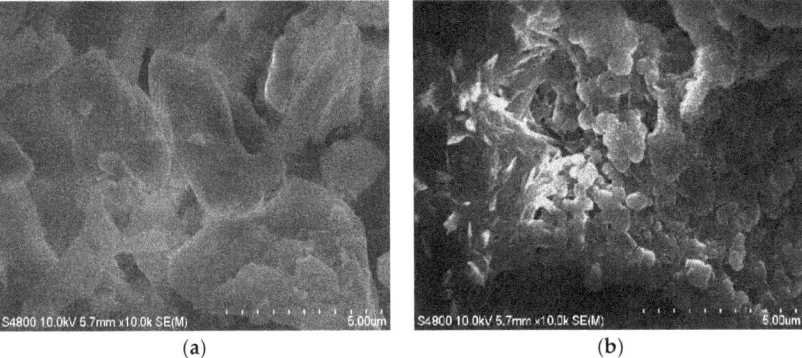

Figure 6. (a) Scanning electron micrographs of *Salix psammophila*; (b) Scanning electron micrographs of the catalyst.

In the catalyst reaction mechanism of Figure 7, the *Salix* carboniferous solid acid catalysts provide H^+, which formed a monomolecular carbocation with the oxygen on the carbonyl group in the oleic acid structure. Then the hydroxyl group on the methanol attacks the carbonyl group on the oleic acid to form a nucleophilic addition reaction. The intermediate is obtained, which is extremely unstable. After reversible rearrangement, the water molecule is lost as a leaving group, and finally methyl oleate is formed.

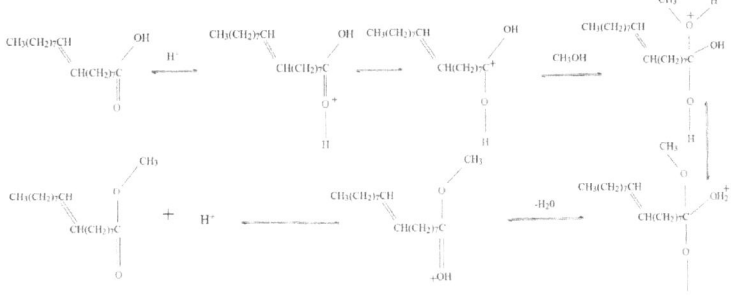

Figure 7. Mechanism of catalyst participation in esterification.

4. Conclusions

Salix carboniferous solid acid catalysts were prepared from waste biomass *Salix psammophila* of Inner Mongolia shrub in one step by using the dehydration and acidity of concentrated sulfuric acid and a low-cost reduction method. The response surface curve analysis was carried out using a large amount of experimental data, and the data model was established. The experiment was combined with the data simulation to further optimize the activity of the *Salix* carboniferous solid acid catalysts.

The one-step carbonation and sulfonation could effectively introduce the sulfonic acid group. Through the test of catalyst recycling, it was concluded that the coating of methyl oleate on catalyst is the main factor affecting the effect of catalytic recycling. Through the combination of experimental optimization and test characterization, the relationship between the structural characteristics of the catalyst and the mechanism of the catalyst in the catalytic reaction was explored, and the obtained catalyst had good activity, stability and application value.

Author Contributions: P.L. conceived, designed, performed the experiments, analyzed the data, and wrote the paper; J.G. taught software to process the data; K.W. revised the manuscript.

Funding: This research was funded by the National Nature Science Foundation of China (21,366,018), Key Research.

Conflicts of Interest: The authors declare no conflicts of interest.

References

1. Walter, B.; Gruson, J.F.; Monnier, G. Motorisation Et Carburants Diesel: Un Large Spectre D'évolutions À Venir Contexte Général Et Thématiques De Recherche. *Oil Gas Sci. Technol.* **2008**, *63*, 387–393.
2. Luque, R.; Lovett, J.C.; Datta, B.; Clancy, J.; Campelo, J.M.; Romero, A.A. Biodiesel as feasible petrol fuel replacement: A multidisciplinary overview. *Energy Environ. Sci.* **2010**, *3*, 1706–1721. [CrossRef]
3. Wilson, K.; Lee, A.F. ChemInform Abstract: Rational Design of Heterogeneous Catalysts for Biodiesel Synthesis. *Catal. Sci. Technol.* **2012**, *2*, 884–897. [CrossRef]
4. Sharma, Y.C.; Singh, B.; Korstad, J. Advancements in solid acid catalysts for ecofriendly and economically viable synthesis of biodiesel. *Biofuels Bioprod. Biorefin.* **2011**, *5*, 69–92. [CrossRef]
5. Lotero, E.; Liu, Y.; Lopez, D.E.; Suwannakarn, K.; Bruce, D.A.; Goodwin, J.G. Synthesis of Biodiesel via Acid Catalysis. *Ind. Eng. Chem. Res.* **2005**, *44*, 5353–5363. [CrossRef]
6. Koh, M.Y. A review of biodiesel production from *Jatropha curcas* L. Oil. *Renew. Sustain. Energy Rev.* **2011**, *15*, 2240–2251. [CrossRef]
7. Luque, R.; Clark, J.H. Biodiesel-Like biofuels from simultaneous. transesterification/esterification of waste oils with a biomass-derived solid acid catalyst. *ChemCatChem* **2015**, *3*, 594–597. [CrossRef]
8. Venkateswarulu, T.C.; Raviteja, C.V.; Prabhaker, K.V.; Babu, D.J.; Reddy, A.R.; Indira, M.; Venkatanarayana, A. A Review on Methods of Transesterification of Oils and Fats in Bio-diesel Formation. *Int. J. ChemTech Res.* **2014**, *6*, 974–4290.
9. Liu, H.A.; Bo, Y. Advances in the research of stillingia oil components as feedstock of biodiesel. *Genom. Appl. Biol.* **2010**, *29*, 402–408.
10. Masakuza, T.; Atsushi, T.; Kondo, J.N.; Okamura, M.; Hayashi, S.; Hara, M. Biodiesel made with sugar catalyst. *Nature* **2005**, *438*, 178–179.
11. Lou, W.Y.; Zong, M.H.; Duan, Z.Q. Efficient production of biodiesel from high free fatty acid-containing waste oils using various carbohydrate-derived solid acid catalysts. *Bioresour. Technol.* **2008**, *99*, 8752–8758. [CrossRef]
12. Zong, M.H.; Duan, Z.Q.; Lou, W.Y.; Smith, T.J.; Wu, H. Preparation of a sugar catalyst and its use for highly efficient production of biodiesel. *Green Chem.* **2007**, *9*, 434–437. [CrossRef]
13. Kastner, J.R.; Miller, J.; Geller, D.P.; Locklin, J.; Keith, L.H.; Johnson, T. Catalytic esterification of fatty acids using solid acid catalysts generated from biochar and activated carbon. *Catal. Today* **2012**, *190*, 122–132. [CrossRef]
14. Mika, L.T.; Cséfalvay, E.; Áron, N. Catalytic Conversion of Carbohydrates to Initial Platform Chemicals: Chemistry and Sustainability. *Chem. Rev.* **2018**, *118*, 505–613. [CrossRef]
15. Van de Vyver, S.; Geboers, J.; Jacobs, P.A.; Sels, B.F. Recent Advances in the Catalytic Conversion of Cellulose. *ChemCatChem* **2011**, *3*, 82–94. [CrossRef]
16. Dusselier, M.; Mascal, M.; Sels, B.F. Selective Catalysis for Renewable Feedstocks and Chemicals. In *Top Chemical Opportunities from Carbohydrate Biomass: A Chemist's View Biorefinery*; Nicholas, K.M., Ed.; Springer International Publishing: Berlin, Germany, 2014; Volume 353, pp. 1–40.
17. Geboers, J.A.; Van De Vyver, S.; Ooms, R.; Op De Beeck, B.; Jacobs, P.A.; Sels, B.F. Chemocatalytic conversion of cellulose: Opportunities, advances and pitfalls. *Catal. Sci. Technol.* **2011**, *1*, 714–726. [CrossRef]
18. Tuhy, .; Samoraj, M.; Michalak, I.; Chojnacka, K. The Application of Biosorption for Production of Micronutrient Fertilizers Based on Waste Biomass. *Appl. Biochem. Biotechnol.* **2014**, *174*, 1376–1392. [CrossRef]
19. Vamvuka, D. Bio-oil, solid and gaseous biofuels from biomass pyrolysis processes—An overview. *Int. J. Energy Res.* **2011**, *35*, 835–862. [CrossRef]
20. Sims, R.E.H. Bioenergy to mitigate for climate change and meet the needs of society, the economy and the environment. *Mitig. Adapt. Strateg. Glob. Chang.* **2003**, *8*, 349–370. [CrossRef]
21. Shu, Q.; Gao, J.; Nawaz, Z.; Liao, Y.; Wang, D.; Wang, J. Synthesis of biodiesel from waste vegetable oil with large amounts of free fatty acids using a carbon-based solid acid catalyst. *Appl. Energy* **2010**, *87*, 2589–2596. [CrossRef]

22. Shu, Q.; Nawaz, Z.; Gao, J.; Liao, Y.; Zhang, Q.; Wang, D.; Wang, J. Synthesis of biodiesel from a model waste oil feedstock using a carbon-based solid acid catalyst: Reaction and separation. *Bioresour. Technol.* **2010**, *101*, 5374–5384. [CrossRef]
23. Dehkhoda, A.M.; West, A.H.; Ellis, N. Biochar based solid acid catalyst for biodiesel production. *Appl. Catal. Gen.* **2010**, *382*, 197–204. [CrossRef]
24. Zhang, X.F.; Wang, K.B.; Yan, X.L.; Zhao, Y.B.; Guo, X.M. Alcoholysis of Salix psammophila liquefaction. *Sci. Technol. Rev.* **2014**, *32*, 37–40.
25. Huang, J.T.; Gao, G.H. Analysis of NMR spectra on liquefied products from Salix psammophila and Caragana intermedia woods. *Chem. Ind. For. Prod.* **2009**, *29*, 101–104.
26. Liu, X.; Wan, Y.; Liu, P.; Zhao, L.; Zou, W. Optimization of process conditions for preparation of activated carbon from waste Salix psammophila and its adsorption behavior on fluoroquinolone antibiotics. *Water Sci. Technol. J. Int. Assoc. Water Pollut. Res.* **2018**, *77*, 2555. [CrossRef]

© 2019 by the authors. Licensee MDPI, Basel, Switzerland. This article is an open access article distributed under the terms and conditions of the Creative Commons Attribution (CC BY) license (http://creativecommons.org/licenses/by/4.0/).

MDPI
St. Alban-Anlage 66
4052 Basel
Switzerland
Tel. +41 61 683 77 34
Fax +41 61 302 89 18
www.mdpi.com

Applied Sciences Editorial Office
E-mail: applsci@mdpi.com
www.mdpi.com/journal/applsci

www.ingramcontent.com/pod-product-compliance
Lightning Source LLC
LaVergne TN
LVHW070505100526
838202LV00014B/1790